FUNDAMENTAL PHYSICS OF FERROELECTRICS 2003

Previous Proceedings in the Series of Workshops on Ferroelectrics

Year	Title	Publisher	ISBN/ISSN
2002	Fundamental Physics of Ferroelectrics 2002	AIP Conf. Proceedings Vol. 626	0-7354-0079-2
2001	Fundamental Physics of Ferroelectrics 2001	AIP Conf. Proceedings Vol. 582	0-7354-0021-0
2000	Fundamental Physics of Ferroelectrics 2000	AIP Conf. Proceedings Vol. 535	1-56396-959-9
1999	Williamsburg Workshop (1999)	J. Phys. Chem. Solids 61 (2), 2000	0022-3697
1998	Williamsburg Workshop (5th)	AIP Conf. Proceedings Vol. 436	1-56396-730-8
1997	Williamsburg Workshop (1997)	Ferroelectrics Vol. 206 (1-4) and Vol. 207 (1-2)	0015-0193

Other Related Titles from AIP Conference Proceedings

678 Lectures on the Physics of Highly Correlated Electron Systems VII: Seventh Training Course in the Physics of Correlated Electron Systems and High-Tc Superconductors
Edited by A. Avella and F. Mancini, August 2003, 0-7354-0147-0

657 Review of Progress in Quantitative Nondestructive Evaluation: Volume 22
Edited by Donald O. Thompson and Dale E. Chimenti, March 2003, 2 vol. hard cover set, CD-ROM included, 0-7354-0117-9

633 Structural and Electronic Properties of Molecular Nanostructures: XVI International Winterschool on Electronic Properties of Novel Materials
Edited by Hans Kuzmany, Jörg Fink, Michael Mehring, and Siegmar Roth, November 2002, 0-7354-0088-1

629 Lectures on the Physics of Highly Correlated Electron Systems VI: Sixth Training Course in the Physics of Correlated Electron Systems and High-Tc Superconductors
Edited by F. Mancini, October 2002, 0-7354-0083-0

590 Nanonetwork Materials: Fullerenes, Nanotubes, and Related Systems, ISNM 2001
Edited by Susumu Saito, Tsuneya Ando, Yoshihiro Iwasa, Koichi Kikuchi, Mototada Kobayashi, and Yahachi Saito, October 2001, 0-7354-0032-6

551 Atomic Physics 17: XVII International Conference on Atomic Physics; ICAP 2000
Edited by Ennio Arimondo, Paolo De Natale, and Massimo Inguscio, February 2001, 1-56396-982-3

To learn more about these titles, or the AIP Conference Proceedings Series, please visit the webpage http://proceedings.aip.org

FUNDAMENTAL PHYSICS OF FERROELECTRICS 2003

Williamsburg, Virginia   2-5 February 2003

EDITORS
Peter K. Davies
University of Pennsylvania
Philadelphia, PA

David J. Singh
Naval Research Laboratory
Washington, DC

SPONSORING ORGANIZATIONS
The Office of Naval Research
The University of Pennsylvania

Melville, New York, 2003
AIP CONFERENCE PROCEEDINGS ■ VOLUME 677

Editors:

Peter K. Davies
Department of Materials Science & Engineering
University of Pennsylvania
3231 Walnut Street
Philadelphia, PA 19104-6272
USA

E-mail: davies@lrsm.upenn.edu

David J. Singh
Code 6390
Naval Research Laboratory
455 Overlook Avenue SW, Bldg 49
Washington, DC 20375
USA

E-mail: singh@dave.nrl.navy.mil

The article on pp. 261-268 was authored by U.S. Goverment employees and is not covered by the below mentioned copyright.

Authorization to photocopy items for internal or personal use, beyond the free copying permitted under the 1978 U.S. Copyright Law (see statement below), is granted by the American Institute of Physics for users registered with the Copyright Clearance Center (CCC) Transactional Reporting Service, provided that the base fee of $20.00 per copy is paid directly to CCC, 222 Rosewood Drive, Danvers, MA 01923. For those organizations that have been granted a photocopy license by CCC, a separate system of payment has been arranged. The fee code for users of the Transactional Reporting Service is: 0-7354-0146-2/03/$20.00.

© 2003 American Institute of Physics

Individual readers of this volume and nonprofit libraries, acting for them, are permitted to make fair use of the material in it, such as copying an article for use in teaching or research. Permission is granted to quote from this volume in scientific work with the customary acknowledgment of the source. To reprint a figure, table, or other excerpt requires the consent of one of the original authors and notification to AIP. Republication or systematic or multiple reproduction of any material in this volume is permitted only under license from AIP. Address inquiries to Office of Rights and Permissions, Suite 1NO1, 2 Huntington Quadrangle, Melville, N.Y. 11747-4502; phone: 516-576-2268; fax: 516-576-2450; e-mail: rights@aip.org.

L.C. Catalog Card No. 2003109026
ISBN 0-7354-0146-2
ISSN 0094-243X
Printed in the United States of America

Contents

Preface ix

**Pressure as a Probe of the Physics of Compositionally-Substituted
Quantum Paraelectrics: $SrTiO_3$** 1
    E. L. Venturini, G. A. Samara, M. Itoh, and W. Kleemann

**Temporal Effects in Dielectric Properties of Some Antiferroelectric
Complex Perovskites** 10
    W.-H. Chan, Z. K. Xu, J. Zhai, H. Chen, and E. V. Colla

**Disorder in $BaTiO_3$ and $SrTiO_3$ and the "Ferroelectric"
Transition in $SrTi^{18}O_3$** 20
    R. Blinc, B. Zalar, A. Lebar, and M. Itoh

Non-Debye Domain Wall Response of $SrTi_{18}O_3$ 26
    W. Kleemann, J. Dec, R. Wang, and M. Itoh

Noise and Aging of Relaxor Ferroelectrics 33
    M. B. Weissman, E. V. Colla, and L. K. Chao

Diffusive Phase Transitions in Ferroelectrics and Antiferroelectrics 41
    S. A. Prosandeev, I. P. Raevski, and U. V. Waghmare

**Temperature Dependence of the Local Structure in Pb Containing
Relaxor Ferroelectrics** 48
    T. Egami, E. Mamontov, W. Dmowski, and S. B. Vakhrushev

Anti-ferrodistortive Nanodomains in PMN Relaxor 55
    A. Tkachuk and H. Chen

Micro-Brillouin Investigations of Relaxor Ferroelectrics 65
    H. Hellwig, R. J. Hemley, and R. E. Cohen

Structure of Nanodomains in Relaxors 74
    S. B. Vakhrushev, A. A. Naberezhnov, B. Dkhil, J.-M. Kiat, V. Shwartsman,
    A. Kholkin, B. Dorner, and A. Ivanov

**Conformal Domain Miniaturization and Adaptive Monoclinic
(Pseudo-orthorhombic) Ferroelectric States** 84
    Y. M. Jin, Y. Wang, A. G. Khachaturyan, J. F. Li, and D. Viehland

**Condensation and Slow Dynamics of Polar Nanoregions
in Lead Relaxors** 98
    D. La-Orauttapong, O. Svitelskiy, and J. Toulouse

**Correlations between the Structure and Dielectric Properties
of $Pb(Sc_{\frac{2}{3}}W_{\frac{1}{3}})O_3$—$Pb(Ti/Zr)O_3$ Relaxors** 108
    P. Juhás, W. Dmowski, I. Grinberg, T. Egami, A. M. Rappe,
    and P. K. Davies

**Cation Ordering in Single Crystals of 1:1 and 1:2 Complex
Perovskite Solid Solutions** 118
    I. P. Raevski, S. A. Prosandeev, S. M. Emelyanov, V. G. Smotrakov,
    V. V. Eremkin, F. I. Savenko, I. N. Zakharchenko, E. S. Gagarina,
    O. A. Bunina, and E. V. Sahkar

**First Principles Investigation of Novel Ferroelectric Perovskite Alloys
Based on A-site Substitution** 124
    S. V. Halilov, M. Fornari, and D. J. Singh

Ab Initio Study of Silver Niobate 130
 I. Grinberg and A. M. Rappe

Off-Center Atomic Displacements in $BaTiO_3$ Quantum Dots 139
 H. Fu and L. Bellaiche

First Principles Calculations of Ionic Vibrational Frequencies
in $PbMg_{\frac{1}{3}}Nb_{\frac{2}{3}}O_3$ 146
 S. A. Prosandeev, E. Cockayne, and B. P. Burton

E-Field and Temperature Dependent Transformation
in <102>-Cut PMN-PT Crystal 152
 C.-S. Tu, L.-W. Huang, R. Chien, and V. H. Schmidt

Polarization Rotation and Monoclinic Phase in Relaxor Ferroelectric
PMN-PT Crystal 160
 V. H. Schmidt, R. Chien, I.-C. Shih, and C.-S. Tu

$PbTiO_3$ at Finite Temperature: An Ab-initio Molecular
Dynamics Study 168
 V. Srinivasan, R. Gebauer, R. Resta, and R. Car

Ferroelectric Instabilities and Self-Consistent Mechanism for
the Isotopic Substitution in KDP 176
 S. Koval, J. Kohanoff, R. L. Migoni, and E. Tosatti

First-Principles Calculations of $K_2 SeO_4$ Dielectrics 186
 R. Caracas and X. Gonze

Point Defects and Physical Properties of Ferroelectrics:
Lithium Niobate 196
 G. Malovichko, V. Grachev, and O. Schirmer

Quantum Chemical Modeling of Electron and Hole Polarons
in ABO_3 Perovskites 204
 R. I. Eglitis, E. A. Kotomin, G. Borstel, and V. S. Vikhnin

Calculations of Perovskite Polar Surface Structures 210
 E. Heifets, R. I. Eglitis, E. A. Kotomin, W. A. Goddard III, and G. Borstel

Extending First Principles Modeling with Crystal Chemistry:
A Bond-Valence Based Classical Potential 220
 V. R. Cooper, I. Grinberg, and A. M. Rappe

Large-Scale Quantum Chemical Modeling of the Phase Transitions
in KTN Solid Solutions 231
 R. I. Eglitis, D. Fuks, S. Dorfman, E. A. Kotomin, G. Borstel,
 and V. A. Trepakov

Progress in Quantum Monte Carlo Calculations of Perovskite
Transition Metal Oxides 241
 L. K. Wagner, P. Sen, and L. Mitas

Phase-Free Quantum Monte Carlo Method: Random Walks Using
General Basis Sets 251
 H. Krakauer and S. Zhang

Site-Specific X-ray Photoelectron Spectroscopy: A New Method to
Measure Partial Density of Valence States 261
 J. C. Woicik

Car-Parrinello Molecular Dynamics in a Finite Homogeneous Electric Field 269
    P. Umari and A. Pasquarello

First-Principles WDA Calculations for Ferroelectric Materials 276
    Z. Wu, R. E. Cohen, and D. J. Singh

Author Index 287

PREFACE

This volume contains papers presented at the Fundamental Physics of Ferroelectrics 2003 Workshop that took place in Williamsburg, VA on February 2-5, 2003. Since the first workshop in 1990, this annual meeting has become a traditional venue for discussion and debate of the physics of the ferroelectric, piezoelectric and dielectric behavior of materials. The reviewed papers appearing in this proceedings volume provide a snapshot of the current status of the field and illustrate the many exciting advances being made in the experimental and theoretical understanding of the physics of these challenging systems.

Ferroelectric materials serve many critical needs in modern technology, with applications as transducers, actuators, dielectrics, and nonvolatile memories. In addition, they present fundamental problems in the behavior of insulators in electric fields, spontaneous polarization, piezoelectricity, phase transitions, extreme sensitivity to temperature, composition, and pressure. Experimental and theoretical advances in the last decade have stimulated a major resurgence of interest in this classic problem of condensed matter physics. The vitality of the field was reflected by the large attendance at this meeting, the 13th in the series of the Williamsburg workshops, with over 85 attendees equally balanced between theory and experiment. Through the formal presentations and extensive informal discussions we all learned from each other and our knowledge of the field was deepened.

We thank the Office of Naval Research for supporting the workshop and the University of Pennsylvania for assistance in defraying the organizational costs of the meeting. Thanks also go to Tempie Hayes the Conference Manager at the Woodlands Conference Center for ensuring that every aspect of the meeting ran smoothly.

Peter K. Davies, David J. Singh.
May 2003.

Pressure As A Probe Of The Physics Of Compositionally-Substituted Quantum Paraelectrics: SrTiO$_3$

E. L. Venturini*, G. A. Samara*, M. Itoh[†], and W. Kleemann[¶]

*Nanostructure and Advanced Materials Chemistry, Sandia National Laboratories, Albuquerque, NM 87185-1421, USA
[†]Materials & Structures Laboratory, Tokyo Institute of Technology, 4259 Nagatsuta, Midori, Yokohama 226-8503, Japan
[¶]Laboratorium für Angewandte Physik, Gerhard-Mercator-Universität Duisburg, D-47048 Duisburg, Germany

Abstract. The influence of pressure on the dielectric properties and phase behavior was investigated for two substituted single crystals of SrTiO$_3$: (a) ^{18}O exchanged SrTi^{18}O$_3$ [STO-18] and (b) lightly doped Sr$_{1-x}$Ca$_x$TiO$_3$ with x = 0.007 [SCT (0.007)]. The ^{18}O atoms in STO-18 reduce both the quantum fluctuations of the TiO$_6$ octahedra and the frequency of the soft phonon mode, leading to a ferroelectric (FE) state via a first-order transition with T$_c$ ~ 24 K at 1 bar. T$_c$ decreases very rapidly under pressure with an initial slope of ~ 20 K/kbar, reaching the quantum displacive limit (T$_c$ = 0 K) near 0.7 kbar. In the case of SCT (0.007), fluctuating polar nanodomains surround the off-center Ca dopants and grow in size as the correlation length for dipolar interactions in the SrTiO$_3$ host increases with decreasing temperature. Ultimately, the fluctuations slow down and the nanodomains "freeze" into a relaxor state with T$_{max}$ ~ 18 K; pressure reduces T$_{max}$ with an initial slope of ~ 35 K/kbar. Both the FE transition in STO-18 and the relaxor state of SCT (0.007) are completely suppressed below 1 kbar and a quantum paraelectric state emerges.

INTRODUCTION

Chemical substitution in the classic quantum paraelectrics (QPE's) SrTiO$_3$ and KTaO$_3$ produces remarkable effects on the dielectric properties and phase behavior of these materials. These effects are subjects of considerable current interest. In this study we investigated the effects of hydrostatic pressure on the properties of two substituted SrTiO$_3$ crystals: (i) a nearly fully (> 97%) ^{18}O exchanged SrTiO$_3$ (STO-18) crystal, and (ii) a lightly doped Sr$_{1-x}$C$_x$TiO$_3$ (x = 0.007) [SCT (0.007)] crystal.

Isotopic substitution in ferroelectric (FE) crystals is known to produce significant changes in properties and has often led to a better understanding of the physics underlying the structural instability. Such substitution at the various sites in ABO$_3$ oxide ferroelectrics has resulted in relatively small changes in transition temperatures.[1] However, a significant advance has been the recent discovery by Itoh et al. that the substitution of ^{18}O for ^{16}O in SrTiO$_3$ induces ferroelectricity with a transition temperature, T$_c$, ≃ 24 K for the fully substituted ^{18}O crystal.[2] SrTiO$_3$ (with naturally occurring ^{16}O) is a classic incipient ferroelectric; its soft (FE) mode frequency decreases with decreasing temperature (T), but is stabilized at the lowest temperatures by quantum fluctuations. Consequently, the crystal does not undergo a FE transition and retains its tetragonal paraelectric (PE) structure down to the lowest temperatures. However, because of the delicate balance between the quantum fluctuations and the dipolar interactions, small perturbations, such as ^{18}O for ^{16}O exchange, can induce FE order. Convincing evidence for the ferroelectricity of $SrTi(^{16}O_{1-x}{}^{18}O_x)_3$ for $x \geq 0.33$ has come from dielectric, hysteresis loop, Raman and birefringence measurements.[2-4]

The $Sr_{1-x}Ca_xTiO_3$ system, in the limit of small x, has attracted much interest. The evidence suggests that the Ca^{2+} ions predominantly occupy off-center positions at the Sr^{2+} sites thereby producing random electric fields and strain coupling to the polarization.[5,6] Extensive dielectric and optical studies were performed on the specific composition (SCT (0.007) by Kleemann and co-workers.[5,7] The results suggest that SCT (0.007) exhibits an inhomogeneous polar low-temperature state with evidence for ordering at two length scales.[5] On the shorter length scale, quasi first-order Raman scattering by soft (TO_1) and hard (TO_2 and TO_4) polar modes gave evidence for Ca-induced FE microregions (FMRs), or polar nanodomains, whose size is determined by temperature-dependent correlation radius, r_c, and the crystal exhibits the dipolar glass response of a relaxor ferroelectric. The longer length scale ordering is deduced from field-dependent linear birefringence measurements and is attributed to the presence of large random-field correlated domains that relax independently and behave like switchable superparaelectric moments whose size increases with increasing temperature.[5]

Because the occurrence of FE transitions in the quantum regime is determined by a very delicate balance between competing interactions, the application of hydrostatic pressure can be expected to strongly influence the phase behavior and the dielectric properties and provide important insight into the physics. This expectation motivated the present work on the above two crystals. Indeed the effects of pressure are found to be very large, the FE phase of STO-18 and relaxor state of SCT (0.007) being completely suppressed by very modest pressures (< 1 kbar). In what follows we present, discuss and compare some of the main results on both crystals. More complete accounts of the work will be presented elsewhere.[6,8]

EXPERIMENTAL DETAILS

Both crystals were $(110)_c$ – oriented thin plates with the large faces parallel to the (110) twin-boundary plane (referenced to the high temperature cubic, c, phase). In this orientation the $SrTiO_3$ crystal remains a single domain upon transforming into its tetragonal phase below 105 K. The dielectric permittivity (both real and loss components) was measured as a function of temperature (300 K), hydrostatic pressure (0-6 kbar) and frequency (10^2 to 10^6 Hz). The ac electric field amplitude was below 1 V/cm to minimize nonlinearities in the dielectric response. Pressure was generated in a conventional high-pressure apparatus using gaseous and solid He as the pressure transmitting medium. The temperature dependence was measured during a slow drift of less than 1 K/minute.

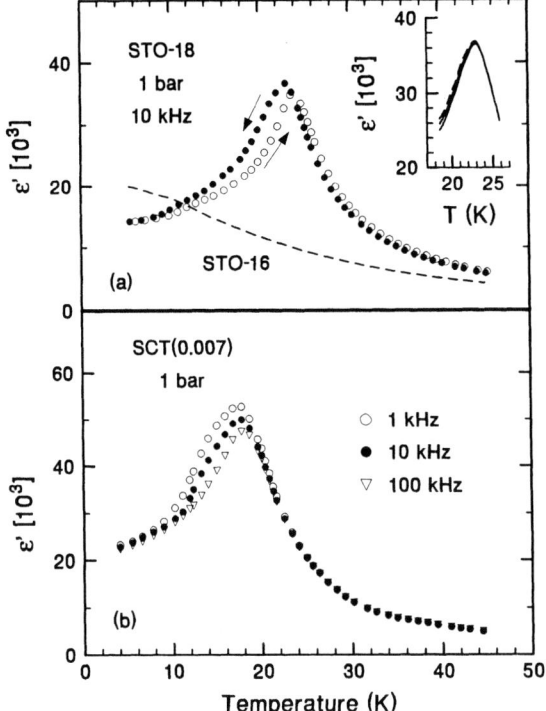

FIGURE 1. Temperature dependencies of the ac dielectric constants of STO-18 and SCT(0.007) at 1 bar. The ε'(T) response of STO-16 is also shown in (a). The inset in (a) contrasts the very weak frequency dispersion of STO-18 at 1, 10 and 100 kHz with that of SCT(0.007) in (b).

RESULTS AND DISCUSSION

The 1 Bar Dielectric Response

The features of the temperature (T) dependence of the real part of the dielectric constant (ϵ') of the STO-18 crystal at 1 bar are shown in Fig. 1a. Data are shown for both heating and cooling cycles at 10^4 Hz. The higher $\epsilon'(T)$ peak on cooling and the thermal hysteresis in the low T FE phase are the suggestive of a first-order phase transition. The data also reveal about a 0.5 K hysteresis in the high T PE phase. We believe this hysteresis (not seen in some other scans) is an artifact having to do with the placement of the thermocouple, which is not allowing it to track the temperature of the sample accurately. Correcting for this T offset by shifting the heating curve down by 0.25 K and cooling curve up by 0.25 K reveals a hysteresis of 1 K in the transition temperature T_c. The dashed line in Fig. 1 shows $\epsilon'(T)$ for STO-16 for comparison.

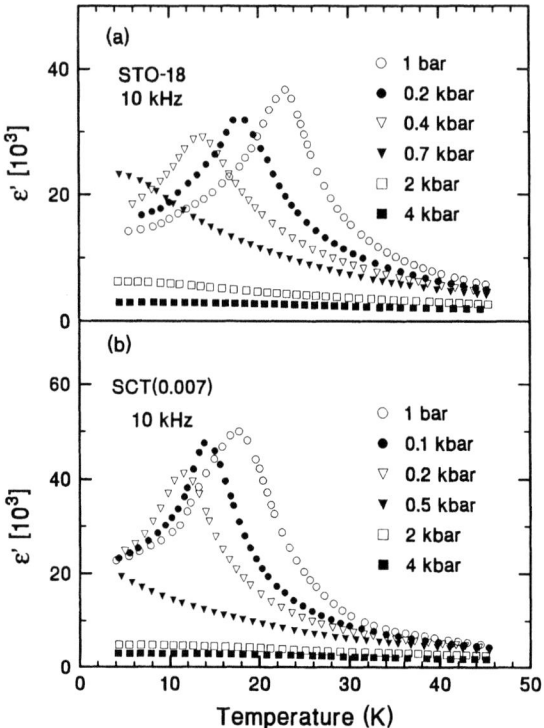

FIGURE 2. Influence of pressure on the ε'(T) responses of STO-18 and SCT(0.007) at 10 kHz.

The first-order nature of the transition has been predicted on the basis of a nonlinear electron-phonon interaction model.[9] However, recent measurements at 1 bar on another STO-18 crystal by Kleemann et al. do not show significant thermal hysteresis in T_c, but exhibit hysteresis in ε'(T) below T_c.[10] Thus, the thermodynamic order of the transition requires further study.

The inset in Fig. 1a shows the frequency dependence of the ε'(T) response. The very weak dispersion in ε'(T) in the FE phase, seen also by Itoh et al. [2], is not unusual in ferroelectrics and does not represent relaxor FE behavior, as will be discussed later. The rounded ε'(T) peak is a largely a manifestation of quantum fluctuations at low T's.

Fig. 1b shows the ε'(T) response of SCT (0.007) at 1 bar. For this crystal there is no thermal hysteresis, but there is strong frequency dispersion in ε'(T) at and below the peak temperature, T_{max}. These features are the signatures of a relaxor state, and they emphasize the qualitative difference in the responses of the two crystals in Fig. 1.

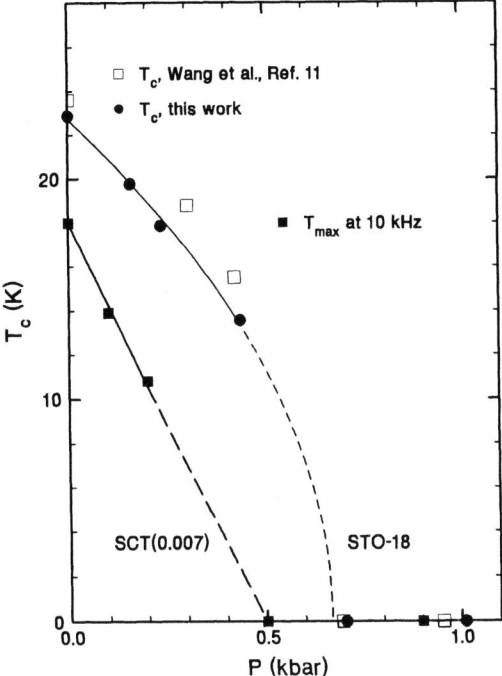

FIGURE 3. Temperature-Pressure phase diagrams of STO-18 and SCT(0.007).

Influence Of Pressure On The Dielectric Responses And Temperature-Pressure Phase Diagrams

Pressure has a very strong influence on the dielectric responses of both crystals. Fig. 2 shows that with increasing pressure there is a rapid shift of the transition temperatures to lower T's followed by the complete suppression of the FE phase of STO-18 below 0.7 kbar and of the relaxor state of SCT (0.007) below 0.5 kbar, and finally the evolution of a QPE state. The QPE state is characterized by a T-independent $\epsilon'(T)$ over a substantial T range at the lowest T's as shown for the 4 kbar data for STO-18 and SCT (0.007). STO-16 exhibits such a state below ~3 K at 1 bar. Accompanying the suppression of the relaxor state of SCT (0.007) is the vanishing of the frequency dispersion in $\epsilon'(T)$.

Similar results were obtained by Wang et al. using a clamp-type high pressure apparatus and a $(100)_c$ – oriented STO-18 crystal that becomes multi-domained below the antiferrodistortive transition at ~110 K.[11] This implies that the pressure response is not significantly affected by the domain structure. Additionally, the pressure dependence of T_c is not significantly affected by the pressure transmitting medium.

Fig. 3 shows important differences in the Temperature-Pressure phase diagrams of the two crystals. For STO-18, T_c decreases with an initial slope $dT_c/dP = -20$ K/kbar. The magnitude of the slope increases with increasing pressure with indication that T_c (dashed line) vanishes with infinite slope, i.e., $dT_c/dP \rightarrow -\infty$ as $T_c \rightarrow 0$ K. This is a strict requirement of the third law of thermodynamics for an equilibrium first- or second-order phase transition and is a reflection of the fact that the entropy difference between the two phases vanishes as $T_c \rightarrow 0$ K. The solid-dashed line for STO-18 in Fig. 3 is a fit of the equation $T_c = A \cdot (1 - P/P_c)^{1/2}$ to the $T_c(P)$ data (solid circles) obtained from $\epsilon'(T)$ versus decreasing T, where $A = 22.7$ K and $P_c = 0.67$ kbar. The form of this equation is predicted from theory for quantum ferroelectrics.[12] The 0.67 kbar value of P_c for this fit (where $T_c = 0$ K) is consistent with the data which show that the sample is in a QPE state at 0.7 kbar (Fig. 2a). Fig. 3 also shows close agreement between our results (solid circles) and Wang et al's data (open squares).

Fig. 2b shows that the SCT (0.007) crystal is in the paraelectric (PE) state at temperatures above 20 K for all pressures. At lower T's and low pressure the $\epsilon'(T)$ response is relaxor-like. Using T_{max} at 10 kHz versus pressure to separate the paraelectric and relaxor states, the relaxor state is confined to the triangular region from 0 to 18 K and between ambient and 0.5 kbar in Fig. 3. The results in Fig. 3 show that T_{max} decreases linearly with pressure at the remarkably large slope $dT_{max}/dP = -35$ K/kbar with the suggestion (dot-dash line) that $T_{max} \rightarrow 0$ K with a finite slope. Thus, the behavior in this case is qualitatively different from that for STO-18. The vanishing of the relaxor state with a finite slope, dT_{max}/dP, is a feature that we have observed earlier for both perovskites and H-bonded relaxors.[12] We believe that this feature is a manifestation of the non-equilibrium nature of the relaxor state, specifically residual configurational entropy in the relaxor state as $T_{max} \rightarrow 0$ K allows for a finite dT_{max}/dP.[12]

An additional feature emerges from the results. Defining the QPE state by the onset of a T-independent ϵ' at the lowest T's, the results in Fig. 2 show that the T-range of this state increases substantially with pressure – a feature that has become well established.[12]

Nature Of The Phase Transition: STO-18

Much of the evidence from the present as well as earlier work by Itoh et al. has indicated that highly substituted STO-18 exhibits on cooling a transition to a normal FE state. From a thermodynamic point of view, the vanishing of Tc with an infinite slope dTc/dx at a critical concentration,[13] or a critical pressure (Fig. 3) is indicative of an equilibrium phase transition, and, furthermore, the thermal hysteresis observed in the present work points to the first-order character of the transition. The evidence also points to the softening of the FE mode in the high T paraelectric phase with increasing ^{18}O substitution as the mechanism for the transition. In STO-16 this mode also softens with decreasing T, but ultimately quantum fluctuations prevent it from softening sufficiently to induce the transition. In STO-18 two effects due to the heavier mass of ^{18}O conspire to induce the transition: (i) additional softening of the FE mode in the tetragonal phase, and (ii) damping of the quantum fluctuations. And it is

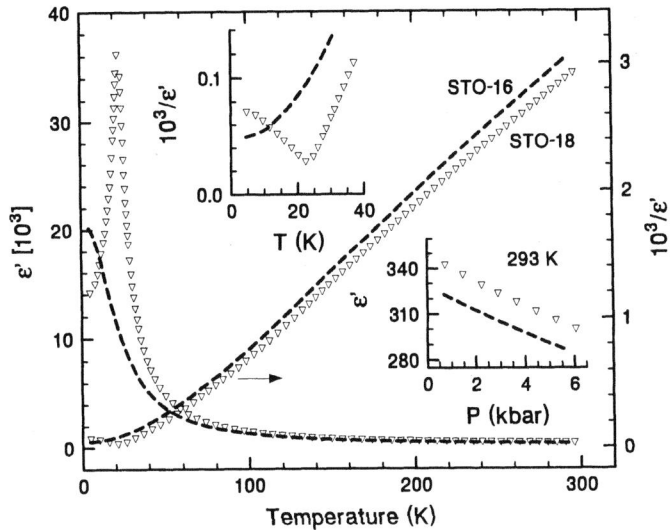

FIGURE 4. The ε'(T) responses of STO-18 and STO-16 at 1 bar over an extended range of T. Also shown is 1/ε'(T). The insets show an expanded view of ε'(T) at low T and the pressure dependence ε'(P) for the two crystals at 293 K.

only in the low T quantum regime where the characteristic energies of the system are so small that the small change in mass produced by ^{18}O substitution can produce such large effects. Thus, e.g., ^{18}O substitution in BaTiO$_3$ raises its FE transition temperature (393 K) by only ~ 0.9 K.[2]

The additional softening of the soft mode frequency due to ^{18}O substitution is reflected in the dielectric response shown in Fig. 4, which is a plot of ε'(T) and 1/ε'(T) at 1 bar over an extended temperature range for our STO-18 crystal as well as for STO-16. The data show that ε' for STO-18 is larger than that for STO-16, and the difference, of course, increases on approaching T$_c$. Recalling that ε' and ω$_s$, the frequency of the soft transverse optic (TO) phonon, are related by the Lyddane-Sachs-Teller relationship so that $\omega_s^2(T) \cdot \varepsilon'(T)$ is constant, the higher value of ε' for STO-18 reflects a lower ω$_s$ for this crystal – an expected effect due to the heavier mass of ^{18}O.

The FE soft mode (Slater mode) in SrTiO$_3$, a long wavelength TO phonon, consists primarily of vibrations of the Ti^{4+} ions against their surrounding oxygen octahedra. Clearly replacing ^{16}O by ^{18}O should reduce the frequency of this mode. Specifically, the ratio of the frequencies is related to the ratio of the effective masses (μ) of the Ti-O$_6$ octahedral units by $\omega_{18}/\omega_{16} = (\mu_{16}/\mu_{18})^{1/2}$. For the Slater mode it is readily shown that $\omega_{18}/\omega_{16} = 0.97$, i.e., a 3% decrease in ω$_s$ on complete ^{18}O substitution in SrTiO$_3$.

Our dielectric data accurately confirm this prediction. This is most clearly seen in the inset in Fig. 4, where ε' is plotted versus pressure at 293 K and the ε' curve for STO-18 is 6% higher than the STO-16 curve. Specifically at 1 bar (0 kbar), ε'_{18} = 348 and ε'_{16} = 328 so that $\varepsilon'_{18}/\varepsilon'_{16}$ = 1.06, i.e., a 6% enhancement which from ε' ∝

($1/\omega_s^2$) implies a 3% decrease in ω_s, as calculated and as recently confirmed by unpublished hyper-Raman scattering data.[14] It is this additional softening, that is further enhanced on approaching T_c (inset Fig. 4) that triggers the FE transition making it difficult for the disordering tendency of the ^{18}O-reduced quantum fluctuations to overcome the ordering tendency of the dipolar interactions. The suppression of T_c and ultimate vanishing of the FE phase with pressure result from the stiffening of ω_s – a well-established result.[15]

Nature Of The "Transition": SCT (0.007)

In the case of SCT (0.007), we envision each Ca^{2+} - induced dipolar entity surrounded by a polarized region of the soft $SrTiO_3$ host lattice, forming a nanodomain whose size is determined by the T-dependent correlation length, r_c, of the host. Because of the soft ferroelectric mode nature of the host, on cooling from high T these nanodomains grow as r_c increases, increasing their dynamic correlations.[6] However, for this dilute sample, these correlations do not become large enough to precipitate a global, long-range ordered FE state at low T. Rather, they exhibit a dynamic slowing down of their orientational motion resulting in the frequency dependent peak in $\epsilon'(T)$ – a dynamic glass-like transition temperature, and the frequency dispersion on the low-T side of the peak. At sufficiently low T all such motion freezes, and the dispersion vanishes as shown below ~ 4K (Figs. 1b and 2b).

The very low value of the ω_s of $SrTiO_3$ at low T's makes the polarizability (~ $1/\omega_s^2$) very large. This polarizability is further enhanced, and ω_s renormalized, by the presence of the polar nanodomains. The correlation length is proportional to the polarizability and, thus, varies as $1/\omega_s^2$. The increase in ω_s with pressure leads to a decrease in r_c and the effect is large near the transition – a result that is now well established for perovskites.[12] The decrease of r_c implies that the size of the nanodomains decreases with pressure. At sufficiently high pressure these domains become small enough so that there is no overlap or correlations among them. At this point the $\epsilon'(T)$ peak and relaxor state vanish, as observed. A more detailed and quantitative account of the physics is presented elsewhere.[6]

CONCLUDING REMARKS

The present results highlight the usefulness of pressure in the study of ferroelectric and relaxor properties. The occurrence of the FE transition in STO-18 and of the relaxor state in SCT (0.007) is determined by a very delicate balance between short- and long-range interactions. Pressure is an excellent variable for delicately tuning the balance between these competing interactions.[12,15] The unusually large effects of pressure for the present two crystals is a reflection of the fact that their "transitions" occur at very low temperatures where the characteristic energies are small, so that small perturbations cause large changes in properties.

ACKNOWLEDGEMENTS

The authors express their appreciation to David Lang for technical support in performing the dielectric measurements and to Dr. Ruiping Wang for her complementary earlier pressure studies on STO-18. The work at Sandia National Laboratories was supported by the Division of Materials Sciences and Engineering, Office of Basic Energy Sciences, U.S. Department of Energy. Sandia is a multiprogram laboratory operated by Sandia Corporation, a Lockheed Martin Company, for the U.S. Department of Energy under Contract No. DE-AC04-94AL85000.

REFERENCES

1. Hidaka, T., *Ferroelectrics* **137**, 291-295 (1992).
2. Itoh, M., Wang, R., Inaguma, Y., Yamaguchi, T., Shan, Y.-J., and Nakamura, T., *Phys. Rev. Lett.* **82**, 3540-3543 (1999).
3. Kasahara, M., Hasebe, H., Wang, R., Itoh, M., and Yagi, Y., *J. Phys. Soc. Jpn.* **70**, 648-651 (2001).
4. Yamanaka, K., Wang, R., Itoh, M., and Iio, K., *J. Phys. Soc. Jpn.* **70**, 3213-3216 (2001).
5. Bianchi, U., Dec., J., Kleemann, W., and Bednorz, J.G., *Phys. Rev. B* **51**, 8737-8746 (1995).
6. Venturini, E.L., Samara, G.A., and Kleemann, W., *Phys. Rev. B* (to appear).
7. Kleemann, W., Albertini, A., Kuss, M., and Lindner, R., *Ferroelectrics* **203**, 57-74 (1997).
8. Venturini, E.L., Samara, G.A., and Itoh, M. (to be published).
9. Bussmann-Holder, A., Büttner, H., and Bishop, A.R., *J. Phys.: Condens. Matter* **12**, L115-L120 (2000).
10. W. Kleemann, private communication (February, 2003).
11. Wang, R., Sakamoto, N., and Itoh, M., *Phys. Rev. B* **62**, R3577-R3580 (2000).
12. Samara, G.A., "Ferroelectricity Revisited – Advances in Materials and Physics," in *Solid State Physics*, edited by H. Ehrenreich and F. Spaepen, Academic Press, New York, 2001, Vol. **56**, pp. 239-483.
13. Wang, R., and Itoh, M., *Phys. Rev. B* **64**, 174104(6) (2001).
14. Itoh, M., private communication (January, 2003).
15. Samara, G.A., and Peercy, P.S., "The Study of Soft-Mode Transitions at High Pressure," in *Solid State Physics*, edited by H. Ehrenreich., F. Spaepen and D. Turnbull, Academic Press, New York, 1981, Vol. **36**, pp. 1-118 and references therein.

Temporal Effects in Dielectric Properties of Some Antiferroelectric Complex Perovskites

Wai-Hung Chan, Zheng Kui Xu, Jiwei Zhai, Haydn Chen

*Department of Physics and Materials Science
City University of Hong Kong, Kowloon, Hong Kong*

Eugene V. Colla

*Department of Physics
University of Illinois at Urbana-Champaign, Urbana, IL 61801, USA*

Abstract. Two complex perovskite antiferroelectric (AFE) systems were studied; they are $Pb_{0.97}La_{0.02}(Zr_{0.6}Sn_{0.3}Ti_{0.1})O_3$ (PLZST) in ceramic form and $Pb_{0.99}Nb_{0.02}(Zr_{0.82}Sn_{0.12}Ti_{0.04})_{0.98}O_3$ (PNZST) in thin film form. Dielectric and transmission electron microscopy (TEM) studies of the PLZST ceramics confirmed the AFE nature of the specimen, but at low temperature the electrical field application could induce a temporary ferroelectric (FE) phase with characteristic life-time dependent on temperature This life-time reaches a value of ~ 0.1s at 248K. Further decrease of temperature leads to a much faster than Arrhenius growth of the recovery time, which eventually exceeds the reasonable laboratory time scale below 210K. We believe that the strongly temperature-dependent kinetics arises from a cooperative freezing of the incommensurate AFE order in the presence of quenched composition disorder. The PNZST films' polarization properties are highly dependent on the film thickness and at the lowest studied limit (~170 nm) the field application in the surface normal direction can induce the FE phase with the time of recovery back to the AFE state being a magnitude of several hours. In this case the filed-induced FE ordering is very asymmetric with respect to the field direction, is thickness dependent and the FE phase can be induced if the time of field application is on scale of a second. The application of the field for a much shorter time (e.g. 1ms) does not significantly affect the AFE properties of the film. We suggest that the substrate/film interface could be responsible for this phenomenon.

INTRODUCTION

The antiferroelectrics are very promising materials for various applications and their fundamental physical properties are still not well understood. The important issue for the soft AFE materials is the relation between ferroelectric (FE and AFE ordering and the stability of the AFE phase. We present results in this study of two AFE systems to demonstrate the trend of the FE ordering after application of the DC bias. Despite some common features observed in both materials the nature of the temporary destruction by the DC bias of the AFE state in PLZST ceramics and in PNZST thin films is absolutely different. As will be made clear from the following discussion in the case of bulk PLZST the induced AFE-FE kinetic transition is based on the development of the intrinsic competition between AFE and FE ordering while for PNZST films properties are linked to the electrode-film interface.

EXPERIMENTAL DETAILS

Ceramic PLZST samples were prepared by the traditional solid-state reaction method [1, 2]. The PNZST thin films of different thickness were prepared by sol-gel processing [3]. Thin films were deposited by spin coating of each layer at 3000 rpm for 30s on the LaNiO$_3$ (LNO) buffered Pt/Ti/SiO$_2$/Si(100) substrates. The LNO layer serves two purposes: one as a conducting oxide bottom electrode and the second as a template layer promoting PNZST growth with highly (100) preferred orientation [3]. Each spin-on PNZST layer was heat treated at 500°C for 5 min. The spin coating and heat-treatment were repeated several times to obtain the expected film thickness. To form the pure perovskite phase a capping layer of 0.8M PbO precursor solution was added at the top of the film before the final annealing at 700°C for 30 minutes. For the electrical measurements the top gold electrodes of 200×200 μm^2 were deposited by DC sputtering. All P-E curves and time decay dependence of the polarization (P(t)) were obtained using a standard Precision Pro Radiant Technologies ferroelectric tester with a triangular (for P-E measurements) or single pulse (for P(t)) waveforms of various time periods. The Agilent 4284A LCR meter was used for dielectric susceptibility measurements. Measurements were done as a function of temperature in the range of 100K-560K, a Delta chamber (for bulk specimens) or a Sign atone Probe station with a temperature-controlled stage (for films) were used.

KINETIC PHASE TRANSFORMATION IN PLZST CERAMICS

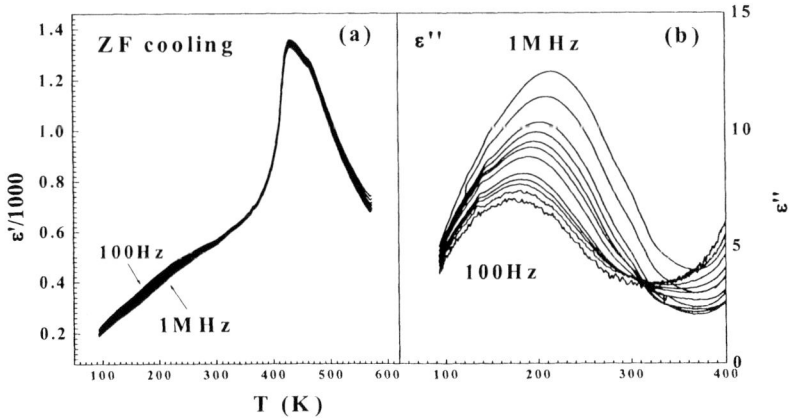

FIGURE 1. The real and imaginary parts of dielectric susceptibility of a bulk PLZST ceramic sample, measured on cooling in frequency range of 100Hz-500kHz.

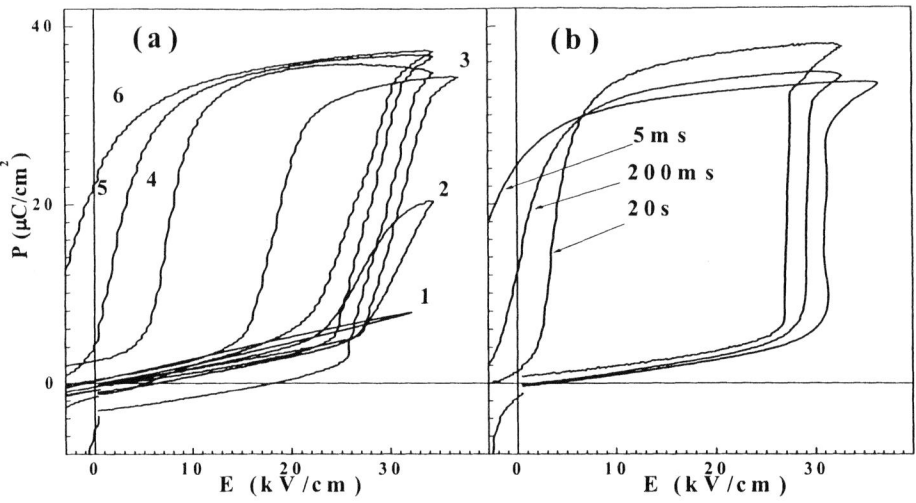

FIGURE 2. The P-E hysteresis loops observed on PLZST at different temperatures: (a) (1 - 493K, 2 - 373K, 3 - 294K, 4 – 255K, 5 – 234K, 6 – 218K). The scanning time for curve 1 was 5ms, for all others - 20s; and (b) at T=243K using different scanning time.

The temperature dependence of the dielectric constants (ε', ε'') is summarized in Fig. 1. The out-of-phase components of $\varepsilon''(T,f)$ reveal frequency dependent broad maximum in low temperature region. We will return to the discussion of $\varepsilon''(T,f)$ behavior later. The dielectric peak position and its magnitude, shown in Fig. 1(a), are in good agreement with literature data [4]. Below about 410K the AFE properties become apparent. This is evident from Fig. 2 where the P-E hysteresis loops measured at various temperatures are presented. The observed hysteresis loops at room temperature (RT) and above are typical for AFE materials and represent field forced parallel dipole alignment and the back switching with reducing field. However below RT we have observed some unusual behaviors of the P-E loops. Besides the frequency dispersion commonly seen in relaxor ferroelectrics, the AFE-FE switching field decreases with decreasing temperature (Fig.2 (a)) and the shape of the hysteresis loop becomes dependent on the time period of the driving voltage. This is seen in Fig. 2(b) where three P-E loops are taken at 243K but with different scanning time. The room temperature switching times have been reported for similar materials [5]; it is in the range of a few microseconds. However we have observed some unusual kinetic behaviors of the P-E loops at lower temperatures. The results presented in Fig. 2(b) point to the drastic increase of the FE-AFE back switching time that was found to be about a few seconds at 243K. To investigate the FE-AFE switching time dependence on temperature we performed the measurements of the polarization as a function of time. Some results are presented in Fig. 3 where we have carried out a series of kinetics experiments by monitoring the decay of the polarization (curve 2 in Fig. 3b) after driving the AFE system to the FE state with a sufficiently large DC pulse that exceeds the critical switching field E_{AFE-FE} (curve 1 in Fig 3(b)). Figure 3(b) shows the polarization decay (P(t)) observed at 236K along with the applied electrical field. A 100 ms wide pulse with amplitude of 45kV/cm was applied to the sample and it took

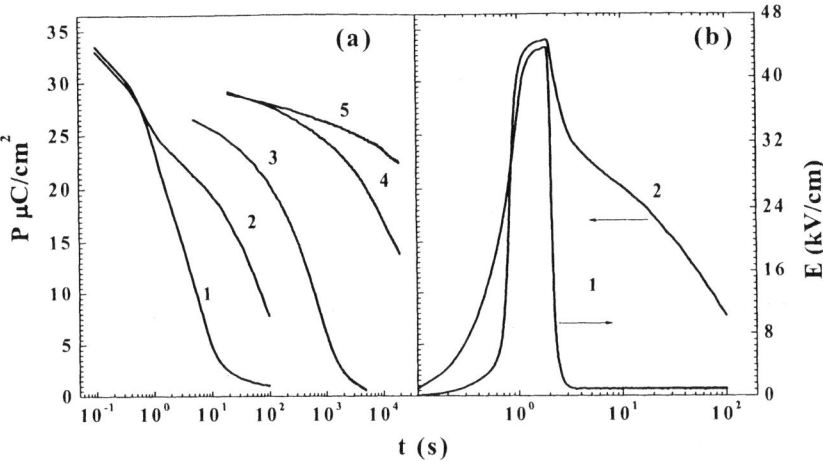

FIGURE 3. (a) The decay of the polarization, after application to the sample of a short pulse, measured at different temperatures: 1 – 245K, 2 – 236K, 3 – 233K, 4 – 223K, 5 – 218K. (b) The shape of pulse pattern: 1– applied to the sample, 2 – polarization response measured at 236K.

several seconds for the poled sample in the FE state to return back to the AFE state. The collection of several P(t) curves obtained at various temperatures is depicted in Fig. 3(a). It is clear that even a small decrease in T creates a huge rise of the recovery time. To analyze the polarization decay we have fitted the experimental data to a stretched exponential function typical of the relaxation process in disordered systems [6,7]:

$$P = P_o \exp\left(-\left(\frac{t}{\tau}\right)^\beta\right) \qquad (1)$$

where P_0 is the initial polarization of the field-induced FE state, τ the characteristic time of the relaxation back to AFE state. Within the temperature range studied, between 218 and 265K, the exponent factor β was found to be 0.73±0.19.

The measured relaxation times at various temperatures, $\tau(T)$, have been fitted (see Fig. 4) to an empirical Vogel-Fulcher relation:

$$\tau = \tau_0 \exp\left[\frac{E_a}{T - T_{VF}}\right] \qquad (2)$$

where τ_0 is the pre-exponential constant, E_a is the activation energy, and T_{VF} the critical temperature. The fitting results revealed the following set of parameters: T_{VF} =136K, E_a = 0.34eV, and τ_0 = 3.1 x 10^{-18}. Similar results were obtained from the

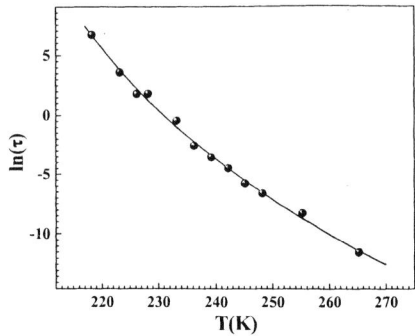

FIGURE 4. Vogel-Fulcher fitting of the time delay data

fitting of the measured frequency f vs. T_{max} (the temperature at which ε'' reaches maximum) data taken from $\varepsilon''(T,f)$ curves set (Fig. 1(b)).

Despite the observed trend to the transition to the FE state our dielectric data provide no evidence of any change of the ground state at the lowest studied temperature (126K); the ground state remains antiferroelectric. This is illustrated in Fig. 5 where two hysteresis loops are presented: (1) the first loop taken from a virgin sample (i.e. the sample was cooled down to 150K in zero field) and (2) a second loop immediately taken after the first. The first P-E loop starts from the origin and has a small slope (dP/dE) typical of AFE behavior. At E~60 kV/cm a step-like transition from the AFE state to the field-induced FE state takes place. Afterward the system displays the "classical" ferroelectric hysteresis loop in (2). This induced FE state is metastable, but for this particular temperature the recovery or the relaxation time τ is

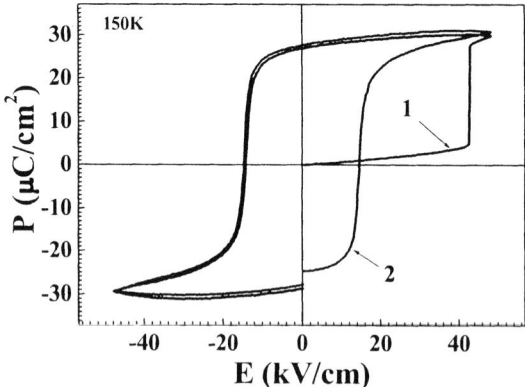

FIGURE 5. Two consecutively taken hysteresis loops at 150 K: Loop 1 – at virgin sample (cooled in zero field); Loop 2 – the second run. The period of driving frequency was 1 s.

far longer than any reasonable laboratory time scale. At higher temperatures we have observed the similar P-E loops (virgin and the second) but the relaxation time becomes shorter.

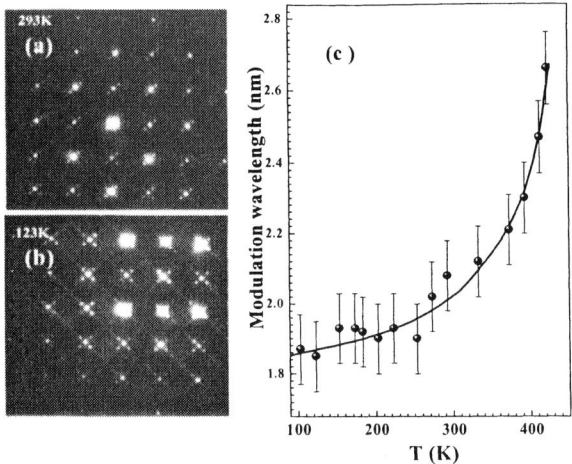

FIGURE 6. SAED patters taken at (a) 293 K and (b) 123 K) from a PLZST sample. Fig. (c) shows the modulation wavelength of the AFE superstructure as a function of temperature

The low temperature TEM study provides the second argument in favor of the AFE nature of the ground state in PLZST. In Fig. 6 the selected area electron diffraction (SAED) patterns are presented. It is clearly seen that the superlattice spots, characteristic of the incommensurate AFE state [8], exist along the diagonal <110> directions in the reciprocal space flanking the fundamental Bragg peaks. They gain strength in intensity at 123K (Fig. 6(b)) when comparing with the RT data (Fig. 6(a)). The modulation wavelength can be determined from the separation distance between the superlattice spots and the Bragg peaks and they are shown in Fig. 6(c) as a function of temperature, which diverges near the transition temperature of ~410K.

TEMPORAL EFFECTS IN PNZST THIN FILMS

The second example of the electrical field-induced temporary AFE-FE transformation was observed on PNZST thin films. The electrical field dependence of the dielectric constant (ε') obtained from the PNZST films (grown on LNO-buffered silicon substrate) of various thicknesses is shown in Fig.7. The measurements were performed by biasing the sample with a bipolar triangular voltage of 650s period and simultaneously measuring the capacitance with an LCR meter. Each panel of graph in Fig.7 represents two consecutive cycles of ε'-E measurements. The solid line depicts the first cycle, while the second by the dashed line. All measurements were made by starting with the positive bias to a level that the AFE-FE switching reached completion termed "poling". Thereafter the DC bias was reversed towards the negative field. Consequently two peaks are present in the positive half period of the scan signifying the AFE-FE forward switching and the FE-AFE back switching. Two more peaks of similar nature appear in the negative half period of the scan.

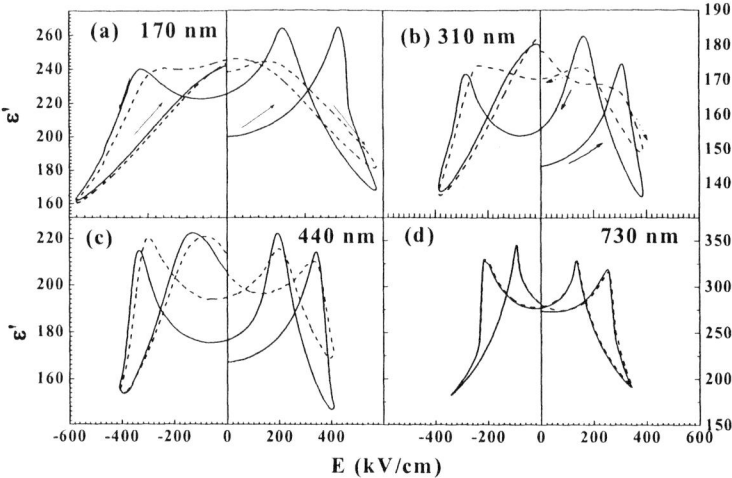

FIGURE 7. ε'-E characteristics of PNZST thin films deposited on LNO substrates with thickness of (a) 170 nm, (b) 310 nm, (c) 440 nm, and (d) 730 nm.

The difference between the first and the second curves for the 730nm-thick film (Fig. 7(d)) is negligible, but with decreasing film thickness the second scan curve shows diminishing AFE features. The FE-AFE back switching becomes less and less pronounced and in the case of 170nm film (Fig. 7(a)) the second curve does not show the peaks corresponding the backward FE-AFE switching. The third and all subsequent scans show nearly identical profiles to the second curve. It is clear from Figures 7 that once the FE state is established after which point the return to the AFE order is retarded or even destroyed for sufficiently thin film. Excursion first to the saturation bias destroys the AFE order but creates a FE order, which does not disappear even after removing the field (no second peak was present while the experiment started with positive bias). This asymmetry in poling properties was confirmed by a special experiment where the scanning started in the negative direction for which no FE-AFE switching was observed even in the first run.

Similar results were obtained in P-E experiments. Shown in Fig. 8 are two P-E hysteresis loops observed on the 440nm-thick PNZST films. The first curve (solid line) corresponds to the fast measurement (1 ms) done on the virgin sample. The second curve (dashed line) was taken after the initial C-V experiment was done and the sample was exposed to the electrical field for more than 10min time. It can be concluded that the time (τ_1) necessary to destroy the AFE ordering by the electrical field is of the order of seconds. The hysteresis loops shown in Fig. 8 were obtained using Radiant Pro equipment with the scanning time of 1ms and all experimental observations confirm that this time is too short to break the AFE alignment in the sample. But after the initial C-V test where poling has taken effect, AFE order is obviously destroyed to the extent that the subsequent P-E curve does not show the characteristic double hysteresis loops.

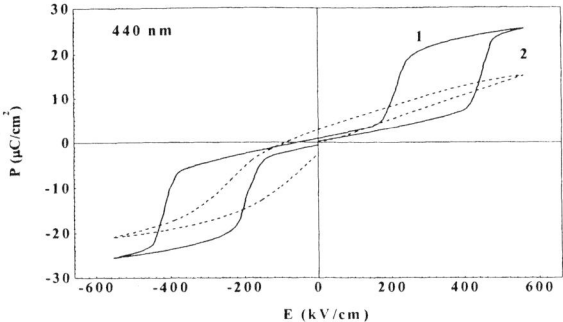

FIGURE 8. Hysteresis loops of a 440 nm thick PNZST thin film in (a) the virgin condition and (b) after electrical field application longer than the time duration of τ_1 (~1s).

The stability of the poled state was monitored by studying the relaxation behavior from the poled FE state back to AFE ordering. This is illustrated in Fig. 9 where the "positive" branches of the double hysteresis loops taken in different time intervals after the first full cycle are displayed. It was found that the electrical field-induced state is metastable and it relaxes back to the original AFE phase over time. Characteristic time (τ_2) of this relaxation is much longer than τ_1. For reference, curve 0 corresponds to the virgin sample (not exposed to negative bias) curves 1-5 represent

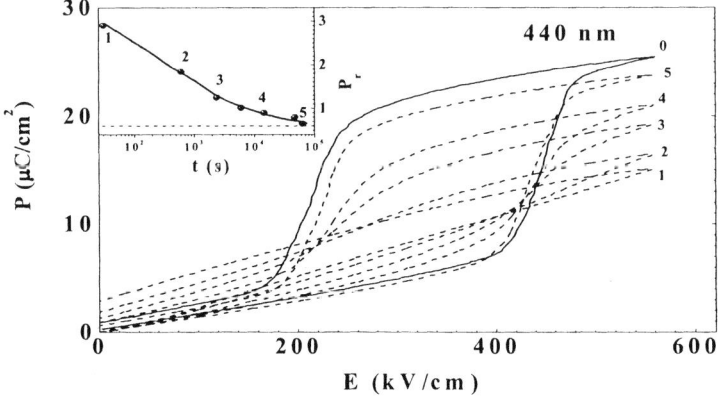

FIGURE 9. Relaxation to the original AFE state after the initial DC field poling is observed by the restoration of hysteresis the P-E loop. The solid line represents measurements done on the virgin sample, whereas dashed lines record data taken at different times after the initial poling cycle. In the insert, the time dependence of the field induced remanent polarization is shown.

the evolution of the hysteresis loop after the sample was subjected to a negative bias for the time longer than τ_1. In the insert to Fig.9 the time dependence of residual polarization (i.e. P(E=0, t)) is shown. The horizontal axis marks the moments when the loops 1-5 were recorded. From Fig. 9 it is evident that the sample eventually returns to its original state. The total recovery time exceeds 10^5 s at room temperature for this particular film thickness. The dashed line in the insert to Fig. 9 corresponds to the initial remanent polarization of the non-poled sample. The sample can be recovered much faster by giving it an annealing at temperature over the critical temperature (~500K) – the temperature above which the system becomes paraelectric.

All experimental results obtained on PNZST films confirm that observed phenomenon of the temporary destruction of the AFE state in PNZST films by the DC field is linked to the properties at the film/substrate interface layer. This comes from the thickness dependence of the AFE stability of after field application. Thicker films are much less susceptible to destruction. The asymmetry in the result of the field application is also an argument in favor of this issue. In order to explain the above experimental results, we have employed a simple diffusion model to explain the possible relaxation processes taking place in the interface layer. As it follows from presented experimental results (original hysteresis loops are rather symmetric: Fig. 7) the only negative bias creates the destruction of the AFE state, and that is why it is very unlikely that there exists strong intrinsic field responsible for self-polarization [9,10]. On the other hand after the application of negative field the P-E curve becomes asymmetric (curve 2 in Fig. 8). From curve 1 in Fig. 2 it is evident that after AFE-FE switching the high polarization phase is stable until the electrical field is over ~200 kV/cm. It means that if after the field application the film looses the AFE features we can expect the existence in the film the induced electrical field of comparable magnitude. The saturation polarization in this case (curve 2 in Fig. 8) is smaller than in virgin sample (curve 1 in Fig. 8) and the new state is expected to be a disordered FE in a strong induced field. Indeed, in the case of the negative bias (the external applied field is in parallel with the induced one) the saturation polarization is bigger and the loop shape (negative branch of curve 2 in Fig. 8) resembles the typical FE hysteresis curve. For positive branch the external field works against the induced intrinsic polarization and the loop shape is much slimmer and not saturated. Both observed processes (AFE phase destruction by the electrical field and the backward recovery to the AFE state) could be characterized by two time constants (τ_1 ~1s and τ_2 ~ 10^3-10^4s). These two temporal effects could be explained in terms of ion diffusion in the interface region.

If D is the diffusivity and L the diffusion distance (in our case L can be assumed equal film thickness) τ_2 can be expressed as

$$\tau_2 \approx \frac{L^2}{D} \tag{3}$$

For diffusion induced by the electrical field, E, τ_1 may be estimated as $\tau_1 \sim \frac{L}{V_f}$, in which V_f is the particles flow velocity that can be estimated from Einstein relation between the diffusivity and mobility [11]: $V_f \approx Eq\mu$, here q is the diffusing particle charge and μ the mobility. Taking into account that $\mu = \dfrac{D}{kT}$ we can find the following expression for τ_1:

$$\tau_1 = \frac{L \cdot kT}{EqD} = \frac{\tau_2 \cdot kT}{qUb}, \qquad (4)$$

where U_b is bias voltage. Using realistic values for $U_b{\sim}30$V and $kT{\sim}25$meV (room temperature) even for $q=1$ the ratio $\dfrac{\tau_2}{\tau_1} \sim 1200$ and this is very close to the observed time constants. Consequently the diffusion driven relaxation effect is a plausible explanation to the observed temporal effect in PNZST thin films.

Conclusions

The presented experimental results for two complex perovskite compositions reveal the instability of the AFE state in respect of the DC field application. In some sense both materials have demonstrated similar properties manifested in temporary destruction of the AFE ordering. Nevertheless very different physical phenomena are responsible for these kinetic effects. In the case of bulk PLZST we have observed the glasslike freezing and this is the intrinsic property of the material, whereas for the PNZST thin film properties are based on the film/substrate interface effects.

Acknowledgments

This research was fully supported by grants from the City University of Hong Kong under the project numbers of 7001237 and 9380015. We are gratefully to Prof. M. B. Weissman of University of Illinois at Urbana-Champaign for fruitful discussions. The technical assistance of Daniel Yau is acknowledged.

References

1. Jaffe, B., Cook, Jr., W.R. and Jaffe, H., *Piezoelectric ceramic*, Academic Press London and NY, 1971, pp. 253-270.
2. Wai-Hung Chan, Haydn Chen, Eugene V. Colla, *Appl. Phys. Lett.* (2003) In print.
3. Zhai, J., Cheung, M. H, Xu, Z.K., Li, X., and Chen, H., *Appl. Phys. Lett.* **81**, 3621, (2002).
4. K. Markowski, S.-E. Park, S. Yoshikawa, and E.L. Cross, *J. Amer. Ceram. Soc.*, **79**, 3297, (1996).
5. W. Y. Pan, C. Q. Dam, Q. M. Zhang, and L. E. Cross, *J. Appl. Phys.* **66**, 6014, (1989).
6. J.C Phillips, *Rep. Prog. Phys.* **59**, 1133, (1996)
7. C.H. Chou, M. H. Tu, and C.L. Wu, *Chin. J. Phys.*, **34**, 143, (1996).
8. Z. Xu, D. Viehland, P. Yang, D.A. Payne, *J. Appl. Phys*, 74, 3406, (1993)
9. I. P. Pronin, E. Yu. Kaptelov, E. A. Tarakanov, T. A. Shaplygina, V. P. Afanasjev, and A. V. Pankrashkin, *Physics of the Solid State*, **44**, 769 (2002).
10. V. P. Afanasjev, A. A. Petrov, I. P. Pronin, E. A. Tarakanov, E. Ju Kaptelov and J. Graul, *J. Phys.: Condens. Matter* **13**, 8755 (2001).
11. C. Kittel, H. Kroemer, Thermal Physics, W.H. Freeman and Company, San Francisco (1980)

Disorder in BaTiO$_3$ and SrTiO$_3$ and the "Ferroelectric" Transition in SrTi^{18}O$_3$

Robert Blinc*, Boštjan Zalar*, Andrija Lebar* and Mitsuru Itoh[†]

*J. Stefan Institute, Jamova 39, 1000 Ljubljana, Slovenia
[†]Materials and Structures Laboratory, Tokyo Institute of Technology, 4259 Nagatsuta, Midori, Yokohama 226-8503, Japan

Abstract. A quadrupole coupling induced ^{47}Ti and ^{49}Ti satellite background is observed in the cubic phases of BaTiO$_3$, SrTiO$_3$ and oxygen 18 enriched SrTi^{18}O$_3$. The satellite background demonstrates that all Ti ions are dynamically disordered between several off-center sites. No splitting or broadening of the central Ti $1/2 \rightarrow -1/2$ NMR transitions is seen on going through the "ferroelectric" transition in SrTi^{18}O$_3$ at $T_c = 25$ K. It seems that in contrast to BaTiO$_3$ the Ti disorder is still present.

I. Introduction

The phase transitions in BaTiO$_3$ and SrTiO$_3$ are generally considered to be classical examples of displacive soft mode type phase transitions [1, 2] describable by anharmonic lattice dynamics [3]. The question of the possible coexistence of a displacive and order-disorder component in the phase transition mechanism is however still open [4, 5, 6, 7]. The problem to be solved is whether the potential for the Ti motion in the cubic paraelectric phase exhibits a minimum at the center of the oxygen cage or whether the Ti ion is disordered between several off-center sites. Since the electric field gradient (EFG) tensor is zero by symmetry at the central position and non-zero at the off-center sites, the above problem could be definitely solved by quadrupole perturbed Ti NMR. Here we report on the results of such a study. Still another problem is the nature of the ferroelectric transition in ^{18}O enriched SrTiO$_3$ around $T_c = 25$ K recently discovered by Itoh [8]. Here we also present the first NMR study of this transition.

II. BaTiO$_3$

In a recent letter we presented the first NMR observation of quadrupole coupling induced ^{47}Ti and ^{49}Ti satellites in the cubic phase of an ultrapure BaTiO$_3$ single crystal above the ferroelectric transition [9]. The presence of this satellite background which exists in addition to the central $-1/2 \leftrightarrow 1/2$ Ti NMR transitions and its angular dependence demonstrate a dynamic local tetragonal breaking of the cubic symmetry. Such a behavior is indeed expected if an order-disorder component is present in the phase transition mechanism. Below T_c the satellite background transforms into well-defined satellite lines. The results show that in the paraelectric phase near T_c all Ti ions are dis-

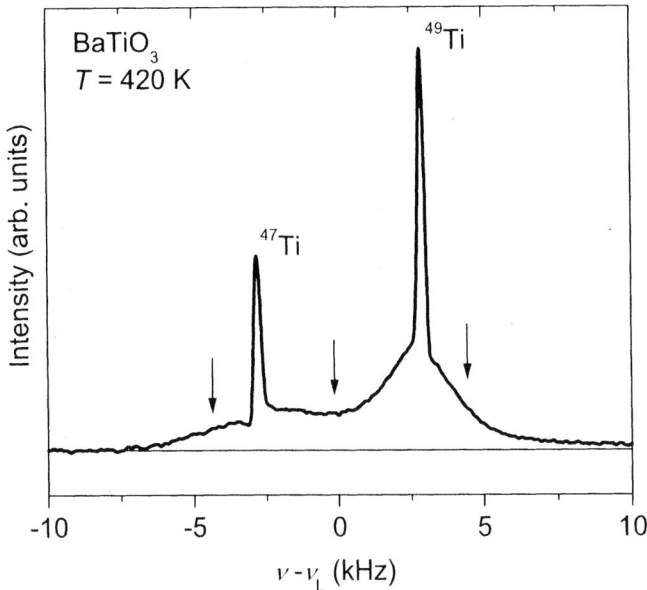

FIGURE 1. Ti NMR spectra of BaTiO$_3$ measured at the orientation **B**∥[001]. The two sharp ^{47,49}Ti $-1/2 \leftrightarrow 1/2$ central lines are superimposed on broad background components originating from unresolved satellite transitions (see arrows). The presence of the satellite background demonstrates that the Ti ions are sitting at off-center sites and not at the central position where quadrupole coupling is zero by symmetry.

ordered between several off-center sites and order into well defined positions below T_c. This agrees with the early view of Slater of a "rattling" Ti ion in BaTiO$_3$ as well as with the original interpretation of the diffuse X-ray scattering [4].

The ^{47}Ti ($I = 5/2$) and ^{49}Ti ($I = 7/2$) NMR spectra and relaxation times were measured in a magnetic field of 9.2 T corresponding to a Larmor frequency of 21.42 MHz for the two Ti isotopes. The ratio of the quadrupole moments of the two isotopes is $^{49}Q/^{47}Q = 0.819$ and the difference in their Larmor frequencies in the cubic phase is about 5.7 kHz. As their natural abundances are relatively low (^{47}Ti: $c = 7.3$ %, ^{49}Ti: $c = 5.5$ %) between 6000 and 20000 scans were collected for the spectra in the paraelectric ($Pm\bar{3}m$) and the tetragonal ferroelectric ($P4mm$) phases, respectively. A two 90°-pulse, four-phase (XX, XY, X-X, X-Y) "exorcycle" pulse sequence was used.

The ^{47,49}Ti NMR spectra of BaTiO$_3$ at 420 K (i.e. for $T > T_c$) are shown in Fig. 1. As expected for the paraelectric cubic phase, the ^{47}Ti and ^{49}Ti NMR spectra of BaTiO$_3$ show two sharp lines belonging to the $-1/2 \leftrightarrow 1/2$ magnetic transitions of the ^{47}Ti and ^{49}Ti isotopes. The two sharp lines are separated by about 5.7 kHz due to the different values of the magnetic moments of the two isotopes. The new and unexpected feature is that each of these two sharp lines "sits" on a broad background component. The integral intensities of the two central sharp lines and the broad components are in the ratio $9/26 = 0.35$ and $8/34 = 0.24$, respectively, as indeed predicted for the ratio of the intensities $|\langle m|I_x|m+1\rangle|^2$ of the central component and the satellites in the ^{47}Ti ($I = 5/2$) and ^{49}Ti ($I = 7/2$) spectra. This also demonstrates that all Ti ions are off-center so that the system is spatially homogeneous and there are no undistorted cubic regions. Below $T_c = 415$ K the broad components disappear and are replaced by well resolved satellites revealing the existence of three physically nonequivalent 90° ferroelectric domains. Instead of two we now have four narrow $-1/2 \leftrightarrow 1/2$ ^{47,49}Ti lines due to 90° degree domains. The main results of the ^{47,49}Ti NMR study of the ferroelectric transition in BaTiO$_3$ at $T_c = 415$ K can be summarized as follows:

(i) On going through $T_c = 415$ K the Ti central $1/2 \rightarrow -1/2$ transition splits - instead of two we now see four central lines.

(ii) From the angular dependence of the satellites we find that the ^{49}Ti quadrupole frequency $v_Q = 3e^2qQ/[h \cdot 2I(2I-1)]$ is

$$v_Q(T > T_c) = 1.3 \text{ kHz} \tag{1}$$

whereas
$$v_Q(T < T_c) = 125 \text{ kHz.} \tag{2}$$

(iii) The splitting of the lines below T_c is due to the appearance of 90° ferroelectric domains.

(iv) The appearance of satellites in the cubic phase demonstrates the existence of off-center Ti sites.

(v) The T_2 spin-spin nuclear magnetization relaxation time values demonstrate that the Ti disorder above T_c is dynamic and not static.

Let us now try to relate the above data to the phase transition mechanism in BaTiO$_3$. According to the hypothesis of Comes et al. [4] - later investigated in detail by Chaves et al. [7] - the Ti ion is at all temperatures above T_c located at one of the eight positions shifted from the center along the body diagonals of the unit cell [10]. In view of the 180° symmetry of the EFG tensor, only four of these EFG tensor orientations are distinguishable by Ti NMR. In the cubic phase all eight sites are equally populated, $n_1 = n_2 = n_3 = n_4 = n_5 = n_6 = n_7 = n_8$. In the tetragonal phase we have in a given 90° domain $n_1 = n_2 = n_3 = n_4 \neq n_5 = n_6 = n_7 = n_8$ so that the Ti ion moves preferentially between the four sites in the x-y plane leading to an effective displacement in the z direction. This is true for one 90° domain. All three 90° domains together then result in effective Ti displacements along the three cubic unit cell edges ($N = 3$) as indeed observed. In each of these 90° domains the Ti EFG tensor is a dynamic average over the four off-center sites located at the body diagonals. In contrast to NMR, XAFS measurements [10] show the presence of rhombohedral deviations from cubic symmetry. The difference is due to the different averaging times of the two methods (typically 10^{-6}s for NMR and 10^{-15}s for XAFS). It should be stressed that both the NMR and the XAFS results can be only understood if one deals with well-defined off-center sites and not only with a shallow Ti potential.

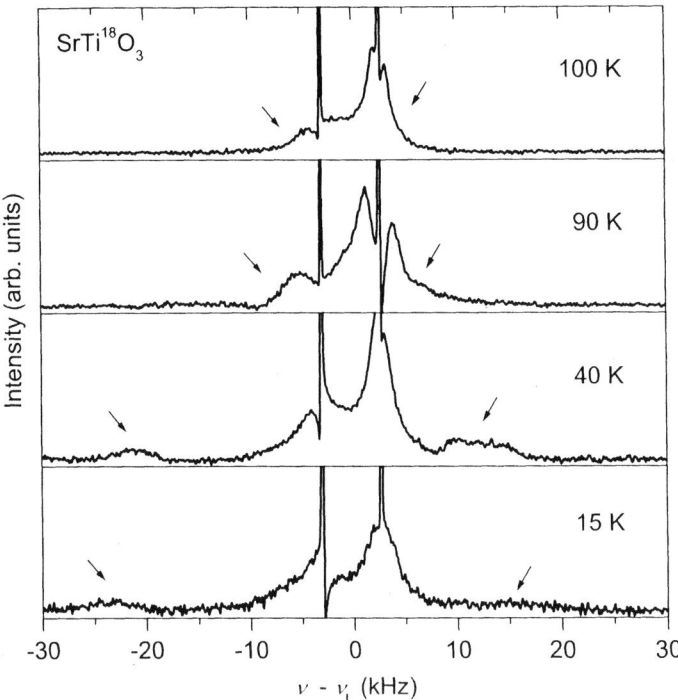

FIGURE 2. Ti NMR spectra in SrTi^{18}O$_3$ at T = 100K, 90K, 40K, and 15K. Resolved satellites are denoted by arrows.

III. SrTi(^{16}O$_{1-x}$^{18}O$_x$)$_3$

SrTi^{16}O$_3$ (STO 16) undergoes an antiferrodistortive phase transition at about T_a = 105K from the cubic $Pm\bar{3}m(O_h^1)$ to the tetragonal I $4/mcm$ (D$_{4h}^{18}$) phase. The material is considered to be a quantum paraelectric where the ferroelectric phase is suppressed by zero-point quantum fluctuations [11].

Itoh et. al. [8] recently succeeded in inducing a presumably ferroelectric phase by substituting ^{18}O for ^{16}O. The maximum in the dielectric constant appears at

$$T_c = 30.4(x - 0.33)^{1/2} \text{ K} \tag{3}$$

i. e. around 25 K for $x = 0.95$.

If there would be a macroscopic symmetry change at T_c connected with long range order and/or the crystal would form well defined 90° (or other) domains, the central $1/2 \rightarrow -1/2$ line would split as at $T_c = 415$ K in BaTiO$_3$. If below T_c there would be polar nanodomains as in relaxors, the central $1/2 \rightarrow -1/2$ line would broaden and

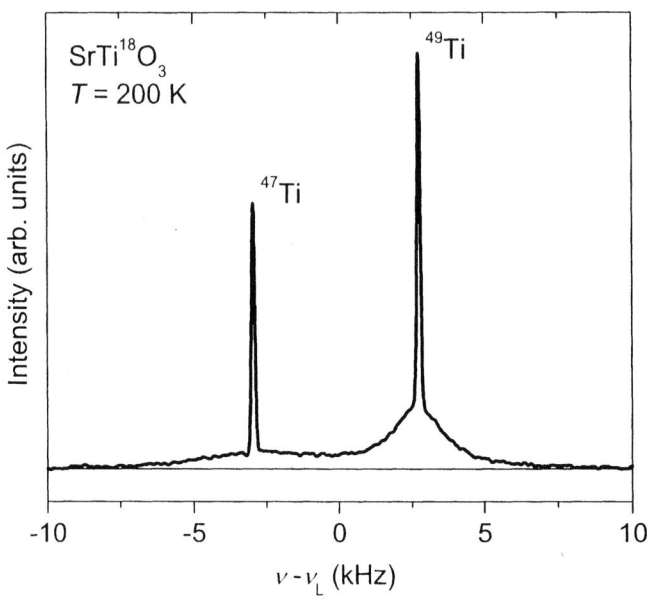

FIGURE 3. Ti spectrum in SrTi^{18}O$_3$ at $T = 200$ K $> T_c$ and orientation **B**∥[001].

show a Gaussian lineshape as in PMN. As it can be seen from Fig. 2 the ^{47}Ti and ^{49}Ti $1/2 \to -1/2$ lines do not split or broaden on going through T_c.

From the obtained ^{47,49}Ti spectra the following conclusions can be made:

(i) The existence of unresolved first order quadrupole satellites in the ^{47,49}Ti NMR spectra in the cubic phase of SrTi^{16}O$_3$ and SrTi^{18}O$_3$ demonstrates that the EFG tensor at the Ti sites is non-zero (Fig. 3) in analogy to the case of BaTiO$_3$.

(ii) This means that the Ti ions in SrTiO$_3$ are not at the centre of the oxygen cage but are disordered between several off-center sites.

(iii) The angular dependence of the second moment M_2 shows that the deviation from cubic symmetry is of tetragonal nature (Fig. 4).

The value of T_2 accounts for the width of the first order satellites and demonstrates that the Ti site disorder is dynamic rather that static.

It seems that the spontaneous polarization in a unit cell is so small (i.e. an order of magnitude smaller than in BaTiO$_3$) that no Ti $1/2 \to -1/2$ line splitting can be observed on cooling through T_c.

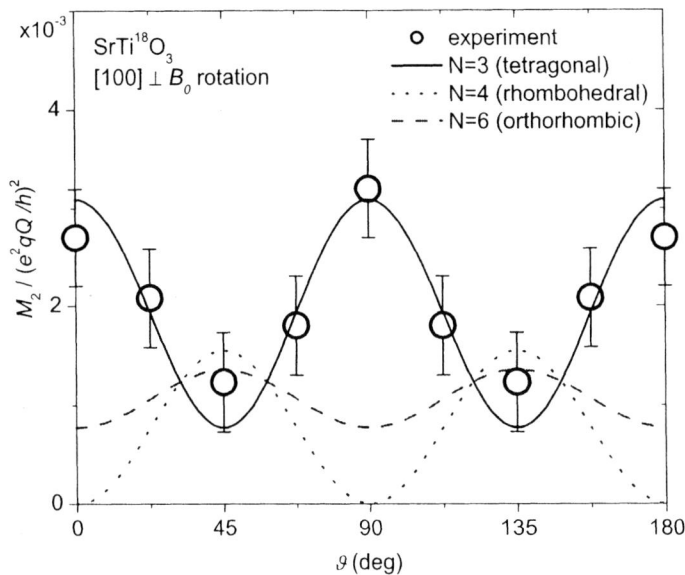

FIGURE 4. Comparison between the experimental and theoretical angular dependencies of the second moment of the unresolved ^{49}Ti satellite background.

REFERENCES

1. Cochran, W., *Phys. Rev. Lett.* **3**, 412-414 (1959).
2. Harada, J., Axe, J. D., and Shirane, G., *Phys. Rev. B* **4**, 155-162 (1971).
3. Migoni, R., Bilz, H., and Bäuerle, D., *Phys. Rev. Lett.* **37**, 1155-1158 (1976).
4. Comes, R., Lambert, M., and Guinier, A., *Solid State Commun.* **6**, 715-719 (1968).
5. Cochran, W., *Phys. Status Solidi* **30**, K157-K160 (1968).
6. Stachiotti, M., Dobry, A., Migoni, R., and Bussmann-Holder, A., *Phys. Rev. B* **47**, 2473-2479 (1993); Bussmann-Holder, A., and Bishop, A. R., *Phys. Rev. B* **56**, 5297-5301 (1997).
7. Chaves, A. S., Barreto, F. C. S., Nogueira, R. A., and Zeks, B., *Phys. Rev. B* **13**, 207-212 (1976).
8. Itoh, M., Wang, R., Inaguma, Y., Yamaguchi, T., Shan, Y. J., and Nakamura, T., *Phys. Rev. Lett.* **82**, 3540-3543 (1999).
9. Zalar, B., Laguta, V., and Blinc, R., *Phys. Rev. Lett.* **90**, 037601-1-037601-4 (2003).
10. Ravel, B., Stern, E. A., Vedrinskii, R. I., and Kraizman, V., *Ferroelectrics* **206-207**, 407-430 (1998).
11. Müller, K. A., and Burkard, H., *Phys. Rev. B* **19**, 3593-3602 (1979).
12. Zhang, L., Kleemann, W., Dec, J., Wang, R., and Itoh, M., *Eur. Phys. J. B* **28**, 163-171 (2002).

Non-Debye Domain Wall Response of SrTi^{18}O$_3$

W. Kleemann^1, J. Dec^{1a}, R. Wang^{2b}, M. Itoh^2

^{1}*Laboratorium für Angewandte Physik, Gerhard-Mercator-Universität, D-47048 Duisburg, Germany;*
^{2}*Materials & Structures Laboratory, Tokyo Institute of Technology, 4259 Nagatsuta, Midori, Yokohama 226-8503, Japan*

Abstract. Two different non-Debye dielectric spectra are observed in different low-frequency regimes in a polydomain relaxor ferroelectric single crystal, SrTi^{18}O$_3$, below its transition temperature, $T_c \approx 23$ K. At $f < 1$ kHz the response is due to irreversible creeplike wall motion, $\chi' \propto \chi'' \propto \omega^{-\beta}$ with $\beta \approx 0.1$ while polydispersive reversible wall segment relaxation with pink noise character is encountered at higher frequencies, $f > 1$ kHz. The crossover between both kinds of dielectric response is attributed to a dynamical phase transition [Nattermann et al., 2001].

SrTiO$_3$ is probably the best-known example of a quantum paraelectric, whose polar instability is suppressed by quantum fluctuations even at lowest temperatures, $T \to 0$ [1] This peculiarity is removed when replacing the ordinary oxygen ions, ^{16}O^{2-}, by the heavier isotope, ^{18}O^{2-}. Itoh et al. [2,3] observed ferroelectric Curie temperatures as high as $T_c \approx 25$K for isotopically modified SrTi^{18}O$_3$ ("STO18"). The nature of the ferroelectric state bears some peculiarities yet. Apart from the high, but broadened peak of the dielectric susceptibility measured at T_c, enhanced susceptibility occurs also in the low-T regime in the presence of external fields [4,5]. It has been argued that this extra response might be due to domain walls. They hint at a random-field induced domain state [6] as a consequence of excess Sr^{2+} vacancies.

In order to reveal more details of the domain state we studied a transverse Ising model complemented by quenched random electric fields which successfully describes both the smeared phase transition and the domain wall response in the low-T domain state [5]. A nonlinear optical study of second-harmonic light scattering clearly revealed the domain state in unpoled STO18 [7]. Analysis of the angular dependence even hints at a triclinic low-T symmetry of the lattice, an assertion which still has to be confirmed by diffraction methods.

This paper is devoted to the low frequency behavior of the dielectric susceptibility in STO18. Domains in ferroelectric crystals are well-known to have a considerable

---

^{a} Present address: Institute of Physics, University of Silesia, Pl-40-007 Katowice, Poland

^{b} Present address: Smart Structure Research Center, National Institute of Advanced Industrial Science and Technology, Tsukuba Central 2, 1-1-1 Umezono, Tsukuba

influence on the complex dielectric susceptibility, $\chi^* = \chi' - i\chi''$, and related quantities [8]. Owing to its mesoscopic character the domain wall susceptibility strongly reflects the structural properties of the crystal lattice. This is most spectacular in crystals with inherent disorder, where the domain walls are subject of stochastic pinning forces and χ^* is highly polydispersive due to a wide distribution $g(\tau)$ of Debye-type response spectra [9,10],

$$\chi^*(\omega) = \chi_\infty + (\chi_s - \chi_\infty)\sum_j g(\tau_j)(1+i\omega\tau_j)^{-1}, \qquad (1)$$

where τ and ω are the relaxation time and angular frequency, respectively.

More generally, the dynamic behavior of domain walls in random media under the influence of a periodic external field gives rise to hysteresis cycles of different shape depending on various external parameters. According to a recent theory of Nattermann et al. [11] on disordered ferroic (ferromagnetic or ferroelectric) materials, the polarization, P, is expected to display a number of different features as a function of T, frequency, $f = \omega/2\pi$, and probing ac field amplitude, E_0. They are described by a series of dynamical phase transitions, whose order parameter $Q = (\omega/2\pi)\oint Pdt$ reflects the shape of the P vs. E loop. When increasing the ac amplitude E_0 the polarization displays four regimes. First, at very low fields, $E_0 < E_\omega$, only "*relaxation*", but no macroscopic motion of the walls should occur at finite frequencies, $f > 0$. Second, within the range $E_\omega < E_0 < E_{t1}$, a thermally activated drift motion ("*creep*") is expected, while above the depinning threshold E_{t1} the athermal "*sliding*" regime is encountered within $E_{t1} < E_0 < E_{t2}$. Finally, for $E_0 > E_{t2}$ a complete reversal of the polarization ("*switching*") occurs in the whole sample in each half of the period, $\tau = 1/f$. It should be noticed that all transition fields, E_ω, E_{t1} and E_{t2}, are expected to depend strongly on both T and f [11].

In this paper we discuss the two different non-Debye responses referring to the field regions $E_0 < E_\omega$ ("*relaxation*") with $Q = 0$ (no hysteresis loop) and $E_\omega < E_0 < E_{t1}$ ("*creep*") with $Q \neq 0$ (slim non-centrosymmetric hysteresis loop) in the low-f ("*LF*") dispersion of STO18 below its transition temperature, $T_c \approx 25$ K. It shows both characteristics in adjacent frequency regimes. While the well-known relaxational $ln(1/f)$ characteristic of relaxing domain wall segments in a weak RF [9,10] applies to "high" frequencies, $f > 1$ kHz, an alternative $(1/f)^\beta$ dependence is observed in the LF regime, $f < 1$ kHz, similarly as recently been reported on the relaxor-type single crystals PbFe$_{1/2}$Nb$_{1/2}$O$_3$ (PFN) [12] and Sr$_{0.603}$Ce$_{0.007}$Ba$_{0.39}$Nb$_2$O$_6$ (SBN:Ce).[13]. In order to understand the latter behavior, we introduce polydispersivity via a broad distribution of wall mobilities, μ_w, which describe the viscous motion of the walls in the creep regime, where they overcome a large number of potential walls due to a high density of pinning defects. As a characteristic of irreversibility the walls stop when switching off the field. Within this concept the rapid individual Debye-type relaxation processes are averaged out on the long-time scale of a creep experiment.

The experiments were done on a single-crystal sample of STO18 (92%) prepared in the same way as described previously [2], with dimensions 0.3x3x7 mm^3 parallel to the cubic axes $x\|[110]_c$, $y\|[-110]_c$ and $z\|[001]_c$, respectively. This geometry warrants the formation of a crystallographic single domain with the tetragonal c axis along the

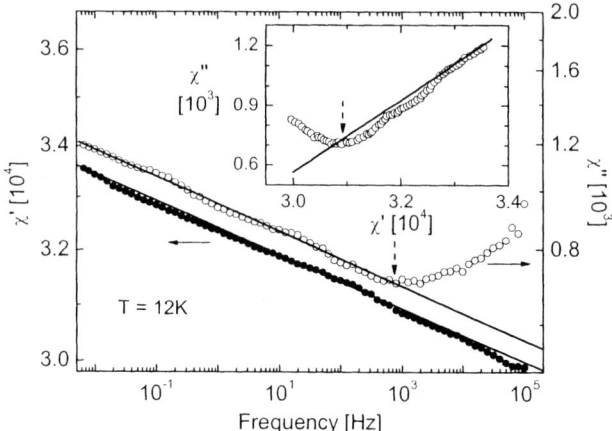

FIGURE 1. Dielectric spectra of χ' and χ'' vs. f (solid and open circles, respectively) and Cole-Cole plots χ'' vs. χ' (inset) of depoled and zero-field cooled STO18 recorded at $T = 12$ K. The vertical broken arrows at $f_{min} = 10^3$ Hz designate the dynamical phase transition between creep (low f) and relaxation (high f). Interpolating solid lines refer to the power laws, Eq. (4), of χ' and χ'' vs. f and to the linear relationship, Eq. (5), of χ'' vs. χ'.

long sample edges when cooling to below the antiferrodistortive transition temperature [1] $T_o = 108$ K [2]. Dielectric response data were taken with probing electric-field amplitudes of 300 V/m applied perpendicular to the c axis. A wide frequency range, $10^{-3} < f < 10^6$ Hz, was supplied by a *Solartron 1260* impedance analyzer with a *1296 dielectric interface*. Different temperatures were chosen both below and above T_c and stabilized to within ± 0.1 K.

Fig. 1 shows the complex susceptibility, χ' and χ'' vs. f, obtained after depoling the sample for 4 hours at $T = 150$ K and subsequent zero-field cooling to $T = 12$ K (solid and open circles, respectively). After this treatment all remnant polarization is safely lost and a statistical distribution of the *RF* inducing impurities is expected. Hence, the theoretically required [6] decorrelation of *RF*s might thus be optimized. The spectra in Fig. 1 illustrate the main features as discussed previously in the case of the dielectric dispersion of zero-field-cooled (*ZFC*) *SBN*:Ce [13]: (i) the imaginary part of the dielectric response increases below $f_{min} \approx 10^3$ Hz (vertical arrow) according to an inverse power law, $\chi'' \propto f^{-\beta}$ (solid line in double-logarithmic presentation), while the real part behaves as $\chi' - \chi'_\infty \propto f^{-\beta}$ (solid line with $\chi'_\infty \equiv \chi'(10^3 Hz) = 29800$) with $\beta \approx 0.15$ in both cases; (ii) a Cole-Cole plot of χ'' vs. χ' in Fig. 1 (inset) is characterized by a straight line at $f < 10^3$ Hz with a slope $\pi\beta/2 = 0.24$ (see Eq. (5) below), which yields, again $\beta = 0.15$; (iii) at $f > f_{min}$, χ'' increases again in a power-law-like fashion, while (iv) the curvature of χ' changes from concave to convex at $f > f_{min}$ and gently bends down in the sense of a polydispersive Debye-relaxation step. The high-f response confirms many of the characteristics predicted by Eq. (1) as discussed previously [13]. This dispersion regime is attributed to polarization

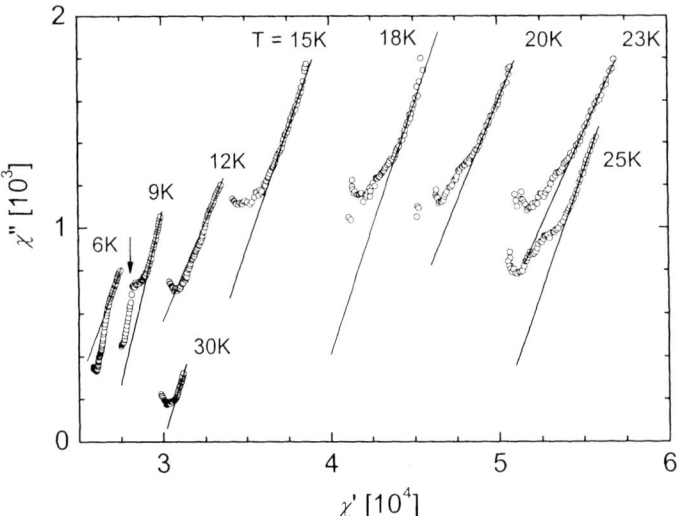

FIGURE 2. Cole-Cole plots χ'' vs. χ' of STO18 obtained at temperatures $6 \leq T \leq 30$ K similarly as in Fig. 1 together with best-fitted linear interpolations according to Eq. (5). An impurity-induced peak-like anomaly of χ'' vs. f at $T = 9$ K is designated by a vertical arrow.

processes due to the reversible motion of domain-wall segments experiencing restoring forces, *i.e.*, relaxation takes place as described by theories involving stochastic *RF* domain wall pinning [9,10]. Contrarily, the low-f response is due to irreversible viscous motion of domain-walls. They experience memory-erasing friction by averaging over numerous pinning centers in a creep process. The latter type of motion becomes possible for at least two reasons: screening of depolarization fields by free charges in the bulk or at the surface and/or pinning of the domain-walls at quenched *RF*s, which is believed to be due to quenched charge disorder in the special case of *SBN*:Ce [14] and is proposed to refer to excess Sr^{2+} vacancies in STO18 [5].

The creep response under the action of an external electric field is readily modeled by considering the average polarization, $P(t) = (2P_s/D)x(t)$, of a regular stripe domain pattern of up and down polarized regions carrying spontaneous polarization, $\pm P_s$, and having an average width D. It arises from a sideways motion of walls perpendicular to the field direction by a distance x. A straightforward calculation [13] reveals a conduction-type *ac* susceptibility

$$\chi_w^*(\omega) = \chi_\infty (1 + 1/i\omega\tau_w), \qquad (2)$$

with $\chi_\infty / \tau_w = (2\mu_w P_s / \varepsilon_0 D)$. χ_∞ refers to the "instantaneous" response due to reversible domain-wall rearrangements occurring on shorter-time scales, $f > f_{min}$ (see above). The "relaxation" time τ_w denotes the time in which the interface contribution to the polarization equals that achieved instantaneously, $\Delta P = \varepsilon_0 \chi_\infty E$.

Since the electric fields used in our experiments ($E_0 = 300$ V/m) are well below the coercive field, $E_c \approx 50$kV/m [2], we have to account for the nonlinearity of v vs. E in the creep regime, where thermal excitation enables viscous motion below the depinning threshold $E_{\text{crit}} \approx E_c$ [15]. Eq. (2) may then be modified phenomenologically [13] by introducing a Cole-Davidson-type exponent $\beta < 1$,

$$\chi_w^*(\omega) = \chi_\infty [1 + 1/(i\omega\tau_{\text{eff}})^\beta], \tag{3}$$

similarly as used in the case of polydispersive Debye-type relaxation [16]. Here τ_{eff} denotes an effective relaxation time. Decomposition of Eq. (3) yields power law dependencies,

$$\chi'(\omega) = \chi_\infty [1 + \cos(\beta\pi/2)/(\omega\tau_{\text{eff}})^\beta] \text{ and } \chi''(\omega) = \chi_\infty \sin(\beta\pi/2)/(\omega\tau_{\text{eff}})^\beta, \tag{4}$$

and a linear relationship

$$\chi''(\omega) = \tan(\beta\pi/2)(\chi'(\omega) - \chi'_\infty), \tag{5}$$

both of which are well supported by our experiments (see Fig. 1 and 2). As can be seen in Fig. 2 from Cole-Cole diagrams obtained at various temperatures between $T = 6$ and 30 K, the low-f slopes vary but slightly as corroborated by a presentation of β vs. T in Fig. 3. Within the accuracy of the measurements an average value $\langle\beta\rangle = 0.15\pm0.02$ is encountered. As expected, the proportionality constant χ_∞ closely follows the temperature behavior of the peak susceptibility, $\chi'_{\text{max}}(T)$ [7]. Closer inspection of Fig. 2 reveals weak peak-like anomalies in the low-f branches of χ'' vs. f at $T < 10$ K (vertical arrow). They are probably due to intrinsic dipolar impurities as stated previously [5].

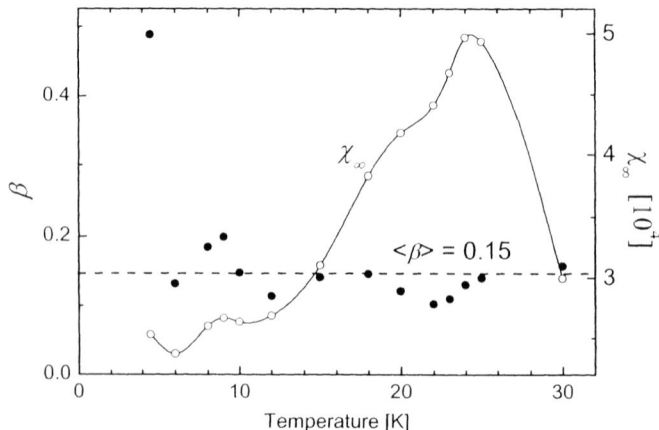

FIGURE 3. Parameters β and χ_∞ extracted from susceptibility data of Fig. 2 according to Eq.(5).

It should be noticed that the monodispersive relation (2) rigorously satisfies the Kramers-Kronig (*KK*) relationships, since $\chi'' \propto 1/\omega$ is a purely conductive contribution due to ohmic-like domain wall sliding and $\chi' = \chi'_\infty$ is constant. This is less obvious for $\beta < 1$, Eq. (3 - 5). However, *KK* analysis starting from $\chi'' \propto \omega^{-\beta}$, Eq. (4), reveals an approximate expression for the real part,

$$(\chi' - \chi_\infty)/\chi_\infty \approx \frac{\sin(\pi\beta/2)}{\pi\beta/2} \frac{1-\beta}{1-\beta/2} (\omega\tau)^{-\beta}, \qquad (6)$$

which comes close to that predicted by Eq. (5) for all values $0 \leq \beta \leq 1$, since $(\pi\beta/2)(1-\beta/2)/(1-\beta) \approx \tan(\pi\beta/2)$. Further attempts have to be taken to describe the creep susceptibility by using the rigorous creep formula [15],

$$v \propto \exp[(-R/k_B T)(E_0/E)^\mu], \qquad (7)$$

where R, E_0 and μ (≈ 1 in the case of *RF*s [17]) are the characteristic energy, critical field and dynamic exponent, respectively. The influence on the effective exponent β of both the non-linearity inherent to Eq. (7) and the interaction between densely spaced domain walls remains to be understood.

In conclusion, the susceptibility due to both irreversible and reversible domain-wall response has been observed in the domain state of zero-field-cooled STO18 in different frequency regimes. Different theoretical descriptions based on creeplike wall motion and polydispersive wall relaxation, respectively, have been employed. Very probably both frequency ranges are linked by a dynamical phase transition from $Q = 0$ to $Q \neq 0$ between the relaxation and creep regimes as predicted [11] for the hysteretic dynamics of ferroic domain-walls. At a given frequency ω and temperature T and below a threshold field amplitude, $E_0 < E_\omega$, the walls do not exhibit any macroscopic motion, since their segments oscillate between metastable states with close energies giving rise to dissipation as described previously. Similarly, a transition is also expected when keeping E_0 fixed, but increasing the frequency to $\omega > \omega_{min}$ as in the present experiments. It remains to be shown that the threshold frequency f_{min} shifts to higher values when increasing the driving field amplitude and that the athermal slide regime appears in the VLF regime with a significant increase of the exponent, $\beta \to 1$. Another task for the future is to transform the theory of dynamic hysteresis, P vs E, into the frequency domain, χ^* vs ω at variant E and T, and to compare it to the present phenomenological theory.

ACKNOWLEDGMENTS

Financial support by DFG (SPP "*Strukturgradienten in Kristallen*") and NATO (*PST.CLG.977409*) is acknowledged.

REFERENCES

[1] Müller, K. A., and Burkard, H., *Phys. Rev. B* **19**, 3593 (1979).
[2] Itoh, M., Wang, R., Inaguma, Y., Yamaguchi, T., Shan, Y.-J., and Nakamura, T., *Phys. Rev. Lett.* **82**, 3540 (1999).
[3] Itoh, M., and Wang, R., *Appl. Phys. Lett.* **76**, 221 (2000).
[4] Wang, R., and Itoh, M., *Phys. Rev. B* **62**, 731 (2000).
[5] Zhang, L., Kleemann, W., Dec, J., Wang, R., and Itoh, M., *Eur. Phys. J. B* **28**, 163 (2002).
[6] Kleemann, W., *Int. J. Mod. Phys. B* **7**, 2469 (1993).
[7] Zhang, L., Kleemann, W., Wang, R., and Itoh, M., *Appl. Phys. Lett.* **81**, 3022 (2002).
[8] Fousek, J., and Janovec, V., *Phys. Stat. Sol.* **13**, 105 (1966).
[9] Ioffe, L. B., and Vinokur, V. M., *J. Phys. C* **20**, 6149 (1987).
[10] Nattermann, T., Shapir, Y., and Vilfan, I., *Phys. Rev. B* **42**, 8577 (1990).
[11] Nattermann, T., Pokrovsky, V., and Vinokur, V. M., *Phys. Rev. Lett.* **87**, 197005 (2001).
[12] Park, Y., *Solid State Commun.* **113**, 379 (2000).
[13] Kleemann, W., Dec, J., Miga, S., Woike, Th., and Pankrath, R., *Phys. Rev. B* **65**, 220101 (2002).
[14] Kleemann, W., Dec, J., Lehnen, P., Blinc, R., Zalar, B., and Pankrath, R., *Europhys. Lett.* **57**, 14 (2002).
[15] Feigel'man, M. V., Geshkenbein, V. B., Larkin, A. I., and Vinokur, V. M., *Phys. Rev. Lett.* **63**, 2303 (1989).
[16] Jonscher, A. K., *Dielectric relaxation in solids* (Chelsea Dielectric Press, London 1983).
[17] Fisher, D. S., *Phys. Rev. Lett.* **56**, 1964 (1986).

Noise and Aging of Relaxor Ferroelectrics

M. B. Weissman, Eugene V. Colla, and Lambert K. Chao

*Department of Physics, University of Illinois at Urbana-Champaign
1110 West Green Street, Urbana, IL 61801-3080*

Abstract. Aging and noise are used to elucidate the types of frozen order formed in the relaxor regime. In PMN, PMN with 10% PT, and PLZT aging deep in the relaxor regime shows several features characteristic of spinglasses, including multiple independent susceptibility 'holes' formed at different aging temperatures. This effect requires complex cooperative glassy freezing of many local units. However, the field scale required to disrupt the aging is anomalously high compared to spinglasses, if a nanodomain is the unit corresponding to a spin. Barkhausen noise results imply cooperative moment changes involving several nanodomains near T_g, but much smaller steps below T_g. This result suggests that kinetic barriers increase at T_g without increasing coupling among nanodomains. Together, these results suggest that in prototypical relaxors the glassy state is not formed by nanodomains, although it affects their dynamics, but rather by smaller units. We tentatively propose that the canted components of the local polarizations (found in scattering experiments) are analogous to the x-y spins in a reentrant spinglass. The first tested prediction of this new picture is that spinglass-like aging effects would be absent in the uniaxial 'relaxor' SBN. Such effects do vanish well below T_g in SBN.

INTRODUCTION

A variety of materials upon cooling develop local ferroelectric order which then freezes into an overall disordered pattern.[1] The term 'relaxor' is often applied to all such materials for which a frequency-dependent temperature of the susceptibility peak, $T_P(f)$, indicates that the freezing is kinetic rather than purely thermodynamic. In some cases, however, the frequency dependence is strong enough to fit an Arrhenius law, indicating that no particularly interesting cooperative effects are driving the freezing.(e.g. [2]) In other cases, including the prototypical material $Pb(Mg_{1/3}Nb_{2/3})O_3$ (PMN, e.g. [3]) and the Ising material $Sr_xBa_{1-x}Nb_2O_6$ (SBN, e.g. [4]), the frequency dependence of $T_p(f)$ is too weak to fit Arrhenius kinetics, indicating that some interesting thermodynamics underlie those relaxor effects. However, even in these cases there is no guarantee that the different relaxor materials have similar physics, since cooperative kinetic freezing effects are rather generic. Hence there is a need for new probes particularly sensitive to the nature of the frozen state. Aging effects [5] and non-equilibrium noise [6] both provide such probes.

One of the simplest unanswered questions concerns the size scale of any cooperative order among nanodomains. Glassy order parameters in general do not show up in conventional scattering experiments. Barkhausen noise [7] is found when materials change their polarization via reorienting domains or domain-wall motions.

The magnitude of Barkhausen noise depends not only on how much the polarization changes but on the size of the steps of the changes, and thus provides a measure of dynamical coherence beyond the size scale over which the static structure has some conventional order parameter. For a given polarization change, the net Barkhausen variance in the polarization should scale linearly with the size of the discrete dipole reorientations. If a large cluster of nanodomains reorients as a unit, Barkhausen noise should detect that change even if the nanodomains within the cluster have random-looking relative orientations.

Aging occurs in most disordered systems, which gradually find lower free-energy states, generally with lower susceptibilities, when allowed to sit under fixed conditions. However, the detailed form of aging is very sensitive to the form of the space of metastable states.[5] Simple collections of two-state systems (e.g. fixed non-interacting Ising domains in local fields) do not show aging. Aging from a simple growth of domains is typically cumulative. In contrast, aging of a spinglass at one temperature, T_A, has little effect on the susceptibility $\chi(T)$ at lower T, but creates a long-term memory that can be read out in $\chi(T)$ later upon warming back to T_A. Such effects require a very complicated collection of metastabilities, and are usually believed to arise from hierarchical state spaces.

In this paper we give an overview of several aging and noise experiments, including some preliminary unpublished results. We suggest a new model for the relaxor freezing in PMN and its relatives, based on an analogy to the reentrant x-y spinglass transition [8-10], in qualitative contrast to the frozen state in SBN, which is suspected of following the random-field Ising model (RFIM).[4]

EXPERIMENTS

The Barkhausen experiments require special techniques because the step size is very small compared to those of simple ferroelectric domain materials. We have designed a balanced-bridge measurement system using two nominally identical capacitors from the same sample, designed to cancel the huge applied signal and significant systematic non-linear responses, leaving the small random non-linear responses.[11] To avoid spurious signals, it is also extremely important to use a very quiet undistorted sine-wave applied signal, unusually well synchronized with the A/D conversion. Even with these precautions, the lowest few harmonics of the driving frequency may contain non-random components from imperfect matching of the two samples.

Aging experiments, on the other hand, require only standard techniques, with some patience and good temperature control. One standard protocol is to age for some hours at T_A, smoothly lower T, then smoothly raise T, while continuously measuring complex χ.

RESULTS AND INITIAL ANALYSIS

Figure 1 shows what the Barkhausen spectra look like in PMN- a mixture of broadband aperiodic noise and random harmonics of the driving frequency. Both

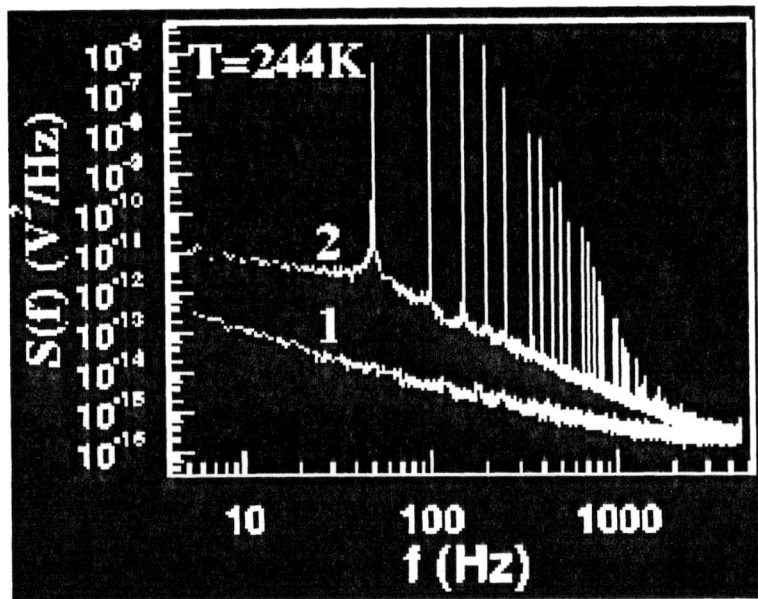

FIGURE 1. The voltage spectral density across a balanced bridge PMN capacitor at 244 K is shown, with an applied 50 Hz ac field of 130 V/cn (trace 2) and with no applied field (trace 1). Harmonics of 60 Hz have been removed. From [11].

types of components are expected theoretically in most models, and arise in some simple simulations we have made of domains in random environments with thermal noise. Aperiodic response to periodic driving arises in part directly from thermal jitter in switching times and also from thermal fluctuations of the metastable states, affecting the switching fields. The magnitude of the noise increases as (very roughly) the cube of the applied field, even though the response is nearly linear. These results suggest that there are large clusters which can reorient coherently, but only if the applied field exceeds not only the random field but also a coercive field, i.e. kinetic barrier. The barriers must grow a bit more rapidly with cluster size than do the moments, or else the large clusters would flip in small fields. The distribution of cluster sizes is strongly constrained by the strong non-linearity of the noise and the near-linearity of the response. However, we do not yet have a good understanding of how the kinetic barriers depend on the size of the reorienting cluster.

Fig. 2 shows magnitudes of components of the Barkhausen noise, normalized to reflect not the total susceptibility but rather the dipole step size under the particular driving condition, taken vs. T. Several features are worth noting. The largest step sizes (at fixed amplitude E_{AC}) are found near T_G. Although absolute calibration of the step sizes is a bit uncertain, the typical dipole moment step size is comparable to ten times the moment of a nanodomain, indicating coherent clusters of some 100 randomly oriented nanodomains. However, this size is not very strongly T-dependent above T_G and shrinks very abruptly below T_G. Very similar results have now been found in PMN doped with up to 10% $PbTiO_3$ (PT), i.e. in material with somewhat

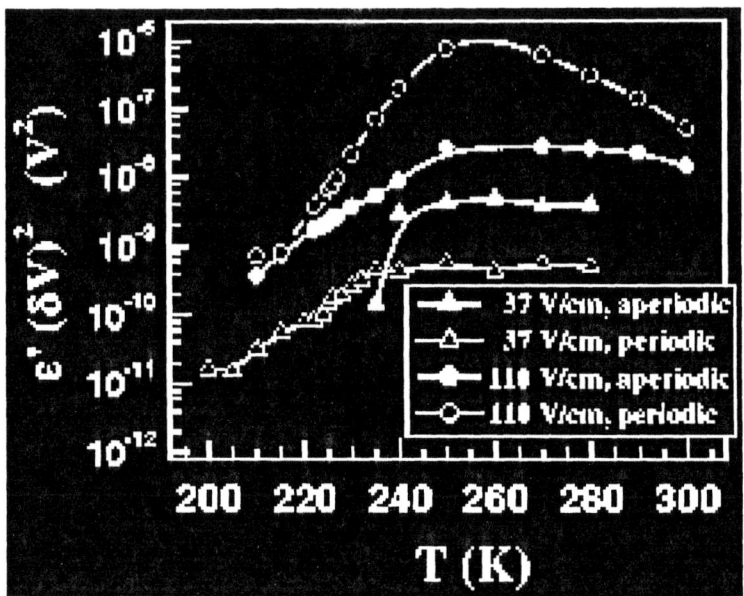

FIGURE 2. The excess noise power above background in the octave from 512 Hz to 1024 Hz from data like that in Fig. 1 is shown as a function of T for two different field magnitudes. The harmonic and anharmonic components are given separately. The values are weighted by $\varepsilon'(T)$ to give a quantity proportional to the Barkhausen step size, which is too small to directly observe. From [11]

stronger ferroelectric correlations. We believe that this sharp drop in the typical step size indicates that the freezing is caused primarily by growth of barriers and not increased clustering of nanodomains.

Aging experiments produce a variety of results in the vicinity of T_G, indicating a complicated range of aging mechanisms, especially in PLZT, for which a atomic rearrangements are likely to be involved.[12,13] Well below T_G, however, PMN, PMN-10% PT, and PLZT all show very similar behavior, very reminiscent of spinglasses. The most striking such behavior is that aging at T_A produces a 'hole' in $\chi''(T_A)$ which does not affect $\chi''(T)$ for $T<0.9\ T_A$. On subsequent warming, the hole in $\chi''(T_A)$ is not affected appreciably by subsequent aging at $T<0.9\ T_A$, although it does depend on aging just below T_A and is quickly erased by aging above about 1.05 T_A. Fig. 3 illustrates one such hole-aging. One confirmed implication is that multiple holes can coexist after aging at a descending sequence of T's. Deep in the relaxor regime, the aging of $\chi''(f)$ depends almost entirely on the product $f\tau$ where τ is the aging time rather than on the separate factors f and τ- another spinglass-like property.[5] The aging can be mostly reset by field jumps big enough to change the dipole energy of a single nanodomain by about kT. In contrast, the necessary field jump to reset a spinglass changes the dipole energy of a single spin by only about kT/300.[14]

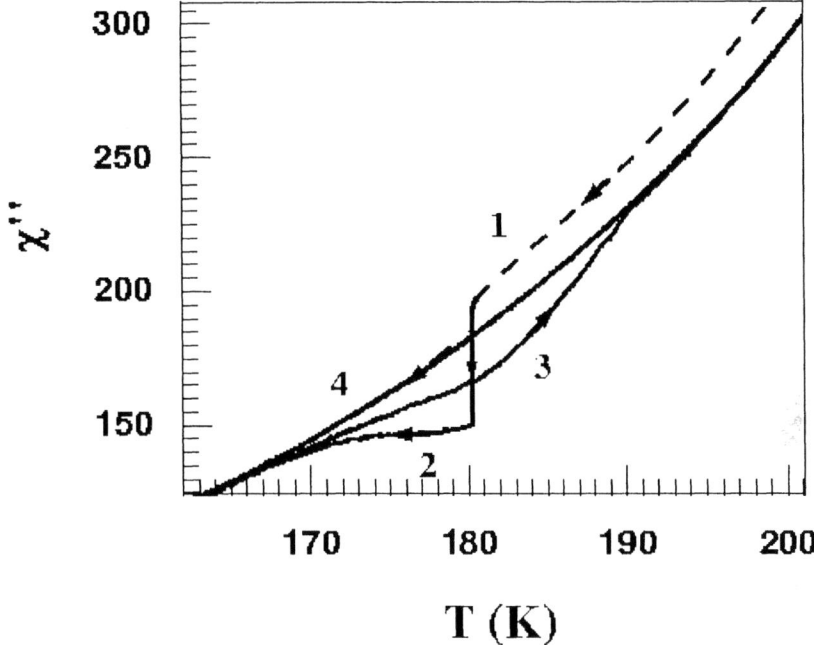

FIGURE 3. The aging of χ''(100Hz) is shown in PMN. Segment 1 was taken on cooling at about 25K/min from 350K, the aging drop at 180K took 64 hours, segment 2 shows cooling at about 1K/min (to 140K) followed be segment 3, heating at about 1 K/min. Segment 3 shows the characteristic hole around the aging temperature. Curve 4 is a reference curve taken with about 1 K/min cooling. From [12].

When the PMN is doped with about 30% PT, forming material with more extended ferroelectric domains [15], the aging no longer resembles that of spinglasses.[13] It becomes cumulative, as expected for ordinary domain growth.

For theoretical reasons discussed below, we suspected that the spinglass-like hole-aging would not be present in the uniaxial SBN. Although near the freezing temperature, some hole-aging was found, it was absent deep in the frozen phase. Instead, as illustrated in Fig. 4, the aging was entirely cumulative, its effects not confined to T near T_A, and there was no abrupt erasure of the aging on heating above T_A.

DISCUSSION

Several new lines of evidence, suggestive but not yet conclusive, point toward a new picture of the relaxor state in the prototypical material, PMN, and its closest relatives. First, we can eliminate some overly simplified pictures. The complicated aging effects, capable of sustaining multiple memories of prior history, are not consistent with any picture of separate domains freezing in quenched random fields. The frozen units must have many coupled degrees of freedom supporting a complicated space of

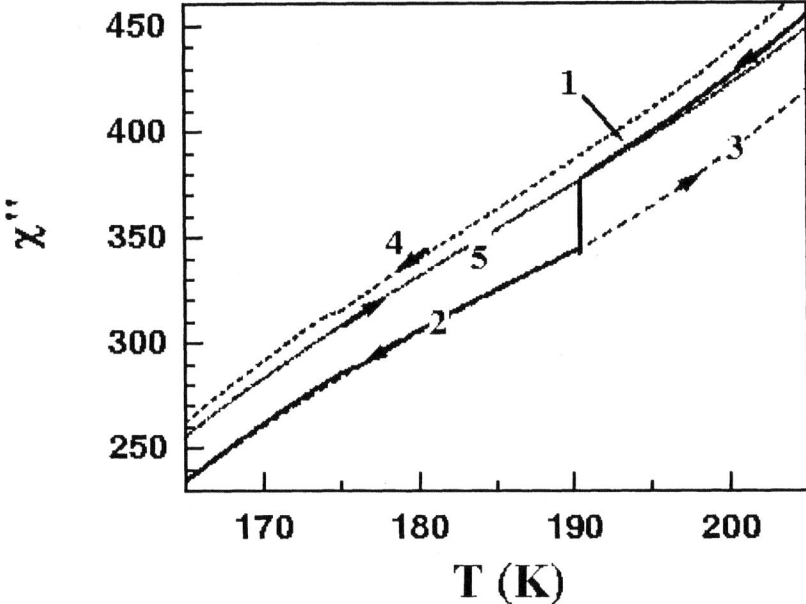

FIGURE 4. Aging of χ"(50Hz) of an SBN sample (60% Sn, 40% Ba, with 1% La doping). All the temperature ramps were at roughly 3K/min. The aging time at 190K was 16 hours. Cooling segment 2 extended to 140K before starting heating segment 3. Segments 4 and 5 were taken under approximately similar conditions without the pause at 190K. The irregular shape of the sample precludes absolute determination of the scale for χ to better than about a factor of 1.5.

metastable states. A direct analogy between the interacting nanodomains and the interacting spins which form canonical spinglasses cannot hold, because the corresponding field scale for disrupting the aging is so much larger in the relaxor case. One immediately suspects that the dipole moments of the constituents of the relaxor glass should then be much smaller than the moments of the nanodomains. There is some anomalous low-temperature heat capacity in some relaxors, although there is some dispute as to its form.[16,17] If it is associated with the same glassy degrees of freedom involved in the relaxor freezing, the characteristic units involved must be much more concentrated, and hence much smaller, than nanodomains. The Barkhausen noise indicates that the largest units involved in Barkhausen steps consist of a few tens of nanodomains, in contrast to spinglass experiments which show that even equilibrium noise (involving units small enough to fluctuate thermally) involves coherent units of 10^4 spins.[18]

If we take seriously the idea that the glassy transition occurs among smaller units, an obvious candidate is supplied by scattering experiments, which show in PMN local moments along (100) type directions, canted with respect to the (111) type nanodomain moments.[19] These canted moments may have a transition analogous to the x-y spinglass transition in reentrant spinglasses.[8] This transition leaves the static (local) ferromagnetic order essentially unchanged, while slowing the dynamics of the ferromagnetism [9,10] due to coupling to the glassy modes in an environment which provides local symmetry breaking. In the experimental reentrant spinglasses, the ferromagnetism also is typically not long-range. In the relaxor, we would expect that barriers to any nanodomain rotations would grow, consistent with the Barkhausen results. Likewise, there would be a genuine spinglass state, supporting the striking and distinctive aging effects. However, coupling between that state and the high-moment nanodomains would give a low characteristic field scale for disrupting those aging effects. There would, however, be one important qualitative difference between the magnetic and dielectric systems, since the material disorder cannot break time-reversal symmetry and thus cannot make random magnetic fields, while random electric fields (or equivalent vector fields) are certainly prominent in the relaxors.

So far these retrodictions are supported by one prediction- the absence of spinglass like aging deep in the frozen state of Ising-like SBN, for which there are no canted moments. We have not yet performed Barkhausen experiments on SBN, but very preliminary simulation results from our colleague R. White indicate that the effective Barkhausen step size does not shrink upon freezing in an RFIM, again in sharp contrast to the data from PMN relaxors. Further theoretical work to understand the implications of different models for both Barkhausen noise and aging should help constrain the models substantially.

ACKNOWLEDGEMENTS

This work was funded by NSF DMR 99-81869. D. D. Viehland supplied the SBN crystal.

REFERENCES

1. Cross, L. E., *Ferroelectrics* **76**, 241-267 (1987).
2. Toulouse, J., B. E. Vugmeister, and R. Pattnaik, *Phys. Rev. Lett.* **73**, 3467-70 (1994).
3. Colla, E. V., S. M. Gupta, and D. Viehland, *J. Appl. Phys.* **85**, 362-367 (1999).
4. Kleemann, W., J. Dec, P. Lehnen, R. Blinc, B. Zalar, and R. Pankrath, *Europhys. Lett.* **57**, 14-19 (2002).
5. Hammann, J., E. Vincent, V. Dupuis, M. Alba, M. Ocio, and J.-P. Bouchaud, Comparative review of aging properties in spin glasses and other disordered materials, in *Frontiers in Magnetism*, edited by H. T. Y. Miyako, S. Miyashita, Phys. Soc., Japan, Kyoto, 1999, pp. 206-211.
6. Weissman, M. B., *Annu. Rev. Mater. Sci.* **26**, 395-429 (1996).
7. Rudyak, V. M., *Uspekhi Fizicheskikh Nauk* **101**, 429-62 (1970).
8. Gabay, M. and G. Toulouse, *Phys. Rev. Lett.* **47**, 201 (1981).
9. Ryan, D. H., J. M. D. Coey, E. Batalla, Z. Altounian, and J. O. Strom-Olsen, *Phys. Rev. B* **35**, 8630-38 (1987).
10. Senoussi, S., S. Hadjoudj, and R. Fourmeaux, *Phys. Rev. Lett.* **61**, 1013-16 (1988).
11. Colla, E. V., L. K. Chao, and M. B. Weissman, *Phys. Rev. Lett.* **88**, 017601/1-4. (2002).
12. Colla, E. V., L. K. Chao, M. B. Weissman, and D. D. Viehland, *Phys. Rev. Lett.* **85**, 3033-3036 (2000).
13. Colla, E. V., L. K. Chao, and M. B. Weissman, *Phys. Rev. B* **63**, 134107/1-10 (2001).
14. Fenimore, P. W. and M. B. Weissman, *J. Appl. Phys.* **76**, 6192-4 (1994).
15. Choi, S. W., J. M. Jung, and A. S. Bhalla, *Ferroelectrics* **189**, 27-38 (1996).

16. Hegenbarth, E., *Ferroelectrics* **168,** 25-37 (1995).
17. Gvasaliya, S. N., S. G. Lushnikov, Y. Moriya, H. Kawaji, and T. Atake, *Physica B* **305,** 90-5 (2001).
18. Weissman, M. B., *Rev. Mod. Phys.* **65,** 829-839 (1993).
19. Dkhil, B., J. M. Kiat, G. Calvarin, G. Baldinozzi, S. B. Vakhrushev, and E. Suard, *Phys. Rev. B* **65,** 024101/1-8 (2001).

Diffusive phase transitions in ferroelectrics and antiferroelectrics

S. A. Prosandeev*, I. P. Raevski* and U. V. Waghmare[†]

Physics Department, Rostov State University, 5 Zorge St., 344090 Rostov on Don, Russia
[†]*J Nehru Centre for Advanced Scientific Research, Bangalore, India*

Abstract. In this paper, we present a microscopic model for heterogeneous ferroelectric and an order parameter for relaxor phase. We write a Landau theory based on this model and its application to ferroelectric $PbFe_{1/2}Ta_{1/2}O_3$ (PFT) and antiferroelectric $NaNbO_3$:Gd. We later discuss the coupling between soft mode and domain walls, soft mode and quasi-local vibration and resulting susceptibility function.

A simple model ferroelectric thin film with dead layers was recently analyzed by Bratkovsky and Levanyuk [1] resulting in an analytic solution. Based on this study, we proposed a microscopic model for an inhomogeneous ferroelectric comprised of ferroelectric slabs sandwiched between dielectric interfacial layers [2]. Low energy solutions of these models reveal that the ferroelectric slabs break into alternating domains (Fig. 1) with zero total macroscopic polarization. The size of the domains was shown to depend only on the relative total width of the dielectric and ferroelectric regions in the direction of the field [2]. The nanodomain structure appears cooperatively and its origin lies in the reduction of depolarization field [1, 2].

The alternating polarization domains are accompanied by shear strain owing to the electrostrictive coupling [2]: the displacements inside these domains are the sum of the polarization and uniform shear strain displacements in accordance with neutron scattering data [3]. We calculated diffused scattering intensity [2] on the proposed static domain structure at the wave vector related to the domain structure. We suggest that the relaxor phase corresponds to a finite value of the order parameter describing these alternating domains. It does not definitely mean that the ideal striped domain structure must appear in relaxors: actually, due to disorder, this structure will be modulated by random fields and stresses. The main point of this theory is that, in order to decrease the depolarization field, the polar regions appear cooperatively, and in the order in which the net polarization vanishes; otherwise, there appears a large contribution of the depolarization field into the energy preventing the appearance of lone polar regions if they are not compensated by local quenched fields. Below we will discuss the importance of the quenched fields in relaxors.

The fluctuations of the order parameter introduced above are conjugated with the field which we will call H_0. We will assume the existence of quenched field H_0 (at least, below Burns temperature). We use the notation η for the relaxor order parameter (the magnitude of polarization inside a nanodomain) and write the Landau expansion:

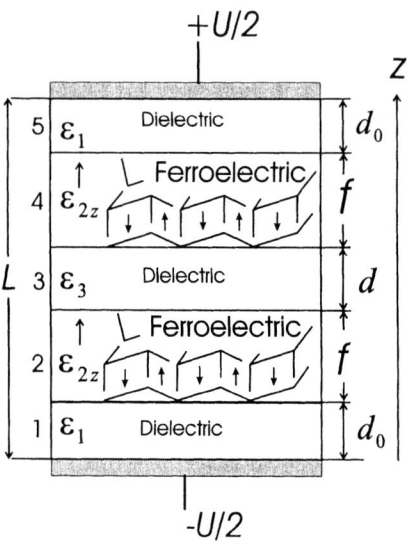

FIGURE 1. A model of inhomogeneous ferroelectrics [2].

$$F = F_0 + \frac{1}{2}\alpha P^2 + \frac{1}{4}\beta P^4 + \frac{1}{6}\gamma P^6 +$$
$$\frac{1}{2}A\eta^2 + \frac{1}{4}B\eta^4 + \frac{1}{6}C\eta^6 - H_0\eta + \frac{1}{2}\lambda P^2\eta^2 \quad (1)$$

where $\alpha = \alpha_0(T - T_{CW})$ and $A = a(T - T_\eta)$.

At zero macroscopic polarization, the equilibrium condition with respect to η is

$$A\eta + B\eta^3 + C\eta^5 - H_0 = 0 \quad (2)$$

At high temperatures η vanishes as $(T - T_\eta)^{-1}$. In this limit, dielectric permittivity

$$\chi = \frac{1}{v(T - T_{CW}) + \lambda \eta^2}, \quad (3)$$

obeys a Curie-Weiss law. At $T = T_\eta$ there is a deviation from this law, and the dielectric permittivity peak is diffused due to the coupling between the new order parameter appearing at the phase transition and frozen conjugated fields.

We used the expression derived in order to describe the diffuseness of the phase transition in relaxor PFT from the para-phase to the relaxor phase with ferroelectric nano-regions. We also took into account a low temperature (glass-type) phase transition which results in a strong decrease of dielectric permittivity below approximately 220 K. We fitted expressions obtained to experimental data (Fig. 2) and got the best fit shown by the solid line. We find that T_η is rather close to the extrapolated high-temperature

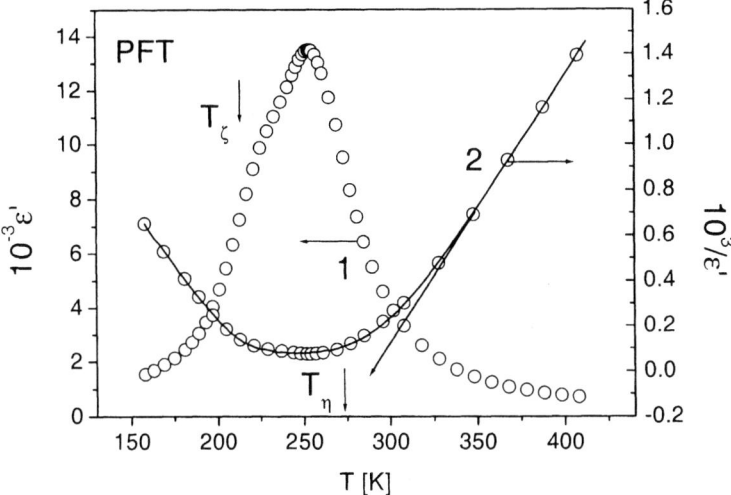

FIGURE 2. Temperature dependencies of ε' (1) and $10^3/\varepsilon'$ (2) measured at 10^6 Hz for PFT crystal. Solid line is the best fit of the theory to experimental data.

Curie-Weiss temperature. This justifies that the ferroelectric relaxor order parameter, η, is connected with local polarization, and can be regarded to a wave of polarization described above.

We use the same expression (3) to treat the diffused phase transition in an antiferroelectric $NaNbO_3$ doped with Gd, where the significance of η now is an antiferrolectric order parameter. The resulting fit to experimental data is rather good shown in Fig. 3. The critical temperatures are indicated in the figure. We also show in Fig. 3 the difference between the inverse dielectric permittivity and high temperature Curie-Weiss law. It is seen that this difference, at low temperatures, behaves linearly with temperature while at the dielectric permittivity peak position it is diffused in accord with the theory.

Experimental data [4] show that the "waterfall phenomenon" exists in the whole temperature interval between the freezing temperature and Burns temperature, implying that the soft mode is strongly damped in the relaxor phase (note that this soft mode is not the uniform ferroelectric). We believe this mechanism to be connected with disorder leading to finite sizes of polar regions (see below), the interaction of the soft mode vibrations with domain walls, and with local dipoles (a mathematical expression is similar to that obtained for the soft mode coupled with microscopic dipoles [5, 6]).

The acoustic mode can have a dip due to its coupling with optical mode [7] (Fig. 4a). We suggest that, due to disorder, the deviation of the optical and acoustic modes from average is different in different regions of the disordered crystal and one has to introduce a distribution function of these deviations (Fig. 4b). We also consider the case when the optical mode has a dip (Fig. 4c) if the interaction with the acoustic mode is not taken into account (the first row in Fig. 4). In this case the optical mode dispersion curve is:

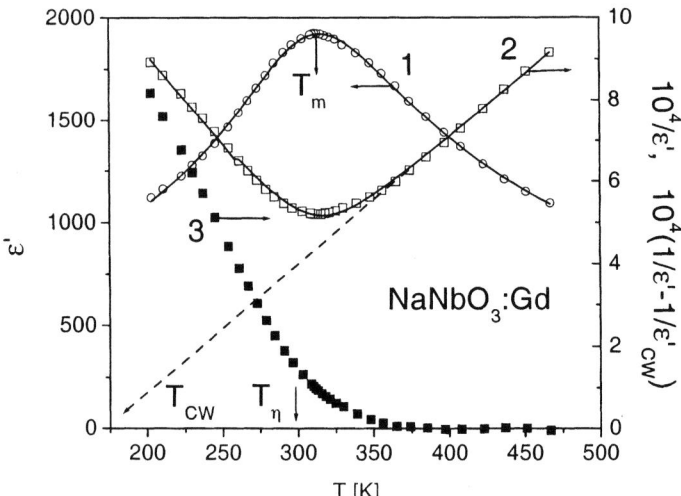

FIGURE 3. Temperature dependencies of ε' (1) and $10^4/\varepsilon'$ (2) measured at 10^5 Hz for 0.88NaNbO$_3$-0.12Gd$_{1/3}$NbO$_3$ crystal, and the difference between the experimental $1/\varepsilon'$ dependence and the Curie-Weiss fit of $1/\varepsilon'$ (3). Solid lines are the best fits of the theory to experimental data.

$\varepsilon = \omega_0^2 - kq^2 + bq^4 + \ldots$. We derived in this case that the correlation function deviates from the Ornstein-Cernike expression by an oscillating factor:

$$\chi(r) \sim \frac{1}{r}\sin(k_0 r)\exp(-k_c r) \tag{4}$$

where k_c is inverse correlation length and k_0 is the wave vector of the susceptibility oscillations. In the k-space susceptibility looks as

$$\chi(q) \sim \frac{a^2}{(q^2 - k_m^2)^2 + a^4} \tag{5}$$

Here $k_m^2 = k_0^2 - k_c^2$ and $a^2 = 2k_0 k_c$. These constants are expressed in terms of parameters in the optical mode dispersion: $\omega_0^2/b = k_m^4 + a^4$ and $k/b = 2k_m^2$. The inverse correlation length k_c can be obtained from:

$$k_c^2 = \frac{1}{2}\left[\sqrt{\frac{\omega_0^2}{b}} - \frac{k}{2b}\right] \tag{6}$$

where, at $\omega_0^2 > k > 0$, k_c is real. Otherwise, it is imaginary, and there appear harmonic beatings.

A general form of the Hamiltonian taking into account the interaction between the acoustic (u) and optical (x) displacements (and a dip in the optical mode) is:

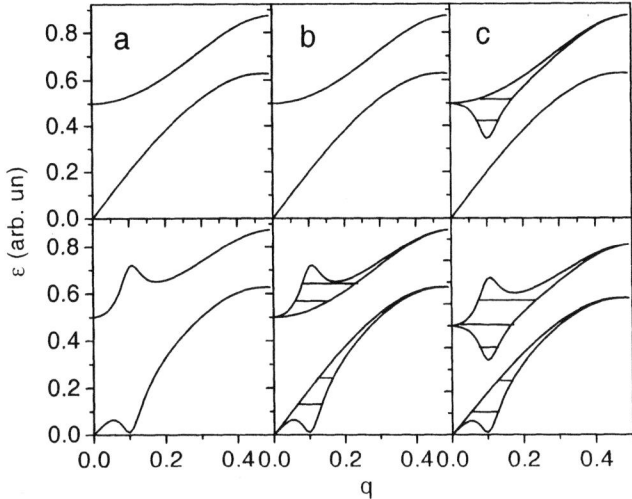

FIGURE 4. Model dispersion curves: a) the condensation of the modes is due to mode-mode coupling, b) the same as (a) but with taking into account disorder, c) the condensation of the modes is due to both the polarization instability and mode-mode coupling; The first row is without the mode-mode coupling, the second row is with the mode-mode coupling; The dashed region arises because of disorder.

$$H_{harm}(q) = \frac{1}{2}u_{-q}A(q)u_q + u_{-q}Vx_q + x_{-q}V^*u_q + \frac{1}{2}x_{-q}B(k)x_q \qquad (7)$$

where, at small q,

$$A(q) = aq^2 + ... \qquad (8)$$
$$B(k) = \omega_0^2 - bq^2 + cq^4 + ... \qquad (9)$$
$$V(q) = v_1 q^2 + v_2 q^4 + iv_3 Pq + ... \qquad (10)$$

Here P is polarization; Some important contributions in the coupling term appear in symmetry broken regions and in the regions where there exists gradient of polarization, that is at the boundaries of the domains or/and polar regions. This coupling leads to a "repulsion" of the acoustic and optical modes as it is shown in Fig. 4, and, at strong q-dependence, a nanodomain-type state can be condensed. At small q, the length of the wave is much larger than the heterogeneity size, and such a wave does not experience strong damping. When q corresponds to the inverse heterogeneity size then damping is the largest (we used this assumption when plotting Fig. 4 rather than using (10)).

We also will discuss the correlation of polarization in the case of a finite volume of the polar region. Let us write dielectric permittivity of a finite uniform volume V:

$$\chi \sim \frac{T}{V^2} \int dV\, dV' \langle P(r)P(r')\rangle \sim \frac{T_{CW}}{\kappa V} min(r^2, r_c^2) \qquad (11)$$

where an Ornstein-Cernike correlator was used, κ being the constant in front of $(\nabla P)^2$ in the Landau fluctuation energy, the integration is over the finite volume V, r is the radius of the volume V, and $r_c \sim (T - T_{CW})^{-1/2}$, in the first approximation. It is seen from this expression that, at high temperatures, susceptibility behaves according to the Curie-Weiss law; when the ferroelectric correlation radius reaches the volume V size then susceptibility saturates. The decrease in polarization at the boundary of a polar region would smoothen this crossover. At temperatures lower than this crossover, the interaction between different polar regions is responsible for the temperature dependence of susceptibility, and a ferroelectric or glass-type phase transition of the order-disorder type takes place due to the correlation of different polar regions. Below the temperature of this phase transition the ferroelectric fluctuations inside the polar regions become harder and a Curie-Weiss temperature dependence is seen again [8] because the correlation radius again becomes lower than the average polar region size.

We finally consider the Hamiltonian consisting of the part describing the long-range ordered polarization $P_\mathbf{q}$ (with a wave vector $\mathbf{q}$) and coupled with it local polar vibrations with local polarization P_l:

$$\begin{aligned} H &= \frac{1}{2}\alpha_\mathbf{q} P_\mathbf{q}^2 + \frac{1}{2}\alpha_l P_l^2 - \alpha_{\mathbf{q}l} P_\mathbf{q} P_l + \frac{1}{4}\beta P_\mathbf{q}^4 + \frac{1}{6}\gamma P_\mathbf{q}^6 \\ &+ \frac{1}{4} B P_l^4 + \frac{1}{2} g P_\mathbf{q}^2 P_l^2 - E P_\mathbf{q} - E P_l \end{aligned} \qquad (12)$$

The susceptibility connected with quasilocal vibrations at zero polarization $P_\mathbf{q}$ can be found by an ordinary procedure [5]:

$$\chi_l = \frac{\chi_{l0}}{1 - \alpha_{\mathbf{q}l}^2 \chi_{\mathbf{q}0}} \qquad (13)$$

where

$$\chi_{l0} = \frac{1}{\alpha_l + 3\beta P_l^2} \qquad (14)$$

$$\chi_{\mathbf{q}0} = \frac{1}{\alpha_\mathbf{q} + g P_l^2} \qquad (15)$$

It is seen that the quasilocal vibrations become unstable when $\alpha_{\mathbf{q}l}^2 \chi_{\mathbf{q}0} < 1$. It implies that the local vibrations satisfying this condition will freeze in and will be arranged in space in accordance with the wave vector $\mathbf{q}$. These results are consistent with resent experimental findings [9] that there is a local transformation at Burns temperature.

Partially supported by RFBR grants 01-03-33119 and 01-02-16029.

REFERENCES

1. Bratkovsky A. M. and Levanyuk A. P., *Phys. Rev. Lett.* **84** 3177; *ibid* **86** 3642 (2001).
2. Raevski I. P., Prosandeev S. A., Waghmare U., Eremkin V. V., Smotrakov V. G., Shuvaeva V. A., cond-mat/0208116.
3. Hirota K., Ye Z. -G., Wakimoto S., Gehring P. M., and Shirane G., *Phys. Rev. B* **65**, 104105 (2002).
4. Gehring P. M., Wakimoto S., Ye Z. -G. and Shirane G., *Phys. Rev. Lett.* **87**, 277601-1 (2001).
5. Prosandeev S. A., Kleemann W. and Dec J., *J Phys: Condens Matter* **13**, 5957 (2001).
6. Prosandeev S. A., Trepakov V. A., Savinov M. E., Jastrabik L. and Kapphan S. E., *J. Phys: Condens. Matter* **13**, 9749 (2001).
7. Yamada Y., Takakura T., cond-mat/0209573.
8. S. Wakimoto, C. Stock, Z.-G. Ye, W. Chen, P. M. Gehring, and G. Shirane, *Phys. Rev. Lett.* **66**, 224102 (2002).
9. Dul'kin E., Raevski I. P., and Emelyanov S. M., *Phys Sol St* **45**, 158 (2003).

Temperature Dependence of the Local Structure in Pb Containing Relaxor Ferroelectrics

T. Egami, E. Mamontov and W. Dmowski

*Department of Materials Science and Engineering,
and Laboratory for Research on the Structure of Matter,
University of Pennsylvania, Philadelphia, PA 19104, USA*

S. B. Vakhrushev

A. F. Ioffe Institute, St. Petersburg, 194021, Russia

Abstract: Temperature dependence of the local structure of relaxor and related ferroelectric oxide systems, such as PMN, PST and PZT, are discussed based upon the results of the pulsed neutron atomic pair-density function (PDF) analysis. Optical data have shown various changes in the local structure taking place much above the relaxor freezing temperature. However, the origins of such changes have long been controversial. We show that the Burns temperature is the local Curie temperature at which the polar nano-regions are formed, and the transition is of the order/disorder type for Pb polarization. Local Pb polarization persists up to temperatures several hundreds of degrees above the Burns temperature where the Raman scattering suggests disappearance of the local distortion in the oxygen environment.

INTRODUCTION

Relaxor ferroelectric oxides, such as $Pb(Mg_{1/3}Nb_{2/3})O_3$ (PMN), show a diffuse and frequency dependent maximum of dielectric susceptibility at Tg near the room temperature. But some features suggestive of local ferroelectric polarization persist well above T_C, as evidenced by optical refractive index [1,2], and Raman intensities [3]. The origin of these features, however, is unclear and has been a subject of controversy [4,5]. The purpose of the present study is to determine the local structure of some ferroelectric oxides containing Pb, including PMN, and discuss the origin of the dielectric activities above T_C in relaxor ferroelectrics. In addition to PMN, we studied chemically ordered $Pb(Sc_{1/2}Ta_{1/2})O_3$ (PST) and $Pb(Zr_{0.52}Ti_{0.48})O_3$ (PZT52/48), both of which show normal ferroelectric transition but are closely related to relaxors, namely disordered PST and PLZT. This study will demonstrate the universality in the temperature dependence of the local environment of Pb in all the compounds studied, the similarity in the nature of the Burns temperature of PMN with the Curie temperature of PST and PZT, the order/disorder character of these transitions, and the possible disappearance of the local distortions of the PbO_{12} group around 1000 K.

LOCAL STRUCTURE OF Pb CONTAINING FERROELECTRICS

Pulsed neutron diffraction experiments were carried out at the SEPD station of the IPNS, Argonne National Laboratory on the powder of PMN, PST and PZT52/48. The PMN sample was provided by Prof. A. Bhalla of the Pennsylvania State University [6], and the powders of PST and PZT were prepared in the laboratory of Prof. P. K. Davies of the University of Pennsylvania [7,8]. The diffraction data were analyzed both by the Rietveld method for the average crystal structure, and the atomic pair-density function (PDF) analysis for the local structure. PZT52/48 is close to the morphotropic phase boundary, and has a monoclinic structure at low temperatures [9].

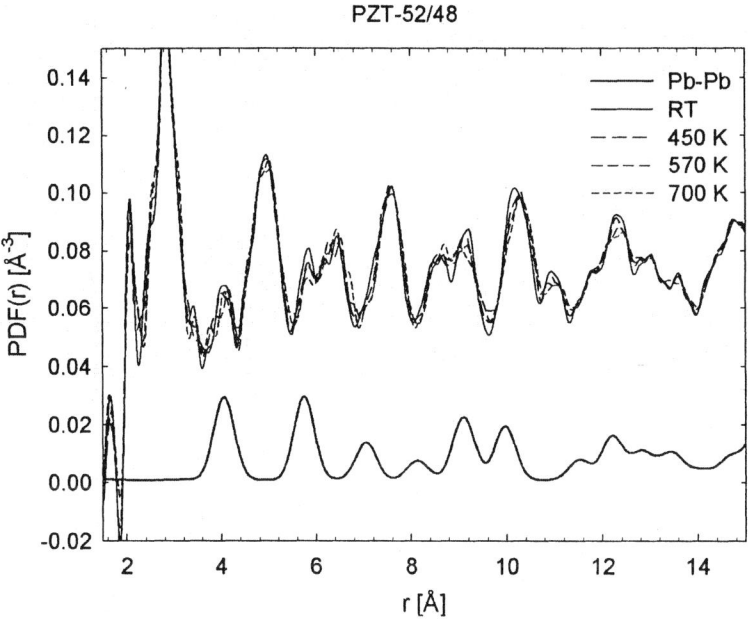

Fig. 1 Change in the PDF of PZT-52/48 with temperature (above), and the Pb-Pb partial PDF (below).

The PDF's of PZT at various temperatures up to 15 Å are shown in Fig. 1. Normally the PDF peaks become lower and wider as temperature is increased due to atomic displacements. It is quite clear that the changes are not uniform, with some peaks showing much stronger temperature dependence than others. The main reason for this non-uniformity is that different PDF peaks represent different atomic correlations. For instance the first positive peak at 2.05 Å represents the Zr-O distance, the second peak at

2.8 Å is due to O-O and Pb-O distances, etc. Since the average structure can be determined by the Rietveld analysis we can calculate the partial, compositionally resolved PDF of the average structure and compare them with the experimental PDF. A careful inspection suggests that large changes in the PDF with temperature coincide with the Pb-Pb separation, as shown in the lower portion of Fig. 1, indicating that what changes most with temperature is the Pb-Pb correlation. The situation is the same for PMN.

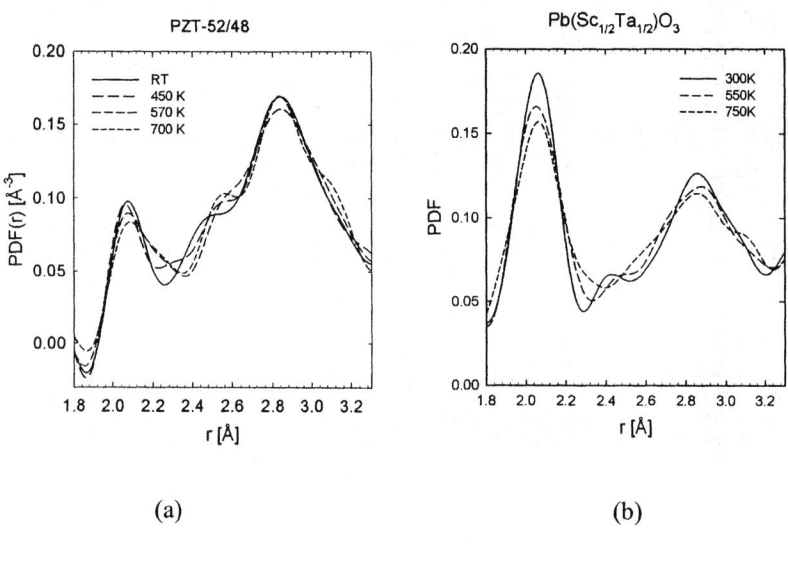

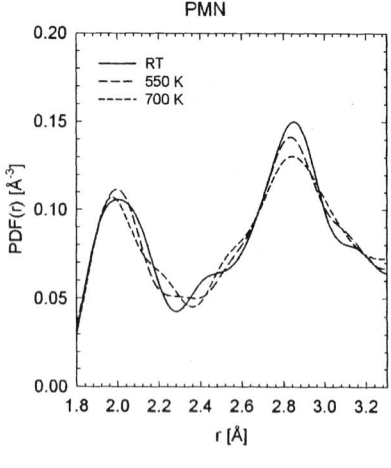

Fig. 2 The PDF of (a) PZT-52/48, (b) PST, and (c) PMN over the first and second peaks. Note the subpeak of the second peak around 2.4 – 2.6 Å moves with temperature.

In addition, the sub-peak of the second peak at 2.4 – 2.6 Å was found to show significant temperature dependence, as shown in Figs. 2 for PZT, PST and PMN. This peak represents short Pb-O distances due to Pb off-centering in the oxygen cage [10], and is clearly seen as an isolated peak in the PDF of PbZrO$_3$ [11]. The subpeak position, R_{SP}, increases in r with temperature. Since the average Pb-O distance is R_{Pb-O} = 2.8 Å, ΔR = $R_{Pb-O} - R_{SP}$ represent the off-centering, and thus is proportional to the local Pb plarization. Fig. 3 shows ΔR as a function of temperature. It is noticed that the data for three samples are very similar, and therefore the temperature dependence of the local Pb polarization must be very similar for these three compounds.

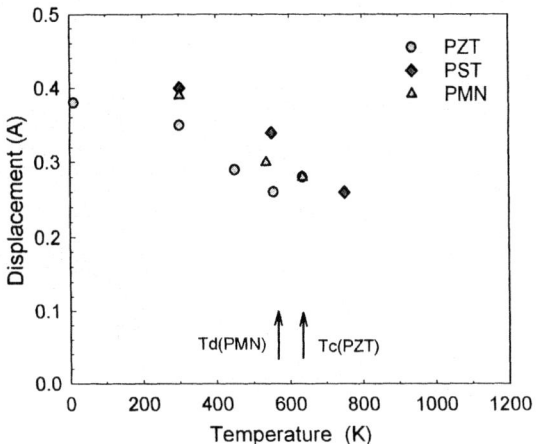

Fig. 3 Temperature dependence of Pb off-centering represented by the change in the short Pb-O bond length from the average (2.8 Å) for PZT, PST and PMN. Burns temperature of PMN, T_d, and the Curie temperature of PZT, T_C, are also shown.

DIELECTRIC TRANSITIONS

The three compounds studied here, PMN, PST and PZT52/48, show quite different dielectric properties. PMN is a relaxor ferroelectric with the freezing temperature, T_g, around 230 K [12]. Both the ordered PST and PZT52/48 are ferroelectric, with T_C of 160 K for PST and 630 K for PZT. Thus it is rather remarkable that all three showed very similar temperature dependence of the local Pb polarization. The most logical way to explain this apparent paradox is that the basic parameters that determine the local Pb polarization, such as the local structure around Pb and the bond strengths, are similar, but the dielectric properties are determined by other factors involving the B-site ions.

Among them PZT is the simplest case, since the value of T_C in PZT changes nearly linearly from PZ to PT [13]. In spite of complications associated with the incompatibility of the Ti and Zr environments [8], apparently there is little effect of frustration on the value of T_C. Thus this value, ~600 K, is a measure of the ferroelectric interactions in the system, a "mean-field value" of T_C, $T_C(MF)$. While macroscopically PST is a regular ferroelectric, the local structural study by the pulsed neutron PDF analysis revealed deep frustration in the system [7]. Even though the average polarization is along <111>, locally Pb wants to be polarized along <100>. However, <100> polarization is frustrated by the size effect of Ta, resulting in non-collinear polarization. Thus chemical disorder of Sc and Ta drives the system to the relaxor state. The sample studied is nominally B-site ordered (Sc/Ta), but ordering is not perfect (87 %), and this small disorder may be enough to frustrate the system. The low value of T_C in PST, therefore, is likely to be the consequence of such frustration, and is suppressed from the mean-field value.

PMN has a T_g value of 230 K. There is convincing evidence that T_g simply represents the spin-glass freezing of polar nano-regions [14,15], while the polar nano-regions persist up to Burns temperature, T_d. The data shown in Fig. 3 clearly demonstrate the similarity of $T_C(MF)$ and T_d, and characterize T_d as the *temperature of local ferroelectric transition* [16]. Because of geometrical frustration related to the chemical disorder of Mg and Nb local alignment of polarization does not extend to long range, and merely forms polar nano-regions. Because of the small size of the nano-regions they remain super-paraelectric down to T_C where they freeze in a similar manner to spin-glasses.

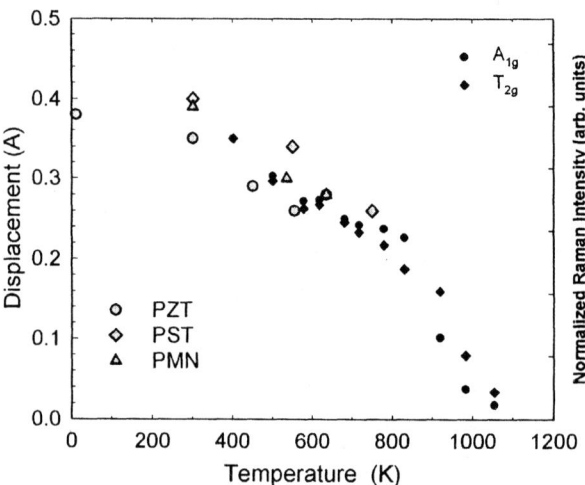

Fig. 4 Pb off-centering (Fig. 3) compared to the Raman intensity normalized by the thermal factor [3].

It is very interesting to note that around $T_C(MF)$ and T_d the Pb polarization is still about 70 % of the saturation value at low temperatures. The data for PMN shows the local Pb polarization persists above T_d. This strongly suggests that the ferroelectric transition in PZT at T_C and the local ferroelectric transition in PMN at T_d are of order/disorder in nature. It is also interesting to compare the data in Fig. 4 with the intensity of the phonon Raman scattering normalized by dividing with the thermal factor (Bose-Einstein factor) as shown in Fig. 4. The two phonon modes (A_{1g} and T_{2g}) are Raman inactive (pure IR) in the average structure, however, they become Raman active below 1000 K, indicating the local symmetry is lower than the average symmetry below that temperature. The similarity of the two data is obvious, and strongly suggests that the local symmetry breaking for the phonons observed by Raman scattering is due to local PbO_{12} distortion, that disappears around $T_L \approx 1000$ K. The temperature dependence of the order parameter shown in Fig.4, however, suggests that the transition at T_L may have a second-order character. Thus this transition may not be a purely local phenomenon, and some collectiveness could be associated with it. One possibility is that it is an antiferroelectric transition with limited correlation length, involving local rotation.

Thus our data clearly demonstrate that the dielectric changes in PMN occur at three distinct temperature ranges representing three different energy scales of interaction. T_L must be related to the anharmonicity of the Pb-O potential in the PbO_{12} cluster. T_d is determined by the inter-ionic ferroelectric interaction, while T_C depends on the size of the polar nano-regions and interactions among them. The similarity of the temperature dependence of the local PbO_{12} distortions among the three compounds, and that of $T_C(MF)$ and T_d suggest that the inter-ionic ferroelectric interactions are similar for these three, in spite of differences in the dielectric properties of the B-site ions. Ti, Nb and Ta can be ferroelectrically active, although Nb in PMN and Ta in PST appear to be only weakly polarized [7,17], and Zr, Mg and Sc are not ferroelectric. Thus the effect of the B-site ions in ferroelectric interaction must be rather weak. Their main role appears to be to produce frustrations through the size effect.

CONCLUSIONS

The evolution of the local structure with temperature was studied for three Pb containing dielectric perovskites. In spite of great differences in the dielectric properties the changes in the Pb environment with temperature are remarkably similar in all three compounds. The dielectric behavior in these compounds is largely determined by Pb polarization, with the B-site ions playing a minor role. The similarity in the temperature dependence of Pb polarization between PZT and PMN helps to identify the Burns temperature as the local ferroelectric transition temperature to form polar nano-regions. Our results also suggest that the local polarization of PbO_{12} persists up to around 1000 K, and ferroelectric transitions in PZT and PMN (local) are of order/disorder type. Thus three distinct energy scales and three distinct interactions associated with them have been clarified and identified.

ACKNOWLEDGEMENTS

This work was supported by the ONR N000-14-01-10860 and the CRDF RP1-2361-ST-02. The IPNS is supported by the U.S. Department of Energy, Division of Materials Sciences, under contract W-31-109-Eng-38.

REFERENCES

1. Burns, G. and Dacol, F. H., *Solid St. Commun.* **48**, 853 (1983).
2. Burns, G and Scott B. A., *Solid St. Commun.* **13**, 423 (1973)
3. Siny, I. G. and Smirnova, T. A., *Ferroelectrics* **90**, 191 (1989).
4. Vakhrushev, S. and Shapiro S., *Phys. Rev. B* **66**, 214101 (2002)
5. Wakimoto, S., Stock, C., Ye, Z.-G., Chen, W., Gehring, P. M. and Shirane G., *Phys. Rev. B* **66**, 214102 (2002)
6. Egami, T, Rosenfeld, H.D., Toby, B.H. and Bhalla, A., *Ferroelectrics* **120**, 11 (1991).
7. Dmowski, W., Akbas, M. K., Davies, P. K. and Egami, T., *J. Phys. Chem. Solids* **61**, 229 (2000).
8. Dmowski, W., Egami, T., Farber, L. and Davies, P.K., *AIP Conf. Proc.* **582**, 33 (2001).
9. Noheda B., Gonzalo J. A., Cross L. E., Guo R., Park S.-E., Cox D. E. and Shirane G., *Phys. Rev. B*, **61**, 8687 (2000).
10. Egami, T., Dmowski, W., Akbas, M. and Davies, P. K., *AIP Conf. Proc.* **436**, 1 (1998).
11. Teslic, S. and Egami, T., *Acta Cryst.* **B54**, 750 (1998).
12. Vakrushev, S., Nabereznov, A., Sinha, S.K., Feng, Y.P. and Egami, T., *J. Phys. Chem. Solids* **57**, 1517 (1996).
13. Jaffe, B., Cook, W. R. and Jaffe, H., *Piezoelectric ceramics*, London, Academic, 1971.
14. Viehland D., Lang S. J., Cross L. E. and Wuttig M. *Phys. Rev. B* **46**, 8003 (1992).
15. Westphal, V., Kleemann, W. and Glinchuk, M. D., *Phys. Rev. Lett.* **68**, 847 (1992).
16. Vakhrushev, S. B. and Okuneva, N. M., *AIP Conf. Proc.* **626**, 117 (2002).
17. Egami, T., *Ferroelectrics* **267**, 101 (2002).

Anti-ferrodistortive Nanodomains in PMN Relaxor

A. Tkachuk*† and Haydn Chen**

*Department of Materials Science and Engineering and Materials Research Laboratory,
University of Illinois at Urbana-Champaign, Urbana, IL 61801
†Current addresses: Advanced Photon Source, Argonne National Laboratory, Argonne, IL 60439
**Department of Physics and Materials Science, City University of Hong Kong, Kowloon

Abstract. Temperature dependent studies of the 1/2(hk0) superlattice reflections (α spots) by synchrotron x-ray scattering measurements were performed in Pb(Mg$_{1/3}$Nb$_{2/3}$)O$_3$ (PMN) and [PbMg$_{1/3}$Nb$_{2/3}$O$_3$]$_{1-x}$ – [PbTiO3]$_x$ (PMN-xPT) with Ti doping $x \leq 0.32$ single crystals. Separation of the α spots from the underlying diffuse scattering background allowed studying them as separate entities for the first time. Structure factor calculations have shown that α spots constitute the presence of a new kind of anti-ferrodistortive nanoregions (AFR) in the form of fluctuations produced by anti-parallel short-range correlated $\langle 110 \rangle$ Pb^{2+} displacements. AFR appear to be different and unrelated to the chemical nanodomains (CND) and ferroelectric polar nanoregions (PNR). Simultaneous presence of AFR and PNR can explain relaxor behavior as a result of competition between randomly occurring ferroelectric and anti-ferroelectric fluctuations. Temperature dependence of the α spots in PMN showed a direct correlation with the freezing phase transition near T$_f$≈220 K.

INTRODUCTION

Pb(Mg$_{1/3}$Nb$_{2/3}$)O$_3$ (PMN) [1] is a relaxor ferroelectric, which among other properties exhibits strong dependence of real χ' and imaginary χ'' parts of dielectric susceptibility on temperature and frequency f of the applied AC electric field. A broad peak in χ' temperature dependence, for example, occurs at 265 K (f=1 kHz), which shifts to T$_f$ ≈220 K as f decreases to zero [1].

Although average crystal structure is indistinguishable from cubic at all temperatures (5-800 K) [2, 3], ferroelectric rhombohedral polar nanoregions (PNR) were postulated to exist in PMN below T$_d$ ≈620 K [4]. Relaxor behavior in PMN is commonly attributed to PNRs, which undergo cooperative freezing near T$_f$ into a glass-like state, reminiscent of magnetic spin glasses, due to random fields produced by underlying chemical and displacement disorder [1, 5, 6, 7]. Spherical-Random-Bond-Random-Field (SRBRF) model is a recent self-consistent theoretical treatment for PMN type relaxors as a special kind of spherical dipole glass [7].

The short-range scale of the nanodomains complicates the direct observation of the PNRs with diffraction techniques. Nevertheless, existence of the PNRs as ferroelectric fluctuations was attributed to the strong temperature dependent neutron diffuse scattering near the Brillouin zone center ($q \ll 0.1$ r.l.u.) below T$_d$ [8, 9]. Correlation radius of these fluctuations was shown to increase from ~50 Å near T$_d$ to ~200 Å saturation value below T$_f$ [8]. Alternatively, the anisotropic x-ray diffuse scattering along $\langle 01\bar{1} \rangle^*$ cubic reciprocal lattice directions was attributed to pure transverse optical (TO) soft modes [10]. Other workers claimed to resolve diffuse scattering interpretation controversy by introducing a phase-shifted soft mode model of PNRs [10]. Moreover, inelastic neutron scattering experiments [11, 12] identified the TO soft mode above T_d, which becomes overdamped at lower temperatures due to condensation of the aforementioned PNRs [12]. Interestingly, subsequent recovery of the Curie-Weiss behavior coincides with the freezing temperature T_f [13]. The presence of a distinct thermodynamic phase transition near T_f into a nonergodic frozen dipolar glass state was suggested in the past based on electroacoustic studies of PMN [14].

An evidence of a new kind of anti-ferrodistortive nanoregions (AFR) in PMN, which can be envisioned as fluctuations different from PNRs, is presented in this work based on temperature dependence and structure factor calculations of 1/2(hk0) short-range order superlattice peaks. The simultaneous presence of AFRs and PNRs is important for understanding PMN relaxor in terms of competing anti-ferroelectric and ferroelectric interactions on the nanometer scale, which is supported by the SRBRF model [7].

EXPERIMENTAL

All scattering studies were performed on PMN single crystals produced by Czochralski and Bridgeman methods. PMN crystals doped with $PbTiO_3$ (PMN-xPT) x≤0.32 were grown by the melted flux method. Crystals were in the form of ⟨001⟩ or ⟨111⟩ oriented platelets having surfaces with linear dimensions no larger than 3-7 mm and thickness ∼1 mm. PMN(111) crystals were sputtered with gold for in-situ x-ray measurements under applied electric field (up to 4 kV/cm). All crystals were of good quality with a mosaic spread no worse than 0.01°, except PMN-0.32PT, which exhibited ∼0.3° mosaic, obtained from x-ray diffraction rocking curve measurements. Dielectric spectroscopy results obtained from the same crystals were published previously [15].

Synchrotron x-ray work was conducted on X-18A beamline at the National Synchrotron Light Source (NSLS), Brookhaven National Laboratory, and at 33-ID beamline at the Advanced Photon Source (APS), Argonne National Laboratory. Both beamlines used focusing mirrors, which also served as high energy harmonic discriminators. Crystals were studied in 10-300 K range inside closed cycle He gas cryostats mounted on 4-circle or 6-circle kappa diffractometers.

Measurements in 300-800 K range were conducted on in-house (CuK_α, 40 kV, 200 mA) Rigaku rotating anode source equipped with focusing pyrolytic graphite monochromator located before the sample mounted inside the evacuated heating stage.

Energy of the incident x-rays at NSLS was chosen to be 10 keV for optimal beamline performance. At APS the x-ray energy in addition to 10 keV was tuned near Nb K (18.99 keV) and Pb L_{III} (13.035 keV). The size of the incident beam was collimated by a pair of slits set to 0.5 × 0.5 mm^2 at APS, 1 × 1 mm^2 at NSLS and 2 × 2 mm^2 for rotating anode. Diffuse scattering measurements at rotating anode required open detector slits and 40 sec/point counting times to achieve reasonable statistics. In contrast, the high flux of the synchrotron sources allowed detector slit size no larger than 1 × 1 mm^2. Typical counting times for diffuse scattering measurements was 1 sec/point at APS and

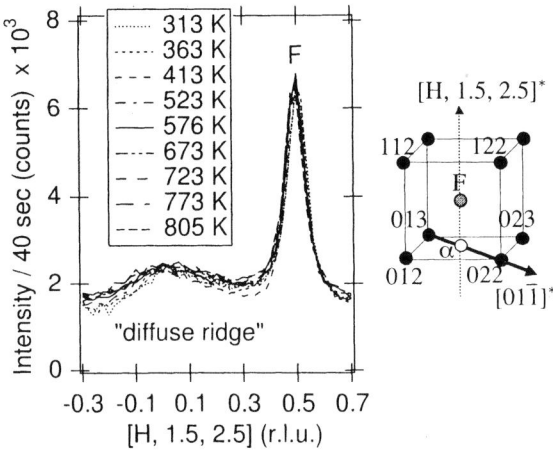

FIGURE 1. Linear [H,1.5,2.5]* scan in the reciprocal unit cell (shown on the right) measured in PMN above T_f at 313-805 K. Fundamental Bragg peaks are at the corners of the cube. Two superlattice peaks are on the face-centered and the body centered positions (H=0 and H=0.5) along the scan direction.

≤10 sec/point at NSLS. High energy harmonics in the incident x-ray beam for all of our experiments were suppressed to the level that did not cause any observable data contamination. Distribution of the scattered intensity in the reciprocal space was measured by fully automated computer-controlled reciprocal space scans. 2D mesh scans allowed for direct point-by-point reciprocal space mapping. The interval between the scan points was ~0.01-0.02 reciprocal lattice units (r.l.u.), where 1 r.l.u. is $2\pi/a \approx 1.55$ Å^{-1} and $a \approx 4.04$ Å is PMN's lattice constant. 2D diffuse scattering measurements were performed 0.1 r.l.u. away from the Bragg peak centers.

RESULTS AND DISCUSSION

Figure 1 depicts linear $[H, 1.5, 2.5]^*$ reciprocal scan measured in PMN above $T_f \approx 220$ K. Its direction in the reciprocal space can be found on the sketch of the reciprocal unit cell, where the corners are fundamental $Pm\bar{3}m$ Bragg peaks. Two Brillouin zone boundary peaks at H=0.5 and H=0 (F and "diffuse ridge") occupy body-centered and face-centered positions in the reciprocal cell, respectively. These peaks are more than ~10^8 times weaker and ~100 times broader than Bragg peaks. Figure 1 shows that both peaks are temperature independent even above $T_d \approx 620$ K, where PNRs start to nucleate [4]. Diffuse scattering distribution in the surrounding reciprocal space, required for unambiguous peak interpretation, was measured using planar 2D mesh scans.

Figure 2 shows reciprocal mesh scans about (022) Bragg peak in PMN at 220 K and in PMN-0.32PT at 300 K in the plane shaded in Fig. 3. Note that the linear H scan in Fig. 1 is also a part of this plane. (022) Bragg peak is outside of vertical scale and only the diffuse scattering tails ≥0.1 (r.l.u.) away from the peak center are visible. Moreover, FWHM of the (022) Bragg peak is ~0.001 (r.l.u.), which is in agreement with its long-

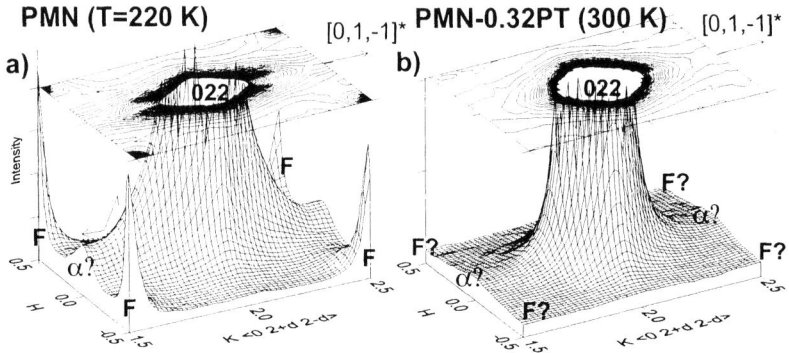

FIGURE 2. Diffuse scattering intensity maps of the shaded area in Fig. 3 near 022 Bragg peak (q>0.1 r.l.u.): (a) PMN at 220 K ; (b) PMN-0.32PT at 300 K. Horizontal axes correspond to [110]* and [011]* reciprocal lattice directions. Diffuse scattering intensity is plotted on the vertical linear scale.

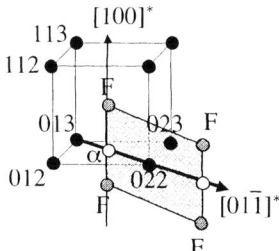

FIGURE 3. Reciprocal unit cell: Bragg peaks (cube corners); F and α spots (body-centered and face-centered positions). Mesh scans in Fig. 2 were measured in the shaded plane.

range order nature. Positions of the gray circles in Fig. 3 correspond to superlattice peaks at four corners of Fig. 2(a), which are referred as F spots in the literature [16]. Origin of the F spots is commonly attributed to chemical nanodomains (CND) produced by $\langle 111 \rangle$ correlated Nb/Mg short-range order [16, 17]. Measured FWHM≈0.75 (r.l.u.) of the F spots in Fig. 2(a) and Fig. 1 corresponds to expected for CND ~50 Å correlation length, which was obtained using the well-known Scherrer equation [18].

Existence of the face-centered 1/2(0kl) superlattice peaks [16, 17, 19, 20, 21], commonly referred as α spots in the literature, is questionable from Fig. 2(a), since diffraction peaks produced by any kind of real space 3D correlations must exhibit a peak cross section in any direction in the reciprocal space [18]. However, there is no clear evidence for any cross section of the α spot (marked with a question mark) along [01$\bar{1}$]* direction (marked with a double ended arrow). It appears that a peak cross section at the position of the α spot in Fig 2(a) along H ([100]*) direction is produced predominantly by the tail of the diffuse scattering ridge extending from the (022) Bragg peak. The FWHM≈0.25 (r.l.u.) of this cross section corresponds to ~15 Å correlation length at 220 K. This same peak cross section is labeled "diffuse ridge" in Fig. 1 when measured above 220 K. These results can explain observation of the α spots in PMN above

~220 K [16, 17, 19, 21] in some cases due to measuring "diffuse ridge" cross sections instead of actual superlattice peaks from 3D correlations. In fact, the first report of the α spot with synchrotron x-rays was also presented using linear [100]* reciprocal scan, which cuts through [01$\bar{1}$]* diffuse scattering ridge [17].

Fig. 2(b) demonstrates that no superlattice peaks of any kind were found in PMN-0.32PT crystal at 300 K from the mesh scan performed with an identical experimental setup. No superlattice reflections were observed for this composition at any temperature down to 10 K. Anisotropy and strength of the diffuse scattering also appears to be much weaker than in the case of pure PMN.

It is well know that diffuse scattering in PMN at $q \ll 0.1$ (r.l.u.) from the Brillouin zone centers exhibits strong temperature dependence [17, 22, 10]. In contrast, temperature independent (see Fig. 1), up to 800 K, diffuse scattering at zone boundaries (q=0.5 r.l.u.) suggests that diffuse scattering observed in different parts of the reciprocal space comes from different origins. Moreover, temperature independent $\langle 1\bar{1}0\rangle$* diffuse scattering ridges were reported to exist in related Pb(Sc$_{1/2}$Nb$_{1/2}$)O$_3$ (PSN) relaxor [23]. The origin of the ridges has been attributed to static or dynamic displacements in {1$\bar{1}$0} planes without (or with weak) correlation between the planes along $\langle 1\bar{1}0\rangle$ directions. Presence of the α spots in PSN was attributed to existence of the linear anti-ferroelectric chains [24]. However, displacement correlation lengths were determined from the α spots without taking into account cross sections of the overlapping $\langle 1\bar{1}0\rangle$ diffuse ridges.

It appears that α spots in PMN can be studied as separate entities from the diffuse ridges only below 220 K [25]. Ambiguity related to the detection and interpretation of the α spots above T$_f$ [16, 17, 19, 21] can be resolved by proper separation of the temperature dependent α spots from temperature independent anisotropic $\langle 01\bar{1}\rangle$* diffuse ridges at zone boundaries [26]. Therefore, only $\langle 01\bar{1}\rangle$* reciprocal scans were used to obtain widths and integrated areas of the α spots needed for structure factor calculations.

Right hand side of Fig. 4 depicts another 2D mesh scan obtained in PMN at 45 K. It was measured 0.1 (r.l.u.) away from four Bragg peaks on the bottom face of the cube in Fig. 3. Note that α spot, which was not pronounced at 220 K along $\langle 01\bar{1}\rangle$* in Fig. 2(a), is now clearly visible along this same direction on top of the diffuse scattering ridge connecting (013) and (022) Bragg peaks. Left hand side of Fig. 4 shows two pairs of linear scans extracted from the similar mesh scans at 300 K and 40 K along two mutually perpendicular [0$\bar{1}$1]* and [011]* directions. Directions of these scans are indicated on the mesh scan with two diagonal arrows labeled by the same type of markers. Figure 4(b) proves existence of the 1/2(035) α spot at 40 K. At the same time Fig. 4(a) reveals a broad cross section of the [0$\bar{1}$1]* diffuse ridge at 300 K (similar to the one in Fig. 1) and also proves the absence of the α spot above T$_f$. Note that data in Fig. 4(a) and (b) are plotted on the same vertical scale, which proves that intensity of the diffuse ridge (solid circles) does not exhibit any pronounced changes below or above T$_f$.

Similar low T measurements, which are presented in Fig. 5, were performed on all six faces of the cube shown in Fig. 3. These data clearly demonstrate that large errors would occur if integrated intensities of the α spots are extracted from the empty triangle scans rather than from the solid circle ones. The relative intensities of the $\langle 01\bar{1}\rangle$* ridges are different and appear to be correlated with the structure factors of the Bragg peaks that they connect. For example, 1/2(145) and 1/2(136) α spots in Fig. 5(e) and (f), which are

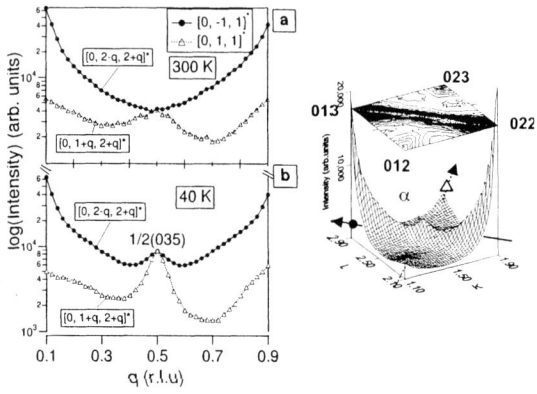

FIGURE 4. Proof of coexistence of the α spot and diffuse ridge in PMN below T_f from reciprocal scans along $[0\bar{1}1]^*$ (solid circles) and $[011]^*$ (empty triangles) measured at: (a) 300 K; (b) 40 K. Directions of the scans are indicated with two arrows on the mesh scan (right hand side of the graph) obtained at 45 K.

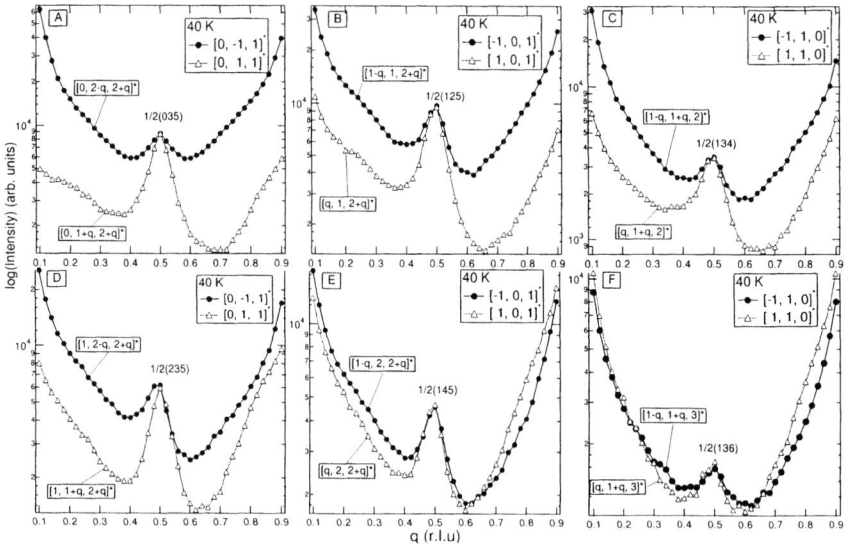

FIGURE 5. Direct proof of the α spot presence below T_f on all six cubic faces in Fig. 3. Directions of the $\langle 01\bar{1}\rangle^*$ diffuse ridges are along the curves plotted with solid circles.

located between relatively weak Bragg peaks, are the least affected by the anisotropy of the diffuse scattering background.

Figure 6(a) presents detailed temperature dependence of the 1/2(035) α spot measured along $\langle 01\bar{1}\rangle^*$ diffuse ridge without electric field. Integrated intensity and FWHM values together with the ones obtained from the 1/2(145) α spot are plotted vs. temperature in Fig. 6(b). Normalized remanent polarization P_r digitized from Ref. [5] is also shown on

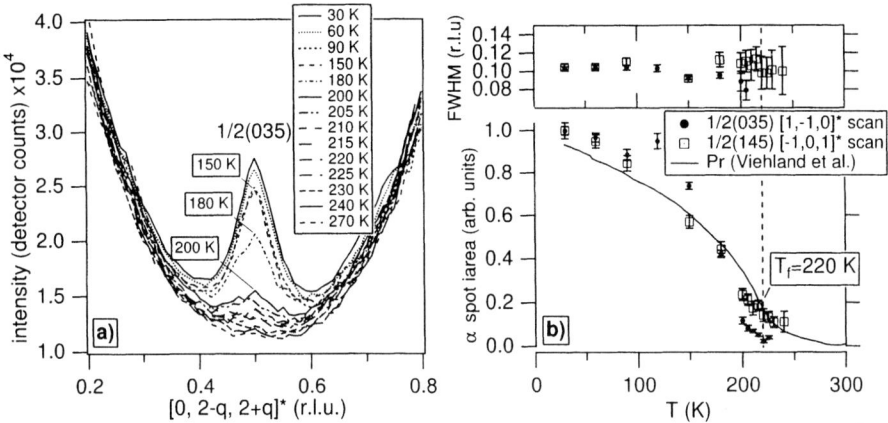

FIGURE 6. (a) Temperature dependence of the 1/2(035) α spot along the ridge in PMN; (b) integrated intensity and FWHM extracted from 1/2(035) and 1/2(145) α spots. Normalized remanent polarization P_r is from Ref.[5] as a guide to the eye.

the same graph as a guide to the eye. Remarkably, significant reduction in the intensity of the α spots on heating occurs near the phenomenological freezing phase transition T_f, which is not accompanied by any kind of Bragg peak splitting [3, 2] in the absence of applied electric field [27, 28]. This does not contradict average cubic structure, since FWHM≈0.1 (r.l.u.) of the α spots in Fig. 6 corresponds to ~30 Å correlation length. It also appears to be temperature independent below T_f within experimental errors.

Similarly, in PMN-0.06PT α spots appear below 265 K [25], which corresponds to the freezing temperature for this composition. However, in PMN-0.1PT only $\langle 01\bar{1}\rangle^*$ diffuse ridges were present at 100 K [26], while F and α spots were not resolvable from the diffuse scattering background for this and higher PT compositions.

Measured at 40 K, integrated intensities of the α spots are represented as white bars in Fig. 7 after performing standard absorption and Lorentz factor data corrections [18]. All the peaks have the same FWHM regardless of their location from the center of the reciprocal space. This indicates that the width of the α spots is primarily defined by the 30 Å size of the corresponding nanodomains. On the contrary, lattice strain peak broadening generally increases away from the center of the reciprocal space [18].

Temperature dependence of the α spots excludes their chemical origins and supports presence of correlated on the nanometer scale atomic displacements along all symmetry equivalent $\langle 011 \rangle$ cubic directions, thereby retaining average cubic symmetry of the whole crystal. According to Glazer classification scheme [29], α spots may result from in-phase oxygen octahedra rotations, common to perovskites. However, structure factor calculations show that 1/2(033) and 1/2(211) reflections underlined in Fig. 7 must be systematically absent [26], which indicates that displacements of other atoms must be involved. For example, Pb^{2+} ions with their lone electron pair are known to be displaced from their equilibrium sites [9, 30]. These displacements will contribute to the structure factor of the α spots only if some fraction of them is correlated in anti-

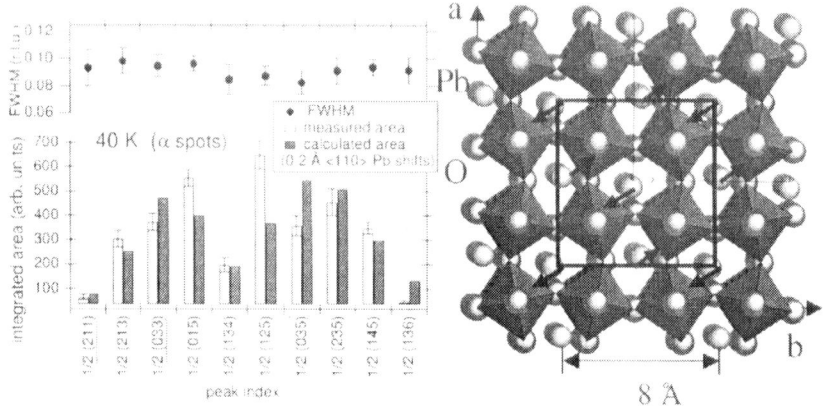

FIGURE 7. Observed and calculated α spot integrated intensities using 0.2 Å Pb^{2+} displacements correlated along equivalent $\langle 011 \rangle$ directions within 30 Å nanodomains. Shown [110] Pb displacements double unit cell along a and b directions. In-phase oxygen octahedra rotations are along the c axis.

parallel fashion that doubles the unit cell in $\langle 011 \rangle$ directions within 30 Å nanodomains. The results of the least squares fit with only two variable parameters are presented as gray bars in Fig. 7. All reflections were fitted at once. One of the fitting parameters was a magnitude of the Pb displacements δ, and another was common for all peaks intensity scaling factor. Debye-Waller factors were taken from the literature [9]. Corresponding Pb displacements with δ=0.2 Å, which gave the best fit, are presented on the right hand side of Fig. 7. Structure factor calculations showed, that this pattern would only contribute to the α spots, which have even third Miller index (l), such as 1/2(134) and 1/2(136) in Fig. 5. Similar patterns with Pb displacements having 30 Å correlation length along other equivalent $\langle 011 \rangle$ directions will contribute to reflections with h or k even indices, accordingly.

Inclusion of Nb displacements or in-phase oxygen octahedra rotations with angles up to 10° only marginally improved the fit. Neutron measurements, which are more sensitive to oxygen, are clearly needed in order to better understand the role of oxygen octahedra rotations. What is important is that oxygen octahedra alone cannot explain α spot structure factor without Pb displacements. Contribution of Pb displacements to the α spots is consistent with differential anomalous factor scattering (DAFS) measurements near Pb L_{III} absorption edge [26]. On the contrary, we were not able to find any evidence for Nb contribution from the DAFS measurement near Nb K edge [26].

Anti-parallel Pb displacements, which give rise to the α spots are expected to compete with ferroelectrically active Pb ions, which are believed to be arranged in a parallel fashion [9, 22]. Because displacements are correlated only on the short-range scale, existence of the Pb displacements in other than $\langle 110 \rangle$ directions is also possible. For example, anti-parallel short-range correlated $\langle 111 \rangle$ Pb Nb displacements can contribute to the structure factor of the F spots [31, 30, 26]. Uncorrelated Pb displacements, on the other hand, will only contribute to the uniform diffuse scattering background.

In a separate experiment we observed expected [27, 28] rhombohedral splitting of the

Bragg peaks in PMN(111) [26] under electric field ≥ 1.8 kV/cm. However, we did not register any changes in either FWHM or intensity of the α spots in both field cooled and zero field cooled regimes below T_f. This fact indicates that changes in the average macrostructure have no effect on either size or number of the nanoregions that give rise to the α spots. On the contrary, the number of PNRs was shown to increase on cooling near T_f from the recent electric-field-induced polarization measurements [32].

CONCLUSIONS

Possible misinterpretation of the α spots above 220 K for cross sections of the temperature independent $\langle 01\bar{1}\rangle^*$ diffuse scattering ridges, was addressed and studied in this work. Separation of the diffuse scattering from the α spot superlattice peaks was achieved along $\langle 01\bar{1}\rangle^*$ by reciprocal space mapping utilizing synchrotron x-ray radiation. Correlation length, obtained from the width of the α spots, is only ~ 30 Å, which defines the average size of producing these peaks nanodomains. These nanodomains are formed by short-range correlated anti-parallel Pb displacements along equivalent $\langle 110 \rangle$ directions with a magnitude of ~ 0.2 Å based upon the structure factor calculations. Fluctuations created by these locally correlated displacements are different from chemical nanodomains (CND) and ferroelectric polar nanoregions (PNR); they constitute a new type of fluctuations with anti-ferroelectric type displacement ordering (AFR) based on anti-parallel nature of the Pb displacements.

Freezing phase transition has been identified in PMN near $T_f \approx 220$ K from the temperature dependence of the integrated intensity of the α spots. Significant enhancement in the intensity of the α spots below T_f we attribute to increase in a total number of AFRs, which average size (~ 30 Å) remains constant down to the lowest measured temperature of 10 K. Nothing can be said about dynamics of these fluctuations, since interaction time between electrons and x-rays during the scattering process is $\sim 10^{-15}$ sec.

Interestingly, temperature dependence of the α spots in related Pb(In$_{1/2}$Nb$_{1/2}$)O$_3$ (PIN) variable order relaxor was shown to be correlated with long-range macroscopic anti-ferroelectric phase transition in the case of the fully ordered PIN [20]. Competition between randomly occurring anti-ferroelectric and ferroelectric fluctuations may be responsible for the relaxor behavior in PMN.

ACKNOWLEDGMENTS

We would like to thank Dr. Zschack from UNICAT ID-33 beamline at Advanced Photon Source (APS), Argonne National Laboratory and Dr. Erlich from MATRIX X-18A beamline at National Synchrotron Light Source (NSLS), Brookhaven National Laboratory for technical assistance during the experiments.

Authors also would like to thank Dr. Colla and Prof. Feigelson for providing good quality PMN single crystals and enlightening discussions.

This research is based upon work supported by the U.S. Department of Energy, Division of Materials Sciences under award No. DEFG02-96ER45439 and by the state of

Illinois IBHE HECA NWU A207 grant, though the Frederick Seitz Materials Research Laboratory at the University of Illinois at Urbana-Champaign.

National Synchrotron Light Source (NSLS) is supported by the U.S. Department of Energy under Contract No. DE-AC02-76CH00016. Use of the Advanced Photon Source was supported by the U.S. Department of Energy, Basic Energy Sciences, Office of Science, under Contract No. W-31-109-Eng-38.

REFERENCES

1. Cross, L. E., *Ferroelectrics*, **151**, 305 (1994).
2. Bonneau, P., Garnier, P., Calvarin, G., Husson, E., Gavarii, J. R., Hewat, A. W., and Morell, A., *Journal of Solid State Chemistry*, **91**, 350 (1991).
3. Mathan, N. D., Husson, E., Calvarin, G., Gavarii, J. R., Hewat, A. W., and Morell, A., *J. Phys.: Condens. Matter*, **3**, 8159 (1991).
4. Burns, G., and Dacol, F. H., *Solid State Commun.*, **48**, 853 (1983).
5. Viehland, D., Li, J.-F., Jang, S. J., Cross, L. E., and Wutting, M., *Phys. Rev. B*, **43**, 8316 (1991).
6. Westphal, V., Kleemann, W., and Glinchuk, M. D., *Phys. Rev. Lett.*, **68**, 847 (1992).
7. Pirc, R., and Blinc, R., *Phys. Rev. B*, **60**, 13470 (1999).
8. Vakhrushev, S. B., Kvyatkovsky, B. E., Naberezhnov, A. A., Okuneva, N. M., and Toperverg, B. P., *Ferroelectrics*, **90**, 173 (1989).
9. Vakhrushev, S. B., Naberezhnov, A. A., Okuneva, N. M., and Savenko, B. N., *Phys. Solid State*, **37**, 1993–1997 (1995).
10. You, H., and Zhang, Q. M., *Phys. Rev. Lett.*, **79**, 3950 (1997).
11. Naberezhnov, A., Vakhrushev, S., Dorner, B., Strauch, D., and Moudden, H., *Eur. Phys. J. B*, **11**, 13 (1999).
12. Gehring, P. M., Wakimoto, S., Ye, Z. G., and Shirane, G., *Phys. Rev. Lett.*, **8727**, 7601 (2001).
13. Wakimoto, S., Stock, C., Birgeneau, R. J., Ye, Z.-G., Chen, W., Buyers, W. J. L., Gehring, P., and Shirane, G., *cond-mat/0112366* (2002).
14. Dorogovtsev, S. N., and Yushin, N. K., *Ferroelectrics*, **112**, 27 (1990).
15. Colla, E. V., Yushin, N. K., and Viehland, D., *J. Appl. Phys.*, **83**, 3298 (1998).
16. Hilton, A. D., Barber, D. J., Randall, C. A., and Shrout, T. R., *J. Mater. Sci.*, **25**, 3461 (1990).
17. Vakhrushev, S., Nabereznov, A., Sinha, S. K., Feng, Y. P., and Egami, T., *J. Phys. Chem. Solids*, **57**, 1517 (1996).
18. Warren, B. E., *X-ray Diffraction*, Dover Publications Inc., New York, 1969.
19. Gosula, V., Tkachuk, A., Chung, K., and Chen, H., *J. Phys. Chem. Solids*, **61**, 221 (2000).
20. Nomura, K., Yasuda, N., Ohwa, H., and Terauchi, H., *J. Phys. Soc. Jpn.*, **66**, 1856 (1997).
21. Miao, S., Zhu, J., Zhang, X. W., and Cheng, Z. Y., *Phys. Rev. B*, **6505**, 2101 (2002).
22. Hirota, K., Ye, Z.-G., Wakimoto, S., Gehring, P. M., and Shirane, G., *Phys. Rev. B*, **65**, 104105 (2002).
23. Malibert, C., Dkhil, B., Kiat, J.-M., Durand, D., Berar, J. F., and d. Bire, A. S., *J. Phys.: Condens. Matter*, **9**, 7485–7500 (1997).
24. Takesue, N., Fujii, Y., Ichihara, M., and Chen, H., *Phys. Rev. Lett.*, **82**, 3709 (1999).
25. Tkachuk, A., Zschack, P., Colla, E., and Chen, H., *AIP American Institute of Physics Conference Proceedings, no*, **582**, 45 (2001).
26. Tkachuk, A., *Synchrotron X-ray Scattering Studies of Local Fluctuations in Lead Magnesium Niobate Relaxor Ferroelectrics*, Ph.D. thesis, University of Illinois at Urbana-Champaign (2002).
27. Calvarin, G., Husson, E., and Ye, Z. G., *Ferroelectrics*, **165**, 349 (1995).
28. Vakhrushev, S. B., Kiat, J.-M., and Dkhil, B., *Solid State Commun.*, **103**, 477 (1997).
29. Glazer, A. M., *Acta Cryst.*, **A 31**, 756 (1975).
30. Egami, T., Dmowski, W., Teslic, S., Davies, P. K., Chen, I. W., and Chen, H., *Ferroelectrics*, **206**, 1 (1998).
31. Zhang, Q. M., You, H., Mulvihill, M. L., and Jang, S. J., *Solid State Commun.*, **97**, 693 (1996).
32. Dkhil, B., and Kiat, J. M., *J. Appl. Phys.*, **90**, 4676 (2001).

Micro-Brillouin investigations of relaxor ferroelectrics

Holger Hellwig*, Russell J. Hemley* and Ronald E. Cohen*

Geophysical Laboratory, Carnegie Institution of Washington, 5251 Broad Branch Road NW, Washington, DC 20015

Abstract. Micro-Brillouin measurements on single crystal lead titanate were carried out at ambient conditions. We used a symmetrical scattering geometry that allows probing of all wave vectors within a given plane. All acoustic modes could be resolved. A central peak that is strongly coupled to the acoustic modes is observed. Detailed analysis of the coupling reveals the importance of the irreducible representation of symmetry E that is closely related to the ferroelectric phase transition at higher temperatures.

INTRODUCTION

Perovskites are of special interest from both an applied and fundamental point of view. One of their prominent features is the variety of phase transitions usually involving a cubic high temperature paraphase, which is centrosymmetric, and often several different phases of lower symmetry. These phases can be polar and give rise to ferroelectric transitions.

Lead titanate ($PbTiO_3$, PT) is a perovskite with a transition from a cubic high temperature phase (space group Pm3m) to a tetragonal polar phase (space group P4mm) at about 766 K [1, 2]. The soft-mode transforms as the three dimensional T_{1u} representation with the basis functions (x, y, z).

Beside Raman frequencies, the light scattering spectrum also contains a central peak. This feature arises from entropy fluctuations close to the phase transition and leads to a strong scattering of light around the Rayleigh line. Raman investigations analyzed the central peak in detail and described it as having the symmetry of the irreducible representation E (point group 4mm) [3]. E is a two-dimensional representation with the basis functions (x, y), (xz, yz), or (R_x, R_y). This mode and the total symmetric representation A_1 correlate to the three dimensional T_{1u} mode in the cubic phase, which represents the symmetry of the ferroelectric soft-mode. The appearance of a central peak is also found in relaxor-type materials and can be described as a common feature for structural phase transitions [4].

The soft-mode has also been studied by neutron inelastic scattering [5] where coupling between an acoustic and an optic mode was observed. Mode coupling between a soft optic mode and an acoustic mode was also detected in $BaTiO_3$ [6] and in PMN [7, 8]. A general overview of mode coupling is given by Wehner and Steigmeier [9]. Scott and coworkers were able to identify mode coupling between the central peak and an acoustic mode in $Ba_2NaNb_5O_{15}$ [10, 11].

Lead titanate is especially interesting since it is an end-member of complex solid solutions like $PbZn_{1/3}Nb_{2/3}O_3 - PbTiO_3$ (PZN-PT), and $PbMg_{1/3}Nb_{2/3}O_3 - PbTiO_3$ (PMN-PT). The PT-poor materials are relaxors that show strong dispersion of the dielectric constant, with temperature dependent maxima. The elastic properties of a material directly reflect the strength of internal bonding and often reveal particular features related to certain types of phase transitions. Knowledge of the elastic properties can also be used to optimize and fine tune materials. Knowledge of the elastic constants, as well as the piezioelectric constants and dielectric constant, is necessary to describe the electromechanical response of a material to applied strains and fields.

Elastic properties can be measured by determining the acoustic sound velocities. Depending on the availability of single crystals of appropriate size, different methods can be employed. Ultrasound techniques require relatively large samples of typically hundreds of microns to millimeter size and are therefore mainly limited to ceramic samples once single crystals of that size are not available [12].

Brillouin has a principal advantage over ultrasonic techniques in the very small sample volumes that can be used. Interaction regions in the order of tens of microns can readily be realized. This offers the possibility of probing single domain samples of very small dimensions. This technique has mainly been utilized in back scattering geometries, where only the longitudinal modes can be probed [13, 14, 15, 16].

Brillouin spectroscopy measures the frequency shift of the scattered light. The scattering geometry translates into a scattering wave vector that normally is on the order of the wave vector of the light wave used to probe the material. In general, this wave vector inside the sample depends on the refractive indices of the sample. The frequency shift and the wave vector give the sound velocity of a particular mode. Since the elastic properties of a piezoelectric material are also affected by the piezoelectric constants, an investigation of the sound velocities makes possible the determination of the elastic and the piezoelectric constants. Both quantities are tensor quantities and are highly dependent on the direction of the wave vector within the material [17]. Measurements along many directions can then be used to fit measured sound velocities to a set of elastic and piezoelectric constants.

The technique requires the measurement to be done on single domain samples. A transition from point group m3m to 4mm implies the possible formation of 6 domains. The three types of domains differ in the orientation of their polar axis, which can be along any 3 of the cubic 4-fold axes, and the other three have their polar direction reversed by 180°. These two types of domains are referred to as 90° and 180° domains, respectively. Because of the anisotropy in point group 4mm, the elastic and piezoelectric properties are different in 90° samples. 180° domain samples show the same elastic behavior, since elastic properties are intrinsically centrosymmetric.

Brillouin measurements on $PbTiO_3$ have been reported and show the general feasibility of the technique for this material [18, 19].

EXPERIMENTAL DETAILS

The sample was a nearly (100) plane parallel plate of lead titanate (PbTiO$_3$) as confirmed by x-ray analysis. The thickness was 70 µm, lateral dimensions were about 200 - 300 µm. Samples were polished and put between two glass plates mounted with silicon grease. No domains were visible by polarized microscopy. We also checked the sample after the Brillouin measurements and found no change in linear optical properties, indicating that the sample did not form 90° domains during the measurements. 180° domains cannot be identified using linear optical techniques. The sample was of slightly greenish color and showed strong anisotropy in the optical absorption, being stronger for the electric field parallel to the polar axis.

Brillouin measurements were performed with a 6-pass tandem Fabry-Perot interferometer [20]. A single mode argon-ion laser operating at 514.5 nm was used together with a custom built micro-optical system, photon-counting system, and a multi-channel scaler card. In order to avoid heating the sample and possible formation of multi-domains induced by high laser power, we used an incident power of about 20-40 mW. The focusing lens and the collecting lens had a focal length of 50 mm. The sample was placed symmetrically with respect to the incoming and collected light such that the difference vector of the incoming and the detected light was in the plane of the sample. The angle between the incoming and the detected light was 80°. The difference vector within the sample represents the scattering vector of the excitation. For this particular geometry, the scattering vector is independent of the refractive indices of the material. By rotating the sample through the face normal as a rotational axis, the scattering vector is rotated in the (100) plane. This geometry also allows all acoustic modes to be probed, since the intensity of the transverse modes is not zero as in back-scattering geometries. Spectra were taken every 10° over a full rotation of the sample. Spectra were collected for 2-15 hours at each single orientation. Spectra at the same angular position were averaged and fitted to the proposed model. The square root of the count rate $\sqrt{N_i} = \sigma_i$ was taken as the uncertainty of each individual count rate N_i. The fitting procedure minimized $\sigma = \frac{1}{F}\sum_{i=1}^{N}(N_i^{obs.} - N_i^{meas.})^2/\sigma_i^2$. F is the number of degrees of freedom, which is the number of data points minus the number of free parameters used to fit the data. Since the data are not completely independent, this scheme does not provide a reliable comparison between models with different numbers of parameters.

We were able to identify all acoustic modes, the longitudinal (LA) and the two transverse modes (TA$_1$ and TA$_2$). Since the transverse modes in the present spectra do not cross, we refer to the mode with a lower frequency as TA$_1$. Another prominent feature present in all the spectra is the central peak, observed as enhanced intensity close to the Rayleigh line. The fitted model contains 3 parameters per band (Gaussian or Lorentzian), an additional parameter for coupling, 2 parameters for the central peak, one parameter for the constant background and another parameter that describes the lineshape convolution; the latter must be applied to the calculated spectrum because the modes that couple to the central peak are Lorentzian.

THEORETICAL ANALYSIS

The measured frequency shift ω and the modulus of the scattering wave vector k are related to the sound velocity v by

$$v = \frac{\omega}{k}. \tag{1}$$

The sound velocity v depends on the elastic constants c_{ijkl}, the piezoelectric constants e_{ijk}, the direction of the scattering wave vector $\mathbf{q}$ (with $\mathbf{q}^2 = 1$), the dielectric properties ε_{ij} and the density $\rho = 7970$ kg/m^3 [19] such that

$$|\Gamma_{ik} - \delta_{ik}\rho v^2| = 0 \tag{2}$$

with

$$\Gamma_{ik} = \left\{ c^E_{ijkl} + \frac{(e_{mij}q_m)(e_{nkl}q_n)}{\varepsilon^S_{rs}q_r q_s} \right\} q_j q_l. \tag{3}$$

δ_{ij} is the Kronecker delta with $\delta_{ij} = 1$ for $i = j$ and $\delta_{ij} = 0$ for $i \neq j$. The superscripts E, D, S and T refer to the boundary conditions of constant electric field, constant electric displacement, constant strain and constant stress, respectively.

In order to calculate the spectra, we express the susceptibilities of each single Lorentzian mode with the central frequency ω_0 and the peak width γ_0 by

$$\chi_0^{-1} = \omega_0^2 - \omega^2 - i\omega\gamma_0 \tag{4}$$

and the central peak with peak width γ_c by

$$\chi_c^{-1} = 1 - i\omega/\gamma_c, \tag{5}$$

where ω is the frequency. To take mode coupling into account, we express the scattering intensity S by

$$S(\omega) = C \cdot [n(\omega) + 1] \cdot \mathrm{Im}\chi(\omega) = C \cdot [n(\omega) + 1] \cdot \mathrm{Im} \sum_{i,j} p_i p_j G_{ij}(\omega) \tag{6}$$

with C being a constant, $n(\omega) = (\exp h\omega/kT - 1)^{-1}$ is the Bose-Einstein function and can be well approximated by $1/\omega$ for frequencies below 1 cm^{-1} and ambient temperature. G_{ij} is given by

$$\begin{pmatrix} \chi_0^{-1} & \Delta^2 + i\omega\Gamma_{mc} \\ \Delta^2 + i\omega\Gamma_{mc} & \chi_c^{-1} \end{pmatrix} \begin{pmatrix} G_{11} & G_{12} \\ G_{21} & G_{22} \end{pmatrix} = \begin{pmatrix} 1 & 0 \\ 0 & 1 \end{pmatrix} \tag{7}$$

The p_i are the mode amplitudes, taken as positive. Δ and Γ_{mc} describe the mode coupling in a general form having a real and imaginary coupling term. The matrix can be extended to include more than one acoustic mode coupling to the central peak. We found the imaginary part sufficient to describe our data and finally restricted the fits to solely negative imaginary coupling. This also avoids ambiguities regarding the results, when spectra can be fit with either pure real or pure imaginary coupling constants (compare Ref. [10]). In this case, mode coupling only introduces one more parameter per mode.

As regards the fitting procedure, mode coupling introduces strong correlations between parameters that are basically independent without mode coupling. This imposes serious problems for the fitting procedure. We used a simplex scheme [21] combined with a variational method that produced randomly new start values in the vicinity of the best results. The variational strength as the magnitude of the variation is varied sinusoidal over 20 iterations. Additionally, the last points in parameter space were saved and used to derive general gradients to have estimates for new points in parameter space. It should be noted that the fits included about 15 free parameters and each spectrum consisted of about 350 data points used for the fit. This ratio is relatively problematic when considering one single spectrum. Therefore initial parameters were also chosen by comparing all spectra and assuming a smooth angular dependence. Different models with the lowest σ were examined. In the event of two σ being nearly identical, the model with fewer parameters was preferred. In cases where the count rate was very high, problems arise from the choice of a proper background function and lineshape. This can lead to very broad peaks with relatively low intensities; these peaks were not considered in the subsequent calculation of the elastic constants. It is assumed that each mode has the same lineshape.

The sound velocities along different directions can be calculated with the data given by Kalinichev et al. [19]. Of special interest here is the contribution of certain constants to the overall effect, denoted by Δ. These contributions can be calculated by a variational method, where the value of each input d_i is varied by $d'_i = 0.999 d_i$ and the total effect on the sound velocity is compared to the original data set from

$$\Delta_i = \frac{v(d_i) - v(d'_i)}{v(d_i)} \times 1000. \qquad (8)$$

The results of these calculations are given in Fig. 1 for the T_1 mode.

Some spectra (compare Fig 2) show features that cannot be explained by a pure superposition of the different modes, the central peak and the background function (consisting of a constant background and the Rayleigh peak at very low wave numbers). While in several spectra some peaks show strong asymmetry, in others there is clear evidence for another peak on the low frequency side of the mode of interest. A strong asymmetry can arise from the finite aperture of the light beam. We used a full angular acceptance angle of 5° for the measurements presented here. This effect would give rise to shoulders on different sides of the peak: for a minimum in the frequency of a particular mode the shoulder would be on the high frequency side, whereas for a maximum the shoulder should be on the low frequency side. The difference between the main peak and its satelite peak is roughly the difference between the mode frequencies over the angular acceptance angle. This effect was not observed in the present spectra. Another possible explanation is the contribution from Raman excitations from thermally excited modes (hot bands) [22]; however, this is unlikely in the very low-frequency (acoustic mode) region. The appearance of distinct peaks on the low frequency side of a particular mode is therefore taken to be of different origin. We focus our attention to the mode coupling mechanism apparent in Fig. 2. Note the strong dip in intensity drops down

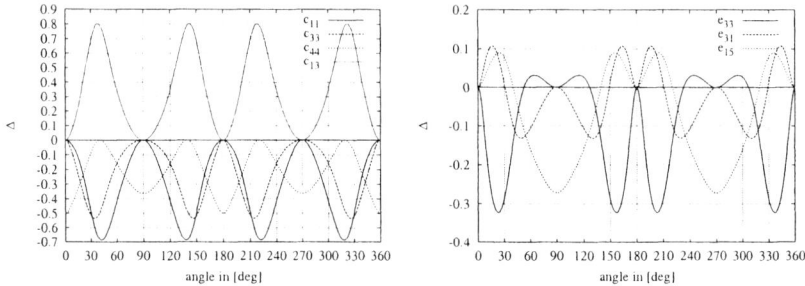

FIGURE 1. Contributions of the different elastic constants (left) and piezoelectric constants (right) to the sound velocities of the lower transverse mode (TA$_1$) within the (100) plane of lead titanate, zero degrees for $\mathbf{q}\|c$ and 90° $\mathbf{q}\|a$. For the definition of Δ see text.

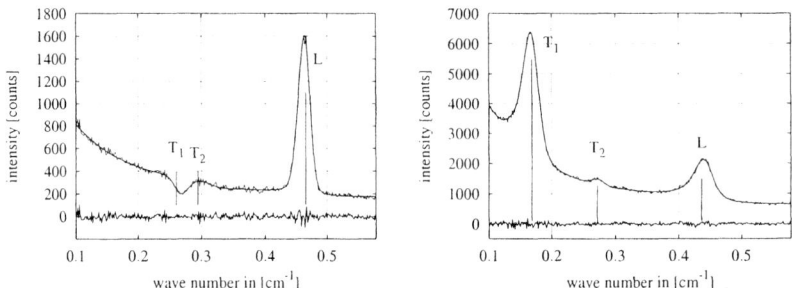

FIGURE 2. Brillouin spectra of lead titanate with $\mathbf{q}$ within (100) (angle between $\mathbf{q}$ and c: 100° and 230° for the left and right figure, respectively). The three different modes (LA, TA$_1$, and TA$_2$) are clearly visible in the right figure. In the left figure, the dip around 0.27 cm^{-1} shows the effect of mode coupling for the lower transverse mode (thin lines: measured curve, thick line: fitted curve with a residual around the zero line; the straight lines indicate the fitted peak position and the labels for the mode assignment are given)

nearly to the constant background level originating from sample independent stray light and dark count in the photon counting system.

RESULTS AND DISCUSSION

Typical spectra are shown in Fig. 2. Besides the enhanced background that originates from the central peak, the longitudinal and transverse peaks are clearly visible. The most prominent features of the effect of mode coupling are the dips in several spectra and strongly asymmetric lineshapes. Fitting the data to a mode coupling model reveals a consistent overall picture for all different directions of $\mathbf{q}$ if the mode coupling is taken as purely imaginary and negative. Although this choice is not unique with respect to the residuals given by σ, we tried to fit all spectra to the same model of modes that couple. In particular, we found that TA$_2$ does not show any significant coupling and LA

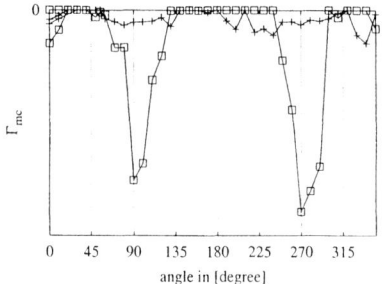

FIGURE 3. Results for the coefficients describing the mode coupling. +: LA mode, squares: TA$_1$ mode. The pronounced peaks at 90° and 270° refer to a wave vector $\mathbf{q}||a$.

couples only weakly, but essentially independently of the wave vector direction $\mathbf{q}$. The strongest coupling was found for TA$_1$ for $\mathbf{q}||a$. These results are illustrated in Fig. 3. The origin of the strong mode coupling can be understood by analyzing the different contributions of the elastic and piezoelectric constants to the sound velocities for the directions measured. The results of these calculations are given in Fig. 1. It can be clearly seen that for $\mathbf{q}$ along a (corresponding to peaks at 90° and 270°) the only contributions are c_{44} and e_{15}. While c_{44} also contributes to a direction with $\mathbf{q}||c$ (corresponding to 0° and 180° in the figure), e_{15} only shows a strong contribution along directions where there is pronounced coupling. The contribution of the piezoelectric constants to TA$_2$ is zero for $\mathbf{q}$ within (100) and the contributions of the piezoelectric constants to the longitudinal effect are strong for e_{33} and e_{31}. It follows that the main component responsible for mode coupling is the piezoelectric constant e_{15}. Since the symmetry of the central peak was described as E (see Ref. [3]), strong coupling is expected for modes of the same symmetry. The symmetry for nearly zone center phonons can easily be analyzed by the basis functions involved in the particular tensor coefficient. The full notation of the short Voigt notation for the piezoelectric constant e_{15} is e_{113}, which relates to an E symmetry. The irreducible representation E is especially interesting since it couples the polar symmetries (x,y) to the quadratic forms (xz,yz). The direct product of E with itself contains the full symmetric representation A$_1$ and can therefore contribute to energy terms. In the case of the other two piezoelectric coefficients e_{33} and e_{31} (equals e_{333} and $e_{311} = e_{322}$ in full notation), we expect coupling with A$_1$ modes as a direct product of two A$_1$ symmetries z and x^2+y^2. This coupling does not seem to be strong.

We have found clear evidence that coupling is a pronounced feature in lead titanate. We identified coupling between the central peak and the LA and TA$_1$ mode for a wave vector within the (100) plane. The coupling terms create a shift of the peak maxima in the spectra. This effect could cause severe problems and might lead to wrong interpretations in other measurements (e.g., neutron scattering) [5, 4, 8]. The fitted elastic and piezoelectric constants are given in Tab. 1 together with results of other Brillouin measurements. In our measurements c_{12} did not contribute strongly to the measured sound velocities and was held fixed to a value of 90 GPa. We also kept the dielectric constants fixed at $K_{11}^S = 102$ and $K_{33}^S = 34$ (with $K_{ii} = \varepsilon_{ii}/\varepsilon_0$).

We extended our investigations to other orientations and found even more pronounced

TABLE 1. Elastic and piezoelectric constants of $PbTiO_3$. The value for c_{12} has not been fitted. c in given in [GPa], e in [C/m^2] with uncertainties given in parenthesis.

	This work	Kalinichev et al. *	Kalinichev et al. †	Li et al. **
c_{11}	237	237(3)	237(3)	235(3)
c_{33}	70	60(10)	90(10)	105(7)
c_{44}	71	69(1)	69(1)	65(1)
c_{66}	97	104(1)	104(1)	104(1)
c_{12}	(90)	90(5)	90(5)	101(5)
c_{13}	77	70(10)	100(10)	99(8)
e_{33}	4.4	5.0	4.1	3.4
e_{31}	2.9	2.1	-0.67	0.98
e_{15}	2.9	4.4	4.8	3.9

* Ref. [19], data set A
† Ref. [19], data set B
** Ref. [18]

mode coupling features for **q** in (101). Another feature that consistently shows up in our spectra is a second peak on the low frequency side that we attribute to a signal arising from regions without piezoelectric coupling. This will be examined in detail elsewhere. Measurements on the relaxor PZN-4%PT are in progress and reveal similar features to the spectra presented here. This material is more sensitive to laser induced formation of domains and shows weaker central peak features. In an 80° scattering geometry we were able to detect all acoustic modes.

ACKNOWLEDGMENTS

We are grateful to D. Rytz and T. Salva for the lead titanate sample. We thank P. Dera for the x-ray analysis of the sample, and H. K. Mao for help and discussions. This work was supported by the Office of Naval Research (grant No. N000140210506).

REFERENCES

1. F. Jona and G. Shirane, *Ferroelectric Crystals*, Pergamon Press, New York, 1962
2. M. L. Lines and A. M. Glass, *Principles and Applications of Ferroelectrics and Related Materials* (Oxford University Press, Oxford, UK, 1977)
3. M. D. Fontana, H. Idrissi, G. E. Kugel, and K. Wojcik: J. Phys.: Condens. Matter **3** (1991) 8695
4. I. G. Siny, S. G. Lushnikov, R. S. Katiyar, and E. A. Rogacheva: Phys. Rev. B **56** (1997) 7962
5. G. Shirane, J. D. Axe, J. Harada, and J. P. Remeika: Phys. Rev. B **2** (1970) 155
6. P. A. Fleury and P. D. Lazay: Phys. Rev. Lett. **26** (1971) 1331
7. A. Naberezhnov, S. Vakhrushev, B. Dorner, D. Strauch, and H. Moudden: Eur. Phys. J. B **11** (1999) 13

8. S. Wakimoto, C. Stock, Z.-G. Ye, W. Chen, P. M. Gehring, and G. Shirane: Phys. Rev. B **66** (2002) 224102
9. R. K. Wehner and E. F. Steigmeier: RCA Reviews **36** (1975) 70
10. J. F. Scott and W. F. Oliver, in: Geometry and Thermodynamics, Ed.: J.-C. Toledano, NATO ASI Series, series B: Physics Vol. 229, PLenum Press, New York (1990) 453
11. W. F. Oliver and J. F. Scott, in: Laser Optics of Condensed Matter, Ed.: J. L. Birman, H. Z. Cummins, and A. A. Kalyanskii (Plenum, New York, 1988) 263
12. ultrasonic velocity and dielectric constant Y. Sasaki, T. Ochiai, Y. Takeuchi, and S. Hayashi: Jpn. J. Appl. Phys. **34** (1995) 4122
13. M. H. Kuok, S. C. Ng, H. J. Fan, M. Iwata, and Y. Ishibashi: Appl. Phys. Lett. **78** (2001) 1727
14. C.-S. Tu and C.-L. Tsai: J. Appl. Phys. **87** (2000) 2327
15. F. M. Jiang and S. Kojima: Jpn. J. Appl. Phys. **39** (2000) 5704
16. H. J. Fan, M. H. Kuok, S. C. Ng, N. Yasuda, H. Ohwa, M. Iwata, H. Orihara, and Y. Ishibashi: J Appl. Phys. **91** (2002) 2262
17. J. F. Nye, *Physical Properties of Crystals*, Clarendon Press, Oxford (1985)
18. Z. Li, M. Grimsditch, X. Xu, and S.-K. Chan: Ferroelectrics **141** (1993) 313
19. A. G. Kalinichev, J. D. Bass, B. N. Sun, and D. A. Payne: J. Mater. Res. **12** (1997) 2623
20. J. R. Sandercock in *Topics in Applied Physics, Light Scattering in Solids III*, Vol.15, edited by M. Cardona and G. Güntherodt, Springer Verlag, Berlin Heidelberg, New York, (1982) 173
21. W. H. Press, S. A. Teukolsky, W. T. Vetterling, and B. P. Flannery: *Numerical Recipes in C*, 2. Edition, Cambridge University Press, Cambridge, UK, 1992
22. C. M. Foster, M. Grimsditch, Z. Li, and V.G. Karpov: Phys. Rev. Lett. **71** (1993) 1258

Structure of Nanodomains in Relaxors

Vakhrushev S. B.*, Naberezhnov A. A.*, Dkhil B.†, Kiat J.-M.**‡,
Shwartsman V.§, Kholkin A§, Dorner B.¶ and Ivanov A.¶

*A. F. Ioffe Physico-Technical Institute, 26 Politekhnicheskaya 194021, St.-Petersburg Russia.
†Laboratoire Structures, Proprietes et Modelisation des Solides, CNRS-UMR 8580 Ecole Centrale Paris, 92295 Chatenay-Malabry Cedex, France
**LSPMS, CNRS-UMR 8580 Ecole Centrale Paris, 92295 Chatenay-Malabry Cedex, France
‡LLB C.E. de Saclay, 91191 Gif-sur-Yvette Cedex, France
§Department of Ceramics and Glass Engineering University of Aveiro 3810-193 Aveiro (Portugal)
¶Institute Laue-Langevin, 38042 Grenoble Cedex 9, France

Abstract. Results of study of the spatial distribution of polarization by scattering (X-ray and neutron) and Piezoresponse Force Microscopy techniques are presented for several relaxors both cubic (PMN and PMNPT solid solutions) and uniaxial (SBN). It is demonstrated that in all cases except pure PMN cooled in zero electric field polar nanodomains are formed. In case of PMN cooled in the applied field nanodomain state is preserved even after subsequent zero field heating to above freezing temperature. AFM measurements of PMNPT10 are used to determine the quantitative characteristics of these nanodomains.

INTRODUCTION

Spatial distribution of the polarization is one of the most important characteristics of any ferroelectric material. Domains and domain walls are well studied in the case of "normal" ordered ferroelectrics. In the case of relaxors the problem, although not less important is much less studied and understood. Many attempts have been made to study the structure of polar nanoregions in relaxors. Mesoscopic structure of cubic and uniaxial relaxors was studied using X-ray [1] and neutron [2] scattering techniques, optical and electron microscopy [3], and recently atomic-force microscopy operating in piezoresonance mode (piezoelectric force microscopy - PFM)[4]. It was shown that the archetypical relaxor $PbMg_{1/3}Nb_{2/3}O_3$ (PMN), carefully annealed and then cooled in zero electric field, the low temperature state can be described as a dipole glass. However application of the electric field or doping with other ferroelectric compounds like $PbTiO_3$ (PMNPT solid solutions) results in the case of strong enough electric fields ($E_{(111)} \geq E_{Th} \approx 2.2$ kV/cm or high enough concentrations of $PbTiO_3$ in the formation of well defined nanodomains. Very little reliable information is available about the mesoscopic structure of pure PMN heated in zero electric field (short-circuited) slightly above the freezing temperature ($T_g \approx 230$ K [1]), after having been field cooled below T_g, or about PMNPT with low lead titanate concentration. However, these situations seem to be the most physically interesting, being related to the mechanism of transformation from the glass to some mixed ferroglass state, and then, finally, to the real ferroelectric phase. A different example is strontium-barium niobate $Sr_xBa_{1-x}Nb_2O_6$ (SBN), which is a uniaxial relaxor. It was proposed long ago [5] and recently confirmed [4], that cooling of SBN in zero field results in the appearance of domain structure. In this paper we present a brief survey of our results related to the nanodomain formation in PMN subjected to cooling in an applied electric field, PMNPT10 (10% of $PbTiO_3$) and SBN.

EXPERIMENTAL

All measurements on PMN and PMNPT solid solutions were performed with samples cut from single crystals grown by the Chochralsky method in the Physics Research Institute of the Rostov-on-Don University. The quality of the single crystals was checked before cutting by rocking curve measurements on a γ-diffractometer. The rocking curves were symmetric with full width at half maximum less than 1'. Neutron scattering measurements were done at the IN3 3-axis spectrometer of Institute Laue-Langevin (Grenoble, France). Measurements on the SBN single crystal were done with the Neutron-3 three-axis neutron spectrometer installed at the WWR-M reactor of PNPI (Gatchina, Russia). X-ray data were collected with the automatic single crystal X-ray diffractometer. $CuK\alpha_1$ radiation from a 18kW rotating anode source was used. An InP monochromator was placed in front of the detector. The topography of polar nanoregions in PMNPT10 was measured with a commercial setup (Multimode, DI) complemented with a lock-in amplifier and function generator. A standard etched Si tip-cantilever system (OTESPA, Olympus) with an average spring constant of 40 N/m (resonance frequency 300 kHz) and a tip radius less than 10 nm was used. Our experimental condition (stiff cantilever with a sharp tip that exerts significant pressure onto the surface of the crystal) corresponds to a so-called strong indentation regime, where the contribution from the piezoelectric vibration to the measured signal is much higher than the parasitic signal due to the electrostatic interaction between cantilever and surface. The imaging was performed under AC voltage (2 V, 50 kHz).

RESULTS AND DISCUSSION.

PMN in Zero Field Heating after Field Cooling regime (ZFHaFC)

The sample was cooled in a field 5.5 kV/cm, applied in (111) direction so that the glassy phase was avoided on cooling (for details see [6]). Measurements were performed on heating of a short-circuited sample (ZFHaFC) for the (111) and ($\bar{1}11$) directions of the reduced wavevector q (parallel and nearly perpendicular to $\vec{E}$. The most essential feature of the observed diffuse scattering was that the shape of the q-dependence of the scattering intensity was not Lorentzian. At relatively large q $I(q)$ followed a power law $q^{-\alpha}$ with α gradually **increasing on heating** from 175K to 330K from 1.75 up to nearly 3. This is strongly different from the case of PMN cooled in zero field when α **increases stepwise on cooling**, from $\approx$ 2.3 at T> T_g to $\approx$ 2.7 at T< T_g. In our previous paper on this subject [6] we followed the approach of [1] and described the obtained results in terms of the fractal model. Here we use instead the Random

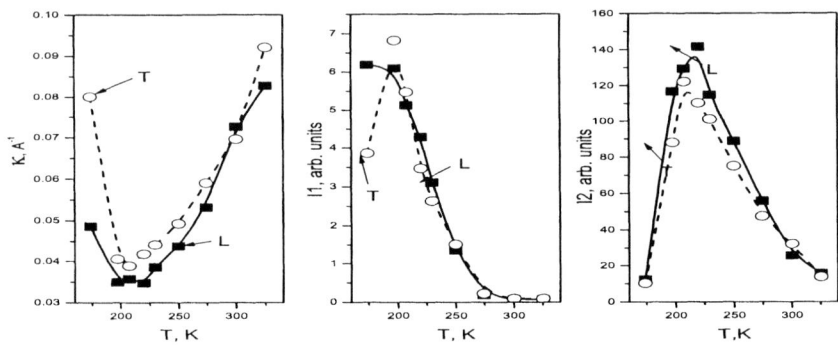

FIGURE 1. PMN ZFHaFC; temperature dependences of the inverse correlation length κ, and of the scattering intensities of components 1 and 2 (see text)

Field Ising Model successfully employed in [7] to describe the neutron scattering in a dilute antiferromagnet. We have decomposed the observed diffuse scattering into two terms [8]: (1) – thermodynamic fluctuations related to the susceptibility with a $q^{2-\eta}$ law at $q > \kappa = 1/R_c$, where R_c is correlation length and (2) – static configuration fluctuations described by a "structure factor" with a $q^{4-\overline{\eta}}$ law with $\overline{\eta} \approx 2\eta$. In the three-dimensional Random Field Ising Model the parameter η is not small ($\eta_{RFIM} \geq 0.5$ [8]). So, depending on the ratio between these two terms the exponent describing the experimentally measured q-dependences of the diffuse scattering intensity can vary from 1.5 to 3.0. For the data fit we have used the expressions from the Tarko-Fisher paper [9] for the Ising model below T_c and in a field. Results of our calculations in the frames of this model are presented in the Fig. 1. One can see, that in the PMN cooled in an electric field the structure factor term is dominating at nearly all temperatures except the lowest. Obtained results are clearly different from the case of "virgin" or annealed PMN crystal in zero field. In the latter case above T_g the diffuse scattering is of nearly Lorentzian shape and can be attributed to the slow critical fluctuations.

All data were consistent with $\overline{\eta} = 2\eta$ and $\eta=0.35$, which is smaller than η_{RFIM} for the model with short-range interactions. The most probable source of the quenched random electric fields in PMN is the random distribution of nonisovalent ions on the B sublattice. Since the short compositional order exists in PMN with correlation length of about 80 Å, [1], the random fields are expected to be correlated at the same distances. Such correlations should affect the critical exponents and $\overline{\eta}$ is supposed to be most strongly affected [10]. Thus despite the fact that we cannot make any quantitative estimate, we feel that the deviation of the experimentally observed η from the RFIM value is a consequence of the chemical order in PMN.

SBN50

Another example of nanodomain formation is $Sr_xBa_{1-x}Nb_2O_6$. We have performed neutron scattering measurements on SBN with x=0.5 (abbreviated later as SBN50). At

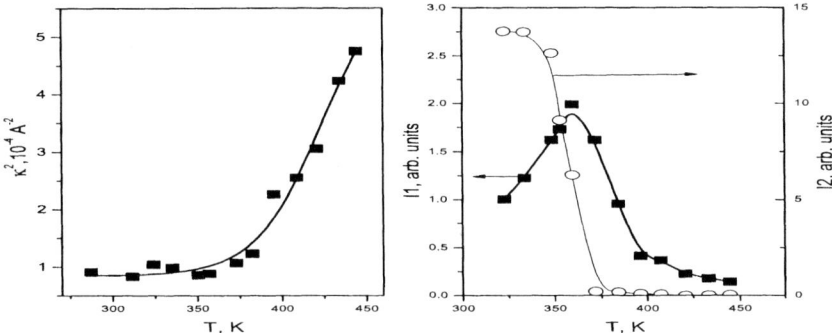

FIGURE 2. Temperature dependences of the inverse correlation length κ, and of the scattering intensities of components 1 and 2 (see text)

temperatures well above freezing temperature ($T_g^{SBN} \approx 390$ K as determined by the pyroelectric measurements [11]) the diffuse scattering was nearly Lorentzian (after resolution corrections). However on passing through T_g^{SBN} the lineshape changed substantially. Similar to the previous case we have decomposed $I(q)$ into two components. We found that a satisfactory fit can be obtained for the mean-field approximation with $I_1 = I_1^{(0)}/(q^2 + \kappa^2)$ and $I_2 = I_2^{(0)}/(q^2 + \kappa^2)^2$. Fitted results are presented in figure 2. One can easily see that the I_2 term is dominant, so there is evidence of the nanodomain nature of the low temperature state in SBN.

PMNPT10

In figure 3 the temperature dependences of the X-ray diffuse scattering intensity extrapolated to $q = 0$ (I_0) and of the width of the Bragg peak are presented for PMNPT10 and PMNPT6 (for comparison). In the case of PMNPT6 the scattering intensity monotonously increases on cooling, while the width of the (300) Bragg peak stays constant and resolution limited. Considering I_0 to be proportional to the static susceptibility, $I_0(T)$ was fitted in terms of the spherical RBRF model [12] and the obtained parameters were found to be close to the limit of glass phase stability. In the case of PMNPT10 the results are essentially different. $I_0(T)$ demonstrates a sharp peak at T_g, and the Bragg peaks broaden. Analyzing the widths of the Bragg peaks as a function of the scattering vector length, the conclusion can be made that this broadening results from the elastic deformation and not from the size effect. (Details will be published elsewhere). According to the phase diagram PMNPT10 is supposed to be at the border between pseudocubic and rhombohedral low-temperature phases. Existing data are contradictory. X-ray re-

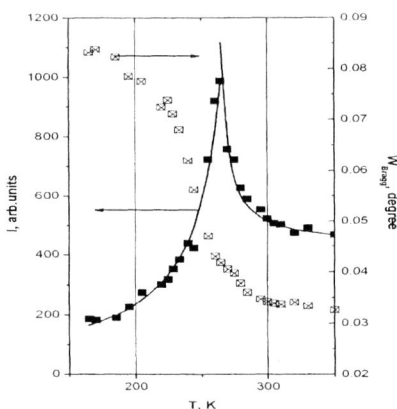 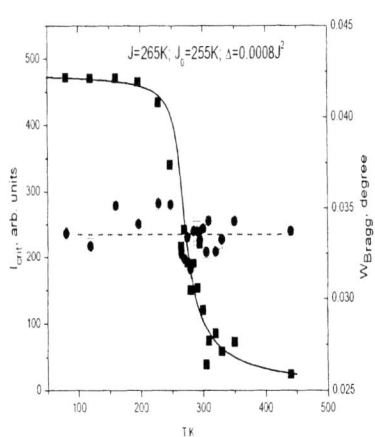

FIGURE 3. Temperature dependences of the diffuse scattering intensity extrapolated to $q = 0$ and the width of the Bragg peak for PMNPT10 and PMNPT6; in case of PMNPT6 parameters of the RBRF model are shown in the figure

sults obtained with ceramic samples clearly indicate rhombohedral distortions,[13] but a recent neutron single-crystal study [14] did not confirm this result. In any case rhombohedral distortions cannot directly affect the width of (h00) Bragg peaks. The most likely explanation is that this broadening is due to the formation of the nanodomains and domain walls. By analogy with the block boundaries in a mosaic crystal, the domain walls should be considered as the defects of the second class,[15] resulting in the destruction of the "true" δ-shaped Bragg reflections and formation of the broadened peaks . At our resolution the size of these domains should be of the order or larger than ≈ 1000 Å to make the finite-size contribution to the total width unobservable.

Piezoresponse Force Microscopy (PFM)

PFM is based on the detection of local vibrations induced by the electric field applied in the immediate vicinity of the sharp conducting tip. Since the local deformation is related to the average piezoelectric coefficient of the deforming volume it is proportional to the local polarization. In a pseudocubic ferroelectric material piezoresponse d_{33} is directly related to the spontaneous or remanent polarization P_S[16]:

$$d_{33} = 2Q\varepsilon_{33}P_S, \qquad (1)$$

where Q is the electrostrictive coefficient and ε_{33} is the dielectric permittivity of the material. In the Figure 4 a PFM image of PMNPT10 at room temperature is presented. White color corresponds to the polarization projection on the (001) direction "up", and black color – "down" relative to the probed surface of the crystal. Polarized regions are clearly seen in the figure. For the quantitative data treatment we have used a correlation function technique earlier used successfully for usual AFM data analysis [17, 18]. We

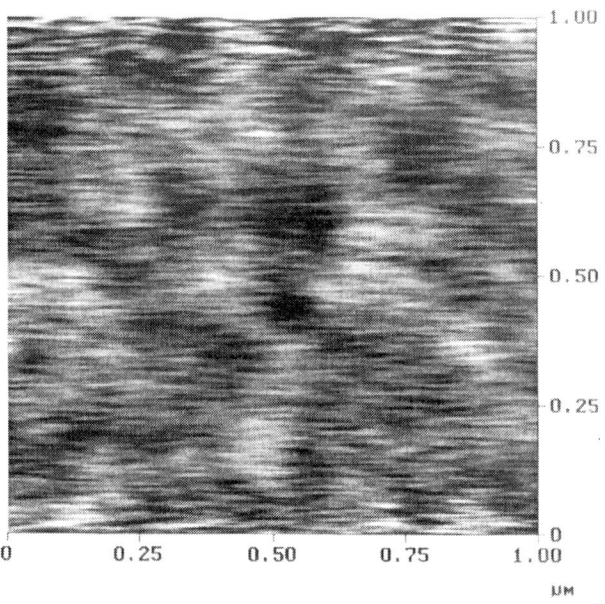

FIGURE 4. PFM image of PMNPT10 at room temperature

have calculated the height-height autocorrelation function:

$$C_h(\vec{r}) = <h(\vec{r})h(0)> \qquad (2)$$

Since the experimentally measured height $h(\vec{r})$ is proportional to the local polarization $P(\vec{r})$, C_h is equivalent to the polarization-polarization correlation function. In the calculations periodic boundary conditions were used. The calculated $C_h(\vec{r})$ is presented in figure 5. Dark regions correspond to negative values of the correlation function (correlation between "up" and "down" polarization). One can clearly observe anisotropy inconsistent with the 4-fold symmetry of a (001) plane. At large r we attribute this anisotropy to a slight miscut of the crystal (of about 5 degrees). At small r the anisotropy can be due to different scanning speed along x- and y-axis. Since lock-in register the deformation signal after some delay, the asymmetry of the correlation image can be just an instrumental effect. In the following analysis we have neglected the anisotropy at relatively large r and so used the angular averaged $C_h(r)$ presented in figure 6 for two independent images, normalized so, that $C_h(0) = 1$. It is easily seen, that the results from these independent images are in good agreement with each other. The shape of the correlation function corresponds to the fairly regular structure of polar nanoregions [18], with a well defined separation distance d, as evidenced by the oscillatory behavior at $r > d$ and existence of a zero-crossing point. For a more detailed analysis we calculated the two-dimensional

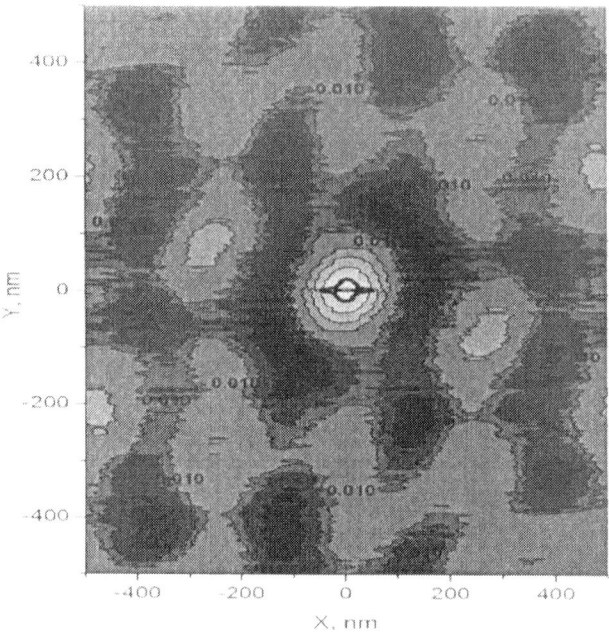

FIGURE 5. Height-height correlation function of the image presented at Fig. 4.

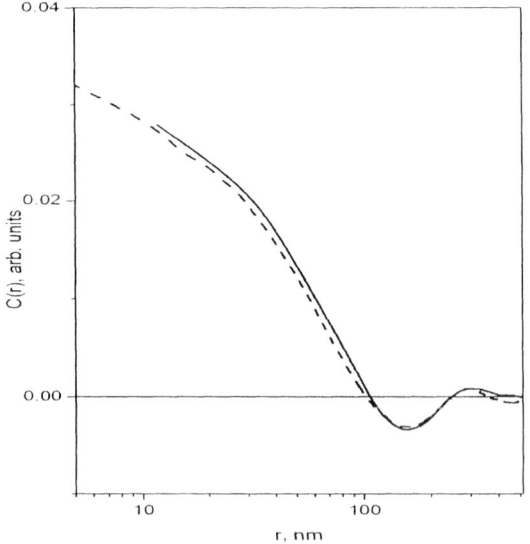

FIGURE 6. Angular averaged height-height correlation functions for two independent images.

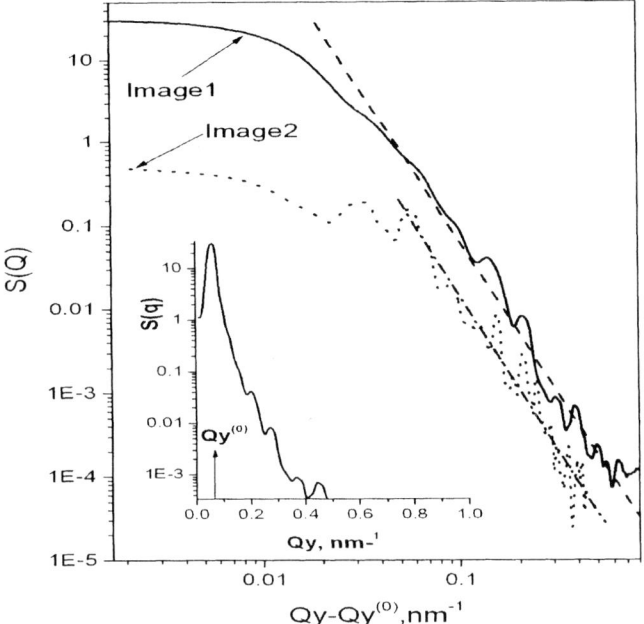

FIGURE 7. Log-log plot of $S_y(q)$ for two independent images, abscissa is corrected for Qy^0; in the inset semilog plot of of $S_y(q)$ for image at Fig. 4, peak at Qy^0 is clearly seen.

Fourier transform of the $C_h(\vec{r})$:

$$S(\vec{q}) = \sum_r C_h(\vec{r}) \exp(i\vec{r}\vec{q}). \qquad (3)$$

$\vec{q}$ is in the plane and so always perpendicular to the analyzed component of the polarization, which is perpendicular to the plane. So $S(\vec{q})$ corresponds exactly to the transverse component of the X-ray or neutron scattering cross-section, usually studied in ferroelectrics. (The longitudinal component is strongly suppressed, see e.g. [19]). Since $C_h(\vec{r})$ is strongly anisotropic at small r, $S(\vec{q})$ is anisotropic at the large q. Instrumental broadening of the image in the X direction made the X component unusable for the analysis. In Fig. 7 the Y components of $S(\vec{q})$ are presented for two images. The clear narrow peak at $q_y^0 = 0.07 \pm 0.01 nm^{-1}$ corresponds to the size 90 ± 15 nm. It is clearly seen in the log-log plots the $S(q)$ follows power law $q^{-\alpha}$ with $\alpha = 3.55 \pm 0.1$ up to $q \approx 0.7\ nm^{-1}$. The decrease of the slope at large q (only partly shown in the figure) is due to the finite experimental resolution, that was in case of image 1 coarser than the pixel size. The shape of $S(q)$ is in conformity with that recently predicted by Prosandeev et al. [20] for the relaxors split into nanodomains. He obtained a Lorentian-squared shifted from

$q = 0$:

$$S(q) \frac{1}{((q-q_h)^2 + \kappa^2)^2}, \qquad (4)$$

where q_h is domain thickness and κ inverse correlation length. In his model based on coupling between the polarization and elastic deformation, the roughness of the domain walls was not included and so at large q, q^{-4} was obtained. In our case the fact that the power exponent is less than 4 can considered as a consequence of the finite width of the domain walls.

SUMMARY

We have briefly considered above the spatial distribution of polarization in relaxors on the nanometer scale. We have shown that in most cases at low temperatures a nanodomain state is formed. This is evidenced by the non-lorentzian shape of the diffuse scattering, that can be separated into the thermodynamic (susceptibility) and configuration fluctuation (structure factor) components. As demonstrated for the case of PMNPT10, creation of the nanodomains results in the arising of long-range strains easily detectable by broadening of the Bragg peaks. PFM measurements of the PMNPT10 single crystal allowed us to observe directly these polar nanodomains. The calculated $S(q)$ follows at large q a power law with index $\alpha = 3.55$, in reasonable agreement with $\alpha = 3.3$ obtained for the structure factor component in PMN after being cooled in the electric field.

ACKNOWLEDGMENTS

It is a pleasure to acknowledge R. Blinc, T. Egami, R. Fisch, M. Glinchuk, V. Sakhnenko and V. Stephanovich for the many useful discussions and various suggestions. Work was supported by the RFBR (grant 02-02-16695), CRDF (Project RP1-2361-ST-02) and Russian program "Neutron Research of Solids".

REFERENCES

1. Vakhrushev, S., Naberezhnov, A., Sinha, S. K., Feng, Y. P., and Egami, T., *J. Phys. Chem. Solids*, **57**, 1517–1523 (1996).
2. Vakhrushev, S. B., Kvyatkovsky, B. E., Naberezhnov, A. A., Okuneva, N. M., and Toperverg, B. P., *Ferroelectrics*, **90**, 173–176 (1989).
3. Miao, S., Zhu, J., Zhang, X., , and Cheng, Z.-Y., *Phys. Rev. B*, **65**, 052101 (2001).
4. Lehnen, P., Kleemann, W., Woike, T., and Pankrath, R., *Phys. Rev. B*, **64**, 224109 (2001).
5. Prokert, F., *Physica Status Solidi B*, **87**, 179 (1978).
6. Chetverikov, Y., Naberezhnov, A., Vakhrushev, S., Dorner, B., and Ivanov, A., *Applied Physics A*, **74 (Suppl1)**, s989 (2002).
7. Slanic, Z., Belanger, D., and Fernandez-Baca, J. A., *J. Phys.: Condens. Matter*, **13**, 1711 (2001).
8. Gofman, M., Adler, J., Aharony, A., Harris, A. B., and Schwartz, M., *Phys. Rev. B*, **53**, 6362 (1996).
9. Tarko, H. B., and Fisher, M. E., *Phys.Rev. B*, **11**, 1217 (1975).

10. Bray, A. J., *J. Phys. C: Solid State Phys.*, **19**, 6225 (1986).
11. Lines, M. E., and Glass, A. M., *Principles and applications of ferroelectrics and antiferroelectrics and related materials*, Clarendon Press, 1977, ISBN 0-19-851286-4.
12. Pirc, R., and Blinc, R., *Phys. Rev. B*, **60**, 13470 (1999).
13. Dkʰıi, B., Kiat, J.-M., Calvarin, G., Baldinozzi, Vakhrushev, S. B., and Suard, E., *Phys. Rev. B*, **65**, 24104 (2001).
14. Gehring, P., Shirane, G., , and Ye, Z.-G., $pmn - 10pt$ is not rhombohedral, These Proceedings (????).
15. Krivoglaz, M. A., *X-ray and neutron diffraction in nonideal crystals*, Springer-Verlag, Berlin, 1996.
16. Eng, L. M., Grafstrom, G., Loppacher, C., Schlaphof, F., Trogisch, S., Roelofs, A., and Waser, R., *Adv. in Solid State Phys.*, Springer-Verlag, Berlin, Heidelberg, 2001, vol. 41, p. 287.
17. Munoz, R. C., Vidal, G., Mulsow, M., Lisoni, J. G., Arenas, C., and Concha, A., *Phys. Rev. B*, **62**, 4686 (2000).
18. Yang, H.-N., Zhao, Y.-P., Chan, A., Lu, T.-M., and Wang, G.-C., *Phys. Rev. B*, **56**, 4224 (1997).
19. Aksenov, V. L., Plakida, N. M., Pontekorvo, D. B., and Stamenkovic, S., *Neutron Scattering by Ferroelectrics*, World Scientific Pub Co, NY, 1990.
20. Prosandeev, S. A., Raevski, I. P., and Waghmare, U. V., Diffusive phase transitions in ferroelectrics and antiferroelectrics (????), these Proceedings.

Conformal Domain Miniaturization and Adaptive Monoclinic (Pseudo-orthorhombic) Ferroelectric States

Y.M. Jin, Yu Wang, and A.G. Khachaturyan

Dept. of Ceramics and Materials Science and Engineering, Rutgers University, Piscataway, NJ 08855

J.F. Li and D. Viehland

Dept. of Materials Science and Engineering, Virginia Tech, Blacksburg, VA 24061

Abstract. Ferroelectric and ferroelastic phases with very low domain wall energies have been shown to form miniaturized microdomain structures. A theory of an adaptive ferroelectric phase has been developed to predict the microdomain-averaged crystal lattice parameters of this structurally inhomogeneous state. The theory is an extension of conventional martensite theory, applied to ferroelectric systems with very low domain wall energies. The cases of ferroelectric microdomains of tetragonal (FE_t) symmetry are considered. It is shown that a nano-scale coherent mixture of microdomains can be interpreted as an adaptive ferroelectric phase, whose microdomain-averaged crystal lattice is monoclinic. The crystal lattice parameters of this monoclinic phase are self-adjusting parameters, which minimize the transformation stress. Self-adjustment is achieved by application of the invariant plane strain (IPS) to the parent cubic lattice, and the value of the self-adjusted parameters constitutes a mixture of the lattice constants of the parent and product phases. Experimental investigations of $Pb(Mg_{1/3}Nb_{2/3})O_3$-$PbTiO_3$ (PMN-PT) and $Pb(Zn_{1/3}Nb_{2/3})O_3$-$PbTiO_3$ (PZN-PT) single crystals confirm many of the predictions of this theory.

INTRODUCTION

The ferroelastic and/or martensite microstructure consists of polydomain plates. Each plate is formed by alternating layers of twin-related domains, where the layers are parallel to the twin plane. The structure of a polydomain martensite plate is schematically shown in Figure 1a. Because the boundary between twin-related domain layers (shown by white and black stripes in Figure 1a) is the twin plane, there is no crystal lattice misfit between layers. The relative thicknesses of the domain layers can be adjusted to establish the macroscopic invariance of the habit plane [1,2], and in so-doing eliminates long-range stress fields generated by crystal lattice misfits [3-5]. This requires the domain-averaged stress-free transformation strain of each plate to be an Invariant Plane Strain (IPS), where the invariant plane is parallel to the habit plane.

Stress-accomodating domain structures typical of martensite should also occur for any displacive transformation, which can form a domain-averaged transformation strain that is an IPS. An example of such a polydomain structure for the tetragonal phase of an ordered CuAu alloy is shown in Figure 1b. The black and white strips in this TEM image are alternating twin-related domains of the tetragonal phase. These stripes assemble into macroplates of the kind shown in Figure 1a, which are in turn arranged into a larger scale pattern.

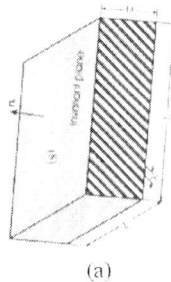 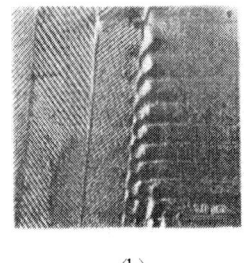

Figure 1. (a) Structure of a polydomain plate consisting of alternating lamellae of two twin-related orientation variants shown by black and white. The boundary between the lamellae is the twin plane. The ratio of the "black" and "white" domain thicknesses provides the macroscopic invariancy of the habit plane. (b) Dark field TEM image of the strain-accommodating domain structure in CuAu alloy [6]. The alternating white and black strips are images of the (110) twin-related tetragonal domains (orientation variants) of the $L1_0$ ordered phase with the alternating directions of the tetragonal axis c. They are arranged in plates similar to that shown in (a). These plates fully fill a sample.

(a)   (b)

Ferroelastics With Very Low Domain Wall Energy: Adaptive Martensite

The crystal lattice parameters of individual martensite domains are those of the low temperature transformed state. However, a unique situation can arise in the case of a very low domain boundary energy: *domains can miniaturize, as the domain size is proportional to* $\sqrt{\gamma}$, *where γ is the domain wall energy*. A thermodynamic and crystallographic theory [7, 8] has been developed for this unique case, and applied to explain unusual features of an intermediate martensitic phase previously observed in both Al-62.5%Ni [9] and Fe-Pd [10] alloys. In this case of miniaturized domains, the crystal lattice parameters of individual domains are no longer those of the low temperature transformed state. Rather, the lattice parameters are a combination of the parent cubic and ferroelastic product phases, as determined by the IPS. As a consequence, the lattice parameters themselves follow the twinning rules. We use the term "adaptive phase" to describe this unique structurally mixed state.

The adaptive phase is a particular (miniaturized) case of conventional martensite with stress-accommodating domains, which can only be expected in situations where the domain wall energy is abnormally small. In this case, the IPS is fully determined by a combination of the lattice parameters of the low symmetry product phase (which forms microdomains) and the high symmetry parent phase. There is no other phase that could have this unique property. Therefore, if the stress-free transformation strain calculated from the crystal lattice parameters of the product and parent phases is the IPS, it can safely be concluded that the martensite phase is the adaptive type. The volume fractions of the various microdomains (i.e., nano-sized orientation variants of conventional martensite) can change in response to applied stress σ and external boundary conditions. Thus, in the adaptive phase, the crystal lattice parameters (which are microdomain-averaged over the plate) are continuously adjusted by σ.

Unlike a conventional martensite phase, the crystal lattice of the adaptive phase can be considered homogeneous only on a scale of ~ 10 nm. At smaller length scales (<10 nm), the structure is inhomogeneous: it is a mixed state, consisting of alternating stress-accommodated coherent structural microdomains (i.e., nano-sized orientation variants of conventional martensite). The spatial distribution (i.e., patterning) of microdomains comprising the adaptive phase is determined by the same condition as that of conventional martensite [3-5] – minimization of the sum of the transformation-induced strain and interfacial domain wall energies. If the adaptive phase is formed from tetragonal microdomains, then the domain structure schematically shown in Figure 1 will be conformally miniaturized. Accordingly, the black and white strips in Figure 1b (which are the differently oriented (110) twin-related tetragonal domains) would become the microdomains of the adaptive phase, and the plates composed of these microdomains schematically shown in Figure 1a would become *macrodomains* of the adaptive phase.

Since the adaptive phase is the particular case of a martensitic domain structure with a very low γ, the physical properties of the adaptive phase will be similar to those of conventional martensite having stress-accommodating domains. The fingerprints of the adaptive martensite phase include: (i) an interdependence of the crystal lattice parameters on those of the parent cubic phase and those forming the microdomains; (ii) shape memory effects; and (iii) a dependence on thermal, stress, and electrical histories (i.e., nonergodicity).

Ferroelectrics With Low Domain Wall Energies: Adaptive Ferroelectrics

Ferroelectric transformations are another important class of materials with displacive crystal lattice rearrangements, where adaptive phases might form. Ferroelectric domains are crystallographically equivalent orientation variants. They are structural domains with polarization. To accommodate crystal lattice misfits, structural domains self-assemble into a martensite-like pattern. The presence of polarization in structural domains will not change the pattern, if neighboring domains are separated by a twin plane boundary (which is the case), and if the twin-related polarization vectors of neighboring domains are directed in a "head-to-tail" fashion. In this case, the gradient of the polarization at the domain boundaries does not generate an electrostatic field (i.e., uncharged domain walls). Accordingly, ferroelectric domains are simultaneously structural ones, provided that the self-assembly pattern eliminates both the strain and electrostatic energies.

Similar to martensitic transformations, a significant reduction in γ will result in domain miniaturization. In such a ferroelectric adaptive phase, both the observed macro-domain polarization and crystal lattice parameters will be microdomain-averaged. The total dipole moment of the adaptive phase is a vector sum of the moments of alternating polar microdomains that compose the macrodomains. It is a weighted sum of the relative thicknesses, ω and $1-\omega$, of the microdomain variant populations. Accordingly, the polarization direction in the adaptive phase will depend on the ratio of the thicknesses of twin-related microdomains. Thus, the polarization direction will not coincide with any fixed high symmetry direction. The observed polarization and crystal lattice parameters will be of the same symmetry, but one with a point group which is reduced to only those symmetry elements common to both the parent cubic and low temperature product phases.

A ferroelectric adaptive phase (consisting of miniaturized domains that are simultaneously polar and structural) should have properties similar to those of a martensitic adaptive phase (with no polarization), plus several additional properties inherent to the presence of polarization in the microdomains. The effect of electric field E and mechanical stress σ on an adaptive ferroelectric should be similar, both resulting in a redistribution of the microdomain variant populations. Under E, the volume fraction of microdomains with a favorably oriented polarization will increase, and that of microdomains which are unfavorably oriented will decrease. This will produce a macroscopic shape change, which will be on the order of the transformation strain. The electric field-induced microdomain rearrangement will affect the crystal lattice parameters of the adaptive phase, resulting in an abnormally high piezoelectricity which is much greater than that of a homogeneous domain state. Because the microdomain-averaged polarization will depend on the ratio of microdomain thicknesses, an E-induced change in this ratio should result in a change in the polarization direction. The direction of the micro-domain averaged polarization will rotate under electric field. No such continuous rotation of the polarization vector is known in a conventional homogeneous ferroelectric domain.

The fingerprints of a ferroelectric adaptive phase are: (i) the IPS is fully determined by a combination of the lattice parameters of the low symmetry product phase (forming microdomains) and the high symmetry parent phase; (ii) the IPS is variabile with E and σ, resulting in abnormally high piezoelectricity; (iii) a gradual rotation of the polarization vector and change in its magnitude upon gradual increase of the electric field; and (iv) shape memory effects, and shape memory polarization. Furthermore, since the domain walls are very mobile due to the very low domain wall energy, the P-E and ε-E hysteresis loops for these materials is expected to be very slim.

These fingerprint properties of adaptive ferroelectrics are in particular interesting in regards to recent reports of new monoclinic ferroelectric phases (FE_m) in oriented poled $Pb(Mg_{1/3}Nb_{2/3})O_3$-$PbTiO_3$ and $Pb(Zn_{1/3}Nb_{2/3})O_3$-$PbTiO_3$ crystals [11-14]. New FE_m phases have been reported near a morphotropic boundary, which separates the phase fields of tetragonal (FE_t) and rhombohedral (FE_r) ferroelectric phases in the phase diagrams. One phase, has the polarization vector confined to the $(010)_c$ plane, while the other has it confined to $(011)_c$. A common feature of these FE_m phases is extremely high electromechanical constants, as the electrostrictive deformation is close to the transformation strain [15-17]. These high electromechanical coefficients have been explained on the basis of a polarization rotation mechanism [11]. It should be emphasized that these pronounced features of the adaptivity are expected due to the easy rearrangement of microdomains under applied E.

In this paper, it will be shown that the special properties of the FE_m phases in PMN-PT and PZN-PT can be explained as a consequence of adaptive phases formed by σ- and E-accommodating microdomains of the FE_t phase. In addition, the crystal lattice parameters of the FE_m (pseudo-orthorhombic) phase have been found to be close to those predicted by the theory of the adaptive phase [18]. Also, the required assumption of a low domain wall energy for adaptive phase formation is in agreement with the fact that the stability field of these FE_m phases is sandwiched in the vicinity of the MPB between the FE_t and FE_r phases. Together, these observations make a strong case in favor of the assumption that the FE_m phases observed near the MPB in PMN-PT and PZN-PT are adaptive phases, with a mixed nano-scale structure.

STRUCTURAL PROPERTIES OF AN ADAPTIVE PHASE CONSISTING OF TETRAGONAL MICRODOMAINS

As mentioned above, the adaptive phase is a particular case of a σ-accommodating domain structure that is conformally miniaturized to the nano-scale level. The theory is based upon abnormally small domain wall energies. Minimization of the sum of the strain and interfacial energies produces structural domains assembled in a polydomain plate, as illustrated in Figure 1a. The typical domain size λ_o is related to the thickness of the polydomain plate D and the domain wall energy γ by the relationship

$$\lambda_o = \beta \sqrt{\frac{\gamma}{\mu \varepsilon_o^2} D} ; \qquad (1)$$

where β is a dimensionless constant, μ is the shear modulus, and ε_o is the twinning strain transforming one domain into its twin-related counterpart [3,4]. It is important that the typical thicknesses of ferroelectric and ferromagnetic domains in a plate of thickness D are also described by a similar relationship that is proportional to $\sqrt{\gamma\,D}$. This is a profound analogy between structural and ferroelectric domains, which is important to adaptive ferroelectric phase formation.

It follows from equation (1) that if $\gamma \to 0$, then $\lambda_o \to 0$ as well. This demonstrates that a reduction of the domain wall energy results in domain miniaturization. If γ is sufficiently small, miniaturization can reach a limit where the domain layer thicknesses λ_o (shown in Figure 1a by white and black strips) approaches the nano- and/or atomic-scale.

As is well known, diffraction from structural domains with twin-related orientations results in a splitting of diffraction spots, along the direction which is perpendicular to the twinning plane. To be able to determine the crystal lattice parameters of miniaturized domains, it is necessary to resolve diffraction spots from individual microdomains of different orientation variants that compose the macro-domain plates shown in Figure 1a. The positions of these diffraction spots would provide the crystal lattice parameters of individual microdomains. Splitting from individual microdomains will only be resolved when the high resolution diffraction condition $H\varepsilon_o\lambda_o \gg 1$, where H is the reciprocal lattice vector of the diffraction spot, is met [19]. If this optical condition is not met (i.e., $H\varepsilon_o\lambda_o < 1$), diffraction will be unable to resolve individual microdomains. The crystal lattice parameters obtained using such "low-resolution" diffraction represent an averaged lattice, obtained from multiple microdomains that compose the macrodomain plates. In this case, groups (or colonies) of microdomains will be perceived as a macrodomain of a structurally homogeneous adaptive phase. The lattice parameters of the adaptive phase are then determined by the positions of these "low-resolution" diffraction spots.

Crystal Lattice Parameters of an Adaptive Phase Consisting of Tetragonal Microdomains

If the adaptive state is composed of tetragonal microdomains, then the microdomains self-assemble to eliminate the transformation-induced stress. Stress-accommodating microdomains are a particular case of a martensitic structure, it can be described by the Wechsler-Lieberman-Read [1] and Bowles-MacKenzie [2] crystallographic theory of the martensite transformation. According to this theory, the domain-averaged stress-free transformation strain is an IPS. The theory of adaptive martensite/ferroelectric phases conformally miniaturizes this geometrical theory to the nano-scale. This conformal miniaturization results in the condition that the lattice parameters are predicted by the IPS.

A fully stress-accommodated macrodomain of the adaptive phase consists of two types of tetragonal microdomains with different orientations. These domains are $(110)_c$ twin-related to avoid the crystal lattice misfits and to form alternating layers along the $(110)_c$ plane. The tetragonal axes of the layers alternate along the $[100]_c$ and $[010]_c$ directions of the cubic parent phase (as shown in Figure 1a). The two types of domains formed from the cubic parent phase by stress-free tetragonal deformations are

$$\varepsilon(1) = \begin{pmatrix} \varepsilon_3 & 0 & 0 \\ 0 & \varepsilon_1 & 0 \\ 0 & 0 & \varepsilon_1 \end{pmatrix} \quad \text{and} \quad \varepsilon(2) = \begin{pmatrix} \varepsilon_1 & 0 & 0 \\ 0 & \varepsilon_3 & 0 \\ 0 & 0 & \varepsilon_1 \end{pmatrix}; \qquad (2)$$

where $\varepsilon_3 = \dfrac{c_t - a_c}{a_c}$ and $\varepsilon_1 = \dfrac{a_t - a_c}{a_c}$ are the stress-free transformation strains, c_t and a_t are the crystal lattice parameters of the stress-free tetragonal martensite, and a_c is the lattice parameter of the high temperature cubic phase. These matrices are presented in the Cartesian coordinate system where the x, y, and z axes are chosen to be parallel to the cubic directions $[100]_c$, $[010]_c$ and $[001]_c$, respectively. This alternating layered structure of domains (which is illustrated in Figure 1) can be (and usually is) interpreted as a structure formed by the periodical $(110)_c$ twinning of the tetragonal phase. The thickness of the $\varepsilon(1)$ domains is $d_1=\omega\lambda$, where λ is the distance between nearest twins and ω is the volume fraction of the martensite plate occupied by the twin with transformation strain $\varepsilon(1)$. The thickness of the lamellar domains of the second orientation variant of the martensite formed by the $\varepsilon(2)$ transformation strain (non-twinned tetragonal phase) is $d_2=(1-\omega)\lambda$.

If the tetragonal strains defined in equation 2 are small (i.e., $|\varepsilon_1|<<1$ and $|\varepsilon_3|<<1$), then the domain-averaged stress-free strain is:

$$<\varepsilon(\omega)> = \omega\,\varepsilon(1)+(1-\omega)\,\varepsilon(2) = \omega\begin{pmatrix}\varepsilon_3 & 0 & 0\\0 & \varepsilon_1 & 0\\0 & 0 & \varepsilon_1\end{pmatrix}+(1-\omega)\begin{pmatrix}\varepsilon_1 & 0 & 0\\0 & \varepsilon_3 & 0\\0 & 0 & \varepsilon_1\end{pmatrix}\begin{pmatrix}\varepsilon_1+(\varepsilon_3-\varepsilon_1)\omega & 0 & 0\\0 & \varepsilon_3-(\varepsilon_3-\varepsilon_1)\omega & 0\\0 & 0 & \varepsilon_1\end{pmatrix}. \qquad (3)$$

To eliminate crystal lattice misfits between variants and thus generation of stress, these two variants [$\varepsilon(1)$ and $\varepsilon(2)$] form alternating twin-related lamellae along the twin plane $(110)_c$. The microdomain-averaged crystal lattice parameters of the adaptive phase, obtained by application of the stress-free strain matrix given in equation 6, have pseudo-orthorhombic symmetry. These lattice parameters referenced to the cubic axes [$\mathbf{a}_1=(a_o,0,0)_c$, $\mathbf{a}_2=(0,a_o,0)_c$, $\mathbf{a}_3=(0,0,a_o)_c$] are

$$a_{ad} = a_t + (c_t-a_t)\omega, \quad b_{ad} = c_t - (c_t-a_t)\omega, \quad c_{ad} = a_t; \qquad (4)$$

where by definition, these lattice parameters are directed along the $<100>_c$ axes. Equation (4) is a fingerprint that the system consists of tetragonal microdomains of two orientation variants with the tetragonality axes along the $[100]_c$ and $[010]_c$ axes. They are the crystallographic fulfillment of the conformal miniaturization of the domain variants that was illustrated in Figure 1a.

However, equation 4 does not eliminate the stress generated by misfits along the habit plane of the macrodomains. To achieve this condition, the system must choice a special value $\omega=\omega_o$ that provides an IPS. Then, complete strain accommodation is achieved between macrodomain plates of the adaptive phase, and the entire specimen exists in a stress-free condition. This is what we designate the condition of "strong" or special invariance, and is the conformal miniaturization of Figure 1b. The strain in equation 4 is the IPS and thus the domain structure is completely stress-accommodated, if $\omega=\omega_o$, where ω_o is:

$$\omega_o = \frac{\varepsilon_1}{\varepsilon_1 - \varepsilon_3} = \frac{a_t - a_c}{a_t - c_t}. \tag{5}$$

Substituting ω_o to (4) provides the lattice parameters of the completely stress-accommodated adaptive phase, produced from the parent cubic lattice parameters by the IPS These parameters are:

$$a_{ad} = a_c, \ b_{ad} = c_t + a_t - a_c, \ c_{ad} = a_t. \tag{6}$$

The ratio of thicknesses of the twin-related tetragonal microdomains in the stress-accommodating state is then $\omega_o/(1-\omega_o) = (a_t - a_c)/(a_c - c_t)$. In fact, the adaptive phase obtained by the averaging over microdomains is monoclinic. However, the monoclinic distortion angle ϕ has been neglected in this pseudo-orthorhombic representation of the crystal lattice parameters of the adaptive phase, as it is proportional to $(\varepsilon_3 - \varepsilon_1)^2$ and thus very small.

It is important to realize that the relations between the crystal lattice parameters of the adaptive, tetragonal, and cubic phases given by equations 6 are a particular case of the relations given by equations 4: equations 6 are valid only for the particular case of a fully accommodated adaptive phase that has zero applied external fields (i.e., $\omega = \omega_o$). Therefore, equation 6 imposes more strong constraints on the crystal lattice parameters of the adaptive phase than equations 4. This strong constraint is lifted under applied field, affecting the geometrical parameter ω, which in turn, is determined by the complex energy balance caused by coupling the applied field and domain structure. In this case, $\omega \neq \omega_o$ and the crystal lattice parameters are determined by equation 7.

Invariants for Adaptive Phases Consisting of Tetragonal Microdomains.

Unlike conventional phases whose crystal lattice constants are intrinsic material constants, the crystal lattice parameters of an adaptive phase are self-adjusting parameters. Self-adjustment provides accommodation of the transformation stress, establishing the invariance condition determined in equation 9.

Changes of E or σ will result in microdomain rearrangement, characterized by the geometrical parameter ω. This results in a violation of complete stress accommodation (or an increase in the transformation-induced strain energy) at the expense of a decrease in the electrostatic or the total strain energies. Nevertheless, the theory of adaptive phases predicts certain invariant relations for the lattice parameters, which result from requiring the multi-layer polytwin structure to be maintained.

Equation (4) describes the situation where the microdomain volume fraction assumes an arbitrary value ω, and consequently does not provide an IPS (i.e., full stress-accommodation). In this case, we have the easily verifiable relations

$$a_{ad} + b_{ad} = a_t + c_t \tag{7a}$$
$$c_{ad} = a_t. \tag{7b}$$

Equations 7 are general: both the crystal lattice parameters of the adaptive phase determined by the partial stress accommodation (see equation 4) and the crystal lattice parameters determined by the full stress accommodation (see equation 6) meet equations 7.

The lattice parameters of the tetragonal phase (a_t and c_t) are intrinsic constants of a conventional phase; they are determined by atomic bonds and are functions of composition and temperature, and are weakly dependent on E and σ. Whereas, the parameters of the adaptive phase (a_{ad}, b_{ad}, c_{ad}) are self-adjusting parameters; they are very sensitive to E and σ, and to the distribution of microdomains. However, as follows from equation 7, the sum of the self-adjustable parameters, a_{ad} and b_{ad}, and the value, c_{ad}, are also intrinsic physical parameters of the conventional tetragonal phase. Therefore, equations 7 provide invariance conditions. Their left-hand sides are invariants that are independent on the applied field, irrespective of the strong dependences of the adaptive phase parameters a_{ad} and b_{ad} on field. In other words, although separately, a_{ad} and b_{ad} can change pronouncedly with T, E and σ; but, taken additively as given in equation 10a, they become invariant.

The invariance relations (7) are easily verifiable by experiment for any intermediate pseudo-orthorhombic phase. It is barely possible that their fulfillment is coincidental at a constant temperature, composition, E and σ. It is practically impossible that their continued fulfillment over the entire temperature, composition, E and σ would be coincidental. Below we will call the conditions of equation 7 the general invariance conditions.

PREDICTIONS AND CONFIRMING OBSERVATIONS OF AN ADAPTIVE PHASE FORMED FROM TETRAGONAL MICRODOMAINS

FE$_m$ Phases in Ferroelectrics Consisting of Tetragonal Microdomains

The invariance conditions are also fulfilled for the recently-discovered FE$_m$ phases of PMN-x%PT and PZN-x%PT. This composition lies on the MPB between FE$_r$ and FE$_t$ phases, where the domain wall energy is expected to be small. If we use the adaptive phase parameters (a_{ad}, c_{ad}, b_{ad}) of equation (6) to designate the FE$_m$ (pseudo-orthorhombic) lattice parameters as ($a_m = a_{ad}$, $b_m = c_{ad}$, $c_m = b_{ad}$), then the special invariance conditions of equations (6) can be rewritten as

$$a_m = a_c; \quad b_m = a_t; \quad c_m = c_t + a_t - a_c. \tag{8a}$$

The parameters corresponding to the general invariancy conditions of equation (4) can be rewritten as

$$a_m = a_t + (c_t - a_t)\omega, \quad b_m = a_t, \quad c_m = c_t - (c_t - a_t)\omega, \tag{8b}$$
$$a_m + c_m = a_t + c_t; \quad b_m = a_t. \tag{8c}$$

Both the special and general invariant conditions given in equations 8a and b should be preserved with changing temperature, if the FE$_m$ phase remains adaptive. Temperature dependent neutron diffraction data has previously been obtained for various PMN-xPT powders [13,14], in the compositional range of 31%<x<37%. These investigations were performed under zero electric field. Thus, any change in the invariance conditions can be attributed solely to the effect of temperature. The data confirms that both the special and general invariance conditions are preserved over a significant temperature range.

Figures 2a-c show the temperature dependence of the lattice constants for PMN-31%PT, PMN-33%PT and PMN-37%PT, respectively. The tetragonal lattice parameters (a_t, c_t) and cubic lattice parameter (a_c) are shown in the figures, in their respective stability ranges. At lower temperatures, the data demonstrate the presence of a FE_m phase with $c_m \neq b_m \neq a_m$. The special "invariance" condition of equation 8a was found to be well maintained. To illustrate that this invariance is preserved, the lattice parameter c_t of the metastable tetragonal phase within the stability field of the monoclinic phase was calculated using the relationship $c_t = c_m - b_m + a_c$, which can be obtained by rearranging equation 8a following the adaptivity theory. The calculated values of c_t are plotted in red, alongside the measured ones of the FE_m phase. It can be seen in Figure 2 that the calculated parameter $c_t = c_m - b_m + a_c$ provides a continuous extension of the lattice parameter c_t measured in the stability field of the FE_t phase, extrapolated into the stability field of the FE_m phase. Also, in accordance with the special invariance conditions, the measured lattice parameter b_m of the FE_m phase is a continuous extension of the lattice parameter a_t measured in the stability range of the FE_t phase. Furthermore, the measured lattice parameter a_m is equal to a_c; however, this parameter does not continuously extend between the cubic and FE_m phases. Rather, it undergoes hibernation in the FE_t phase, spontaneously reappearing in the FE_m phase as the parameter a_m, in accordance with the invariancy condition $a_c = a_m$.

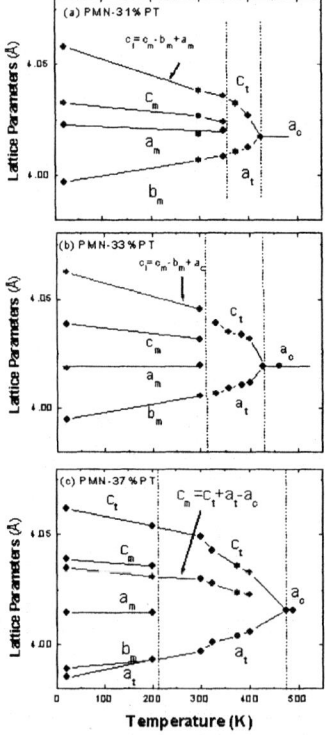

Figure 2. Temperature dependent lattice parameters for (a) PMN-31%PT, (b) PMN-33%PT, and (c) PMN-37%PT. These data were taken by neutron diffraction by Noheda et al. [13,14]. The points and lines in red are the lattice parameters of the tetragonal phase extended into the monoclinic one, which were calculated by equation (6).

The general invariance conditions of equation (8b) were found to be well-preserved for PMN-xPT, over both wide compositional (31%<x<37%) and temperature (20<T<500 K) ranges. Figure 3 shows the general invariance conditions as a function of temperature for the various PMN-xPT compositions; parts (a) and (b) in this figure show $a_m+b_m=a_t+c_t$ and $b_m=c_m$, respectively. [The scale of the y-axes was chosen to have the same relative change in parameter values, as that of the temperature dependent lattice parameters shown in Figure 2. This allows for easy comparisons.] The data clearly demonstrate that the general invariance conditions are obeyed over the entire temperature and compositional range of the FE_m and FE_t phase.

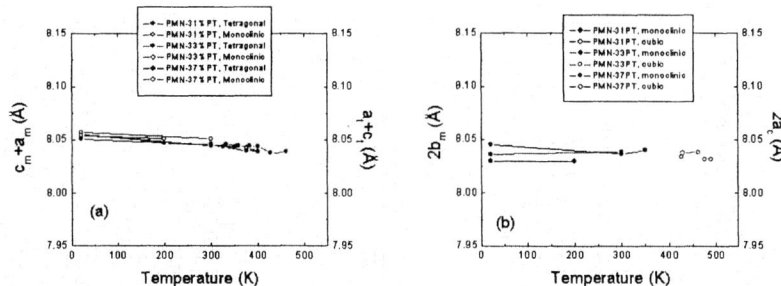

Figure 3. Temperature dependence of the general invariance condition of (a) equation 8b, and (b) equation 8c. Data are shown for PMN-31%PT, PMN-33%PT, and PMN-37%PT. The values of (c_m+a_m) and b_m were calculated from corresponding data in Figures 5a-c using equation 8c.

Application of E violates the special invariance condition of equation 8a, preventing complete stress accommodation. However, the general invariance condition of equation 8b and 8c will still be fulfilled, that is if the FE_m phase remains adaptive. The field dependence of the crystal lattice parameters of PZN-8%PT have been determined by single-crystal neutron diffraction [12,13]. Figure 4a shows data that prescribe the dependence of the crystal lattice parameters of the FE_m (pseudo-orthorhombic) phase on applied E, over the entire FE_m stability range up to the point that the FE_t phase is induced. Figure 4b shows the corresponding plot of the general invariance condition of equation 8b. The results clearly demonstrate that the general invariance condition is rigidly obeyed over the entire FE_m stability range and into the FE_t phase field. The fulfillment of the predicted general invariancy demonstrates that the changes in the lattice parameters with E are due to a redistribution of tetragonal microdomains, which are arranged in $(110)_c$ twin-related muti-layer patterns. The FE_t phase is reached when the crystal has been fully detwined by field, which occurs when when the geometric ratio in equation 4 is $\omega=0$.

Under zero field, for PZN-8%PT, the monoclinic (pseudo-orthorhombic) lattice parameters had the special relationship $a_m=c_m$. When a_m becomes equal to c_m, the symmetry of the lattice increases from monoclinic to orthorhombic. The latter can be seen by using a doubled orthorhombic unit cell in a manner similar to that for orthorhombic $BaTiO_3$ [21,22] rather than a monoclinically distorted primitive cell of the host cubic lattice. We designate this orthorhombic phase as FE_O. In this case, equation 8b still describes this orthorhombic phase as a particular case where the relative volume fractions of the microdomain variants is $\omega=\frac{1}{2}$ and the $(110)_c$

twin-related tetragonal microdomains are (110) atomic layers with equal thickness $a_c/\sqrt{2}$. In fact, investigations of poled (110)-oriented crystals have demonstrated that full remanence can be sustained along the $(110)_c$, in agreement with this possibility [18,20]

The locking of ω into the particular value $\omega=\frac{1}{2}$ corresponding to the formation of the orthorhombic phase can be attributed to interactions between domain walls, which cannot be neglected in the relevant case because the separation distance between the walls is of atomic scale. By and only at this special geometrical value $\omega=\frac{1}{2}$, the symmetry of the system is increased from monoclinic (pseudo-orthorhombic) to orthorhombic. This abrupt increase in symmetry necessitates a singularity in the chemical free energy at $\omega=\frac{1}{2}$. This singular point can be either the free energy minimum (local or global) or the maximum with respect to ω. The minimum should be an infinitesimally narrow "pothole" on the free energy vs ω curve. The observation of this orthorhombic phase confirms that the free energy singularity is a free energy minimum, and that the orthorhombic phase is either stable or metastable.

Since $\omega=\frac{1}{2}$ is not equal to the value ω_o, complete stress accommodation and the special invariance condition are not achieved (equation 8a). However the general invariance condition is well maintained under applied field (see equation 8b). Figure 4 indicates that an applied field unlocks ω from this special value, making $\omega(E) \neq \frac{1}{2}$ in equation 8a. The fact that the parameters of the unlocked phase fulfill the general invariance of equation (8b) demonstrates that both the locked orthorhombic phase FE_O and unlocked monoclinic phase are both described by the adaptive phase theory.

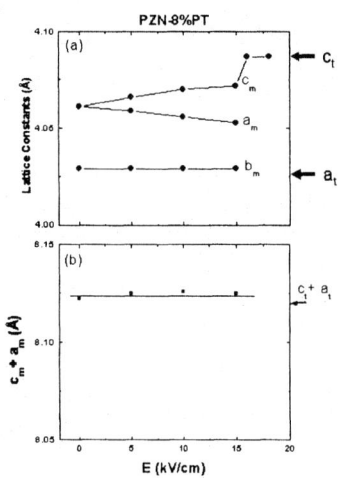

Figure 4. (a) Dependence of the crystal lattice parameters, c_m, a_m, and b_m, of the pseudo-orthorhombic (monoclinic) phase in PZN-8%PT on the applied electric field [12,13]. (b) Dependence of the sum, $a_m + c_m$, on E. The independence of this sum on E is the fulfillment of the invariance condition of equation 8c. The results demonstrate that $(a_m+c_m)=(a_t+c_t)$ as required by the general invariance condition.

The results clearly show that the general invariance conditions are obeyed over a broad range of electric fields (Figure 4b), and temperatures and compositions (Figures 3a and b). This continued fulfillment of the invariance conditions over such a broad stability range unambiguously demonstrates that the FE_m phase is not a homogeneous (unique) phase requiring independent lattice parameters. Rather, it is an adaptive phase which can be described over its entire stability range by adjustable parameters, given by the invariance conditions.

DISCUSSION AND SUMMARY

In this paper, the theory of a single-phase ferroelectric domain structure has been extended to the particular case where the domain wall energy γ is abnormally low. In this case, the sizes of domains proportional to $\sqrt{\gamma}$ (equation 1) is drastically reduced and ferroelectric domains are conformally miniaturized. This is the main assumption of our theory. We believe that this assumption is realistic near a morphotropic phase (MPB) boundary between two ferroelectric perovskite-based phases with different orientations of the polarization vector. This assumption has several significant ramifications.

First, there will be a conformal miniaturization of the domain structure, without a significant change in the topology of the spatial pattern or distribution of variant populations. The domain topology is not affected because miniaturization is driven by the vanishing of the volume-dependent components of the elastic strain and depolarization energies: it is not dependent on the surface energy contributions from domain boundaries. There are two important intrinsic length parameters, which control the size of the domain structure. These are $\frac{\gamma}{\mu \varepsilon_o^2}$ and $\frac{\gamma}{P_o^2}$, which are both proportional to γ and are reduced to the nano- and sub-nano-scale under a condition of drastically small γ.

Second, conformal miniaturization of domains results in a structurally inhomogeneous phase on a scale less than 10 nm. Whereas, on length scales >10 nm, the structure appears homogeneous. Locally (within microdomains), the structure is tetragonal; however the micro-domain averaged lattice is pseudo-orthorhombic. In order to be able resolve the crystal lattice parameters of individual microdomains, the optical condition $H\varepsilon_o\lambda_o \gg 1$ must be met where H=(HKL) is the reciprocal lattice vector of the operational diffraction spit, and (HKL) are the indexes. This estimate can be rewritten as

$$\lambda_o \gg \frac{a_o}{\varepsilon_o \sqrt{H^2 + K^2 + L^2}}. \qquad (9)$$

Using the experimental values of $\varepsilon_o = \varepsilon_3 - \varepsilon_1 \sim 10^{-2}$ and assuming a high reflection index of $\sqrt{H^2 + K^2 + L^2} \sim 10$, we can estimate using equation (13) that microdomains will be resolvable if $\lambda_o \gg 10 a_o$. Diffraction patterns taken under "low resolution" will perceive the lattice parameters as those of the microdomain-averaged state. To perceive that of individual microdomains, the sum must be $H^2 + K^2 + L^2 \gg 10^2$. This estimate indicates that microdomains can reach ten(s) of nanometers and still not be resolved. We designate this structurally mixed state as the adaptive ferroelectric phase. The unique properties of the adaptive ferroelectric phase are due to the fact that it is a macroscopically homogeneous phase, which on the nanometer scale has both an inhomogeneous structure and polarization. Experimentally, transmission electron microscopy investigations of unpoled PMN-PT has revealed the presence of irregularities on this scale [23,24], which have been described as microdomains and tweed. We can conclude that there are inhomogeneities on this scale, whether they are the ones of the adaptive phase remains in questions.

Third, the lattice parameters of the adaptive phase are a mixture of those of the parent cubic and low temperature product ones. They are not intrinsic physical constants, rather they are parameters that are adjusted to achieve stress and depolarization electric field accommodation. Adjustments occur by changes in microdomain variant populations. Under the condition of complete stress accommodation, the crystal lattice rearrangement transforming the paraelectric phase to the adaptive ferroelectric one is the IPS. The invariance conditions (see equations 4 and 6) of the crystal lattice parameters of the adaptive phase and the high symmetry parent phase are the finger-prints of adaptivity. The general invariance condition achieves stress accommodation by eliminating misfits between variants. The special invariance condition does this also, but in addition eliminates misfits along the habit plane of the macrodomain. These invariance conditions imposed on the crystal lattice parameters are so restrictive and special that it is certainly impossible that their continued fulfillment over an entire temperature, concentration, electric field, and stress stability ranges of an intermediate phase is coincidental.

In the case of an adaptive ferroelectric phase consisting of tetragonal microdomains, the microdomain-averaged symmetry is monoclinic (pseudo-orthorhombic). Under the condition of full stress accommodation, crystal lattice parameters of the pseudo-orthorhombic adaptive phase have the following special crystallographic relationship with the cubic and tetragonal ones: $a_{ad}=a_c$, $b_{ad}=c_t+a_t-a_c$, $c_{ad}=a_t$. These special relations are easily verified; and have been for martensitic transformations in Ni-Al [9] and Fe-Pd [10] alloys, and for oriented PMN-PT ferroelectric crystals.

ACKNOWLEDGEMENTS

JUW, YMJ and AGK are gratefully acknowledge the support DMR of the NSF under the grant DMR 9817235. DDV and JFL gratefully acknowledge the support of the Office of Naval Research under grants N000140210340, N000140210126, and MURI N000140110761.

REFERENCES

1. M.S. Wechsler, D.S. Lieberman, and T. A. Read, Trans. Metall. Soc., AIME, **197**, 1503 (1953)
2. J.C. Bowles and J.K. Mackenzie, Acta Metall. **2**, 129, (1954)
3. A. G. Khachaturyan and G.A. Shatalov, Zh. Eksp. Teor. Fiz. **56**, 1037 (1969) (Sov. Phys. JETP **29**, 557 (1969))
4. A.G. Khachaturyan, The Theory of Structural Transformations in Solids, Wiley, 1983, New York
5. A.L. Roytburd, Fiz.Tverd.Tela, 10, 3619 (1968) (Sov.Phys. Solid State, **10**, 2870 (1969))
6. V.I. Syutkina, E.S. Jakovleva, Phys. Status Solids, **21**, 465, 1967.
7. A.G. Khachaturyan, S.M. Shapiro, and S. Semenovskaya, Phys.Rev. B, **43**, 10832 (1991)
8. A.G. Khachaturyan, S.M. Shapiro, and S. Semenovskaya, Mater Trans. JIM, **33**, 278, (1992)
9. S.M. Shapiro, B.X. Yang, G.Shirane, Y. Noda, , and L.E. Tanner, PRL, **62**, 161, (1989)
10. H.Seto, K.Noda, and Y. Yamada, J. Phys.Soc. Jpn., **59**, 965 (1990)
11. B. Noheda, D. Cox, G. Shirane, E. Park, L.E. Cross, and Z. Zhong, Phys. Rev. Lett **86**, 3891 (2001).
12. K. Ohwada, K. Hirota, P. Rehrig, Y. Fujii, and G. Shirane, Phys. Rev. B (arXiv: Cond-mat/0207726 v3, 12 Aug 2002).
13. B. Noheda, J.A. Gonzalo, L.E. Cross, R. Guo, S.E. Park, D.E. Cox, and G. Shirane, Phys. Rev. **B61**, 8687 (2000)
14. B. Noheda, D. Cox, G. Shirane, J. Gonzalo, and L.E. Cross, Appl. Phys. Lett. **74**, 2059 (1999).
15. S. Park and T.R. Shrout, J. Appl. Phys. **82**, 1804 (1997).

16. J. Kuwata, K. Uchino, S. Nomura, Jpn. J. Appl. Phys. **21** (9), 1298 (1982).
17. D. Viehland and J. Powers, Appl. Phys. Lett **78**, 3112-3114 (2001).
18. D. Viehland, J. Appl. Phys. **88**, 4794 (2000).
19. A.G. Khachaturyan, Kristallografia (USSR), **4**, 646, 1959
20. Yu Lu, Q.M. Zhang, and D. Viehland, Appl. Phys. Lett **78**, 3109-3111 (2001).
21. J. Merz, Phys. Rev. **76**, 1221 (1949).
22. H. Kay and P. Vousden, Philos. Mag. **40**, 1019 (1949).
23. D. Viehland, Myung-Chul Kim, Z. Xu, and Jie-Fang Li, Appl. Phys. Let. 67, 2471-2473 (1995).
24. Z. Xu, M.C. Kim, Jie-Fang Li, and Dwight Viehland, Phil. Mag. A 74, 395-406 (1996).

Condensation and Slow Dynamics of Polar Nanoregions in Lead Relaxors

D. La-Orauttapong, O. Svitelskiy, and J. Toulouse

Physics Department, Lehigh University, Bethlehem, PA 18015

Abstract. It is now well established that the unique properties of relaxor ferroelectrics are due to the presence of polar nanoregions (PNR's). We present recent results from Neutron and Raman scattering of single crystals of PZN, PZN-xPT, and PMN. Both sets of measurements provide information on the condensation of the PNR's and on their slow dynamics, directly through the Central Peak and, indirectly, through their coupling to transverse phonons. A comparative analysis of these results allows identification of three stages in the evolution of the PNR's with decreasing temperature: a purely dynamic stage, a quasi-static stage with reorientational motion and a frozen stage. A model is proposed, based on a prior study of KTN, which explains the special behavior of the transverse phonons (TO and TA) in terms of their mutual coupling through the rotations of the PNR's.

Introduction

For many years, relaxor materials (mostly, lead-based $Pb(R_{1/3}Nb_{2/3})O_3$, where R = Mg^{2+} or Zn^{2+} PMN and PZN respectively), have been a focus of research in the ferroelectrics community. Growing attention has been given to their industrially-promising solid solutions with $PbTiO_3$ (PT) [1, 2, 3], especially at concentrations near the morphotropic phase boundary (MPB), where their remarkable piezoelectric and electrostrictive properties are enhanced [3, 4, 5]. However, the fundamental origin of these properties remains a puzzle, and the development of their low-temperature state is still not well understood.

The difficulties in understanding originate from the high complexity of these materials, characterized by chemical, compositional and orientational disorder that coexists with the presence of short-range order [6, 7]. As relaxors are cooled from high temperature, the major structural changes are preceded by the nucleation of polar nanoregions (PNR's) [8, 9]. These PNR's are the consequence of ion off-centering which is commom in many perovskites. By analogy with $KTa_{1-x}Nb_xO_3$ (KTN) at sufficient Nb concentrations [10], one could expect that their development would lead to the appearance of a long-range order. However, unlike KTN where Nb^{5+} is the only off-centered positive ion, lead relaxors present a much more difficult case [11, 12, 13, 14, 15, 16, 17]. It is often assumed[18, 19, 20, 21] that Nb^{5+}, due to its position in the cell and small radius, acts as the main ferroelectric agent. But, the PNR's in the lead compounds, develop in the presence of random fields, that prevent the formation of long-range order [21, 22]. The frustrating effects of these fields can be compensated by an electric field applied in a $\langle 111 \rangle$ direction [23, 24].

The complexity of the lead relaxor materials has resulted in the co-existence of several mutually-excluding interpretations of its light scattering spectrum (for review see [10, 25]). Not only is the phonon assignment of particular lines in question, but also the very existence of first-order light scattering in a cubic crystal remains unaccounted. The absence of a Raman analogy to the "waterfall" phenomenon[26, 27, 28, 29] is also puzzling. First-principle calculations of the lattice modes[30], can be very helpful in answering these questions.

In order to shed some light on the development of the low-temperature phase and to resolve the above mentioned contradictions, we decided to carry out a detailed study of the relaxor behavior in PZN, PZN-xPT (x = 4.5 and 9%), and PMN single crystals (for growth technology see [31, 32, 33]) using neutron and light scattering spectroscopy. The purpose of this article is to present a brief summary of the work completed [34, 35, 36, 10].

Neutron scattering studies and their results

The neutron scattering experiments were carried out on BT9, HB-1, and 4F-2 triple-axis spectrometers at the NIST Center for Neutron Research (NCNR), at the High Flux Isotope Reactor (HFIR) of Oak Ridge National Laboratory, and at the Orphée reactor of the Laboratoire Léon Brillouin (LLB), respectively. The spectrometer was operated in *the final neutron energy* E_f *fixed* at 14.7 meV (λ = 2.36 Å) at NCNR, at 13.6 meV (λ = 2.45 Å) at HFIR, and in *the final neutron wavevector* k_f *fixed* with either at 1.97 Å^{-1} (8.04 meV, λ = 3.19 Å) or at 1.64 Å^{-1} (5.57 meV, λ=3.83 Å) at LLB. The (002) reflection of a highly oriented pyrolytic graphite (HOPG) crystal was used to monochromate and analyze the incident and scattered neutron beams. To suppress contamination by higher order neutrons, a HOPG filter was installed in the scattered beam. The crystal was mounted onto an aluminum sample holder or a boron nitride and oriented with either in the [100]-[011] or [100]-[010] scattering planes.

Constant-$\vec{Q}$ scans were used to collect data by holding the momentum transfer $\vec{Q} = \vec{k}_i - \vec{k}_f$ fixed, while scanning the energy transfer $\hbar\omega = E_i - E_f$. Using this scan, the central peak ($\hbar\omega$=0) (CP) and the transverse acoustic (TA) phonon mode were obtained upon cooling.

The central peak is a consequence of the relaxational motion in the crystal. So when this motion is fast, the CP is broad and its intensity is small. With decreasing temperature, the life time of the clusters increases and their reorientational motion is slowing down. This should lead to the growth of the central peak as shown in Fig. 1 (a). This figure show the CP spectra are shown taken at $\vec{Q}$ = (2,0.35,0) at 900 K and 375 K ($\vec{Q} = \vec{q} + \vec{G}$, where q is the momentum transfer relative to the $\vec{G}$ = (2,0,0) Bragg point, measured along the [010] symmetry direction). The CP spectra (solid lines) were fitted to a delta function.

The temperature dependencies of the CP intensity of PZN, 4.5%, and 9%PT at $\vec{Q}$ = (2,0.35,0) are presented in Fig. 1 (b). The CP intensity initially increases with decreasing temperature until $T \sim T^*$, and goes through a minimum at the transition

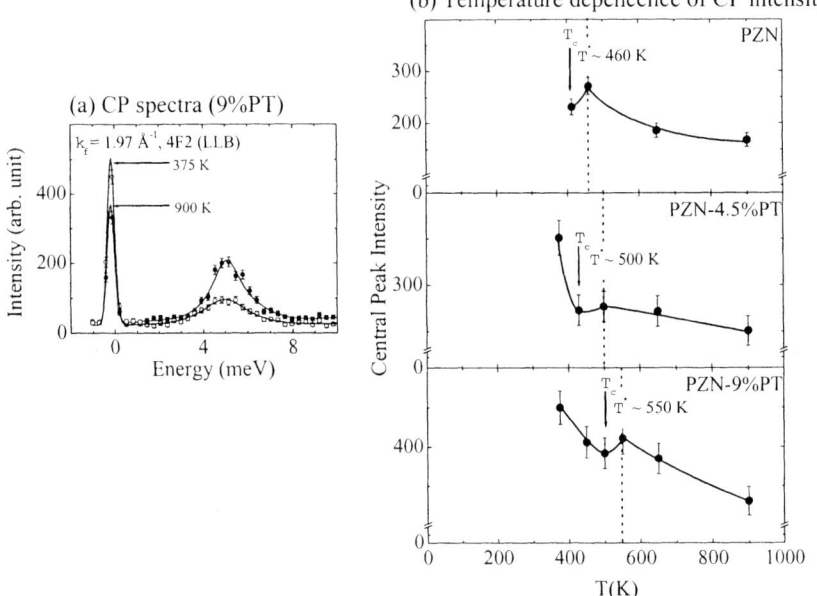

FIGURE 1. (a) Central peak spectra of PZN-9%PT at $\vec{Q} = (2,0.35,0)$ at 900 K and 375 K (b) Temperature dependence of the central peak intenisty at $\vec{Q} = (2,0.35,0)$ in PZN, 4.5%PT, and 9%PT, showing the condensation temperature T^*

before increasing at lower temperatures. Since the central peak is attributed to the relaxation of the precursor ferroelectric clusters[37, 38], the presence of a maximum provides supportive evidence for a temperature T^* at which the polar regions start to condense. The condensation temperatures (PZN : $T^* \sim 460$ K, 4.5%PT : $T^* \sim 500$ K, and 9%PT : $T^* \sim 550$ K) found are in agreement with our previous neutron *elastic* diffuse scattering studies [34, 35]. It is important to note that, with addition of PT, the PNR's condense at a higher temperature above the transition than in pure PZN. This fact is due to stronger correlations between the PNR's in the presence of PT.

When investigating the polarization dynamics of relaxors, it is also important to examine the TA phonons, since these should couple to the reorientation of the localized strain fields that are known to accompany the polar regions [39]. One might expect to observe increased damping of the TA phonons when the polar regions condense. In Fig. 2 (a) we show the TA phonon spectra of 4.5%PT crystal measured at the scattering vector

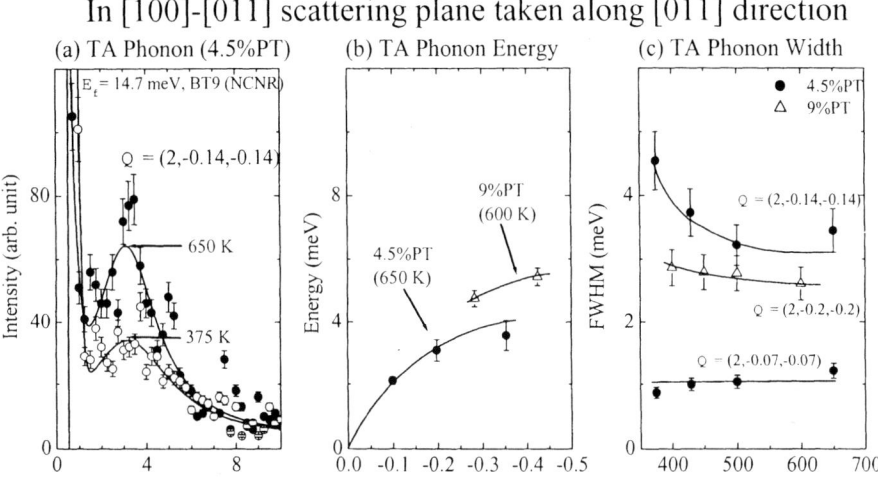

FIGURE 2. (a) TA phonons of 4.5%PT taken at the scattering vector $\vec{Q}$ = (2,-0.14,-0.14) at 650 K and 375 K (b) The dispersion curve of the TA mode for 4.5%PT (650 K) and for 9%PT (600 K) (c) Temperature dependence of the FWHM of the TA mode in 4.5%PT and 9%PT

$\vec{Q}$ = (2,-0.14,-0.14) (or $q \sim 0.20$ rlu)[1] at 650 K and 375 K. The TA spectra (solid lines) were fitted to a Lorentzian function. These profiles suggest that the TA mode damping increases with decreasing temperature. The peak position ($\hbar\omega$) of the scattered neutron intensity as a function of $|\vec{q}|$ is shown in Fig. 2 (b). This figure shows that the frequency of the TA phonon in 9%PT is higher than in 4.5%PT. However, the width or damping of the phonon is smaller for the 9%PT than for the 4.5%PT one as shown in Fig. 2 (c). This observation suggests that with increasing PT concentration, the TA damping decreases but the TA phonon frequency increases.

Our measurements of PZN-xPT, the TA phonon damping is seen to increase, starting at temperatures far above the transition, at a large q first and at a smaller q later with decreasing temperature. In other words, we find that, at a given temperature, the larger the q, the higher the damping. In fact, we do expect such a trend from the coupling to smaller polar regions at higher temperatures and to larger and slower ones at lower temperatures. As seen from Fig. 2 (c) for 4.5%PT, the phonon damping begins to increase at ~ 500 K (or T^*) and $\vec{Q}$ = (2,-0.14,-0.14) (or $|\vec{q}| \sim 0.20$ rlu), which corresponds to about 5 unit cells ($2\pi/q$). This result is consistent with the size of the polar regions derived from our neutron elastic diffuse scattering data [35]. Such an agreement provides evidence that the increase in TA phonon damping is connected to the appearance of the polar regions. In 9%PT, the TA phonon damping is significant lower than in 4.5%PT but shows a similar trend. It is important to note that the measured TA

---

[1] 1 reciprocal lattice unit (rlu) along $[0\bar{1}\bar{1}] = \sqrt{2}\times$ rlu $= \sqrt{2}$ ($2\pi/a$) = 2.19 Å^{-1}, where $a \sim 4.05$ Å.

phonon corresponds to the C_{44} elastic modulus, which can couple to the reorientations of a strain field with rhombohedral symmetry between different [111] directions. The higher frequency of the TA phonon in 9%PT than in 4.5%PT indicates that C_{44} is higher in 9%PT. Both observations, lower damping and higher TA phonon frequency, suggest that, with increasing PT, the polar regions are less able to reorient and the lattice becomes more rigid. In PMN and PMN-20%PT[28, 29], the TA phonon starts to broaden at at temperatures far above the transition or near the wavevector, q_{wf}, at which the TO phonon has been reported to disappear ("waterfall") [26, 27].

Raman scattering studies and their results

We have investigated several lead ferroelectric relaxor crystals. In this paper we briefly report on the results obtained on PMN <100>-cut single crystalline sample (for details see [36]). Such a sample represents the simplest case and may serve as a model.

The scattering was excited by propagating in a $\langle 100 \rangle$ direction, 514.5 nm light from a 200 mW Ar^+-ion laser, focused to a 0.1 mm spot. The scattered light was collected at an angle of 90° with respect to the incident beam (i.e., in $\langle 010 \rangle$ direction) by a double-grating spectrometer. For most of the measurements, the slits were opened to 1.7 cm^{-1}. However, in order to acquire more precise data in the central peak region, at temperatures close to the maximum of the dielectric constant ($100 < T < 350$ K), the slits were narrowed to 0.5 cm^{-1}. Each polarization of the scattered light, $\langle x|zz|y \rangle$ (VV) and $\langle x|zx|y \rangle$ (VH), was measured separately. In order to exclude differences in sensitivity of the monochromator to different polarizations of the light, a circular polarizer was used in front of the entrance slit. For control purposes, we also took measurements without polarization analysis. Finally, to protect the photomultiplier from the strong Raleigh scattering, the spectral region from -4 to +4 cm^{-1} was excluded from the scans. The data were collected in the temperature range from 1000 to 100 K. The cooling rate was 0.5-1 K/min. Every 50-20 K the temperature was stabilized and the spectrum recorded.

The measured spectra were consistent with those from Refs.[40] and [41] and shown in [36]. In the high temperature region, a typical spectrum consists of two strong lines centered approximately at 45 cm^{-1} and 780 cm^{-1} and of three broad bands between them. The line at 45 cm^{-1} exhibits a triplet structure. Lowering temperature leads to the splitting of the broad bands into a number of narrower lines and to the appearance of new lines.

To analyze the data, we decomposed the measured spectra using a multiple peak fitting procedure. Satisfactory fits could be achieved with the assumption that the central peak has a Lorentzian shape and that each of the other peaks is described by the spectral response function of damped harmonic oscillator, modified by a population factor:

$$\Phi_i \sim \frac{\Gamma_i f_{0i}^2 f}{(f^2 - f_{0i}^2)^2 + \Gamma_i^2 f_{0i}^2} F(f,T), \qquad (1)$$

where Γ_i and f_{0i} are the damping constant and the mode frequency and the Bose population factor is given by:

$$F(f,T) = \begin{cases} n(f)+1, \text{ for Stokes part} \\ n(f), \text{ for anti-Stokes part} \end{cases}, \qquad (2)$$

where
$$n(f) = (exp(hf/kT) - 1)^{-1}.$$

As all of the peaks are much better resolved at low temperatures, the fitting procedure was started at the low-temperature end of the data set (at 110 K) and, the evolution of the peaks was then followed with increasing temperature (*i.e.*, in the opposite order of the measurements). At the same time, the number of peaks necessary to achieve a reasonably good fit was minimized. The control data set (measured without polarization analysis) was used to calibrate the positions and widths of the weak and poorly resolved peaks from the VV and VH data sets. Since a large number of parameters is involved, the results of a particular fit may depend on their initial values. To stabilize the results, the best-fit values of the parameters obtained at one temperature were used as initial values for the fit at the next temperature. In this manner, several sets of fits were obtained and analyzed. It is remarkable that, in all of them, the major parameters showed the same trends. Below, we show the most interesting results obtained from the analysis of the central peak and tripet line located at 45 cm^{-1}. For clarity, we describe the observed phenomena from high to low temperatures, following the same order as in measurements (unless stated otherwise).

The temperature dependencies of the fitting parameters for the central peak in the PMN crystal (CP) are presented in Fig. 3 (left-hand side). Circles correspond to the VV and triangles to the VH component of the peak. The existence of the CP is a direct consequence of the lattice fluctuation relaxations, which are very sensitive to the restrictions imposed by the low-symmetry clusters. If the relaxations are fast [38], the CP is low-intense and broad, whereas their slowing causes growth and narrowing of the peak.

We should point out a striking similarity of the temperature behavior of the CP in $PbMg_{1/3}Nb_{2/3}O_3$ (Fig. 3 (left-hand side)) and in $KTa_{0.85}Nb_{0.15}O_3$ (right-hand side) crystals. We have shown[10] that the temperature behavior of the CP in KTN can be explained by a model involving the collective relaxational motion of off-centered Nb ions and its progressive restriction with decreasing temperature. In the cubic phase, Nb ions are allowed to reorient amongst eight equivalent <111> directions. The appearance of the PNR's, followed by a sequence of phase transitions down to a rhombohedral $R3m$ phase, limits the ion motion to four, two and, finally, locks it in only one site. This model is in agreement with the neutron scattering studies of similar systems [42]. The similarity of the CP behavior in PMN and KTN, suggests that the temperature evolution of the polar clusters in $PbMg_{1/3}Nb_{2/3}O_3$ passes through similar stages as those in $KTa_{0.85}Nb_{0.15}O_3$ (KTN). However, it is not accompanied by the appearance of long-range order.

Starting from high temperature, the first important feature is the strong and narrow scattering in the VV geometry, accompanied by relatively weak scattering in the VH geometry. This indicates the presence of a symmetric slow relaxational motion, most likely involving 180° reorientations of ions. Starting from $\sim$ 900 K, the cessation of this motion causes a decrease in intensity of the CP, which reaches a minimum near the

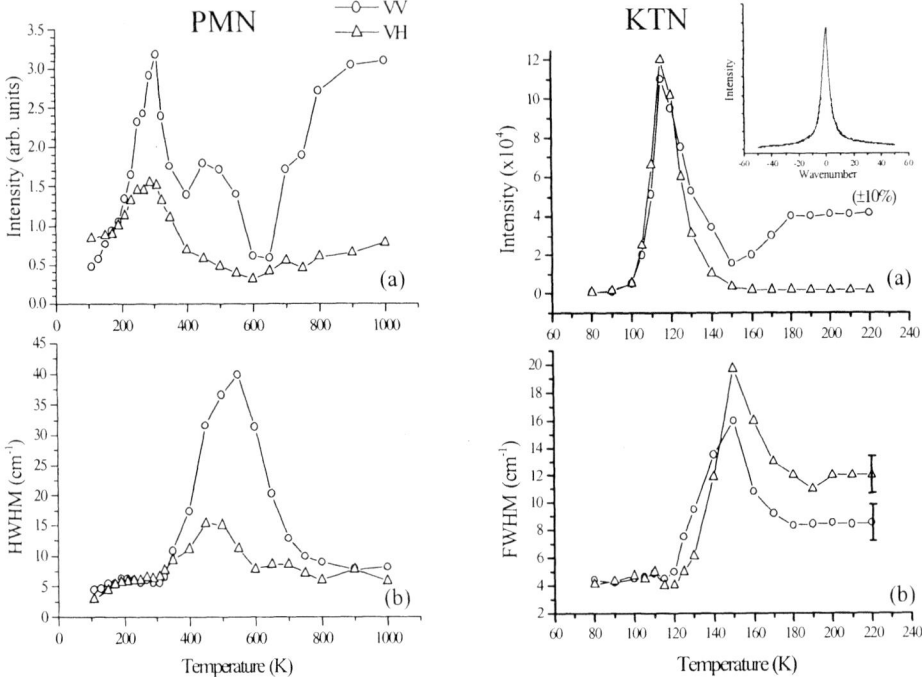

FIGURE 3. Temperature dependencies of the intensity (a) and half-width (b) of the Lorentzian approximation for the central peak in VV (circles) and VH (triangles) geometries of experiment. Left-hand side shows results for PMN crystal, right-hand side compares them with those for KTN crystal [10]. Insert demonstrates quality of the fit.

Burns temperature $T_d \approx 620$ K. The prohibition of 180° reorientations indicates the loss of inversion symmetry in the lattice, caused by onset the distinguishability between Mg and Nb occupied sites and the formation, in the 1:1 ordered areas, of a superstructure with average $Fm3m$ symmetry [43, 44]. It also imposes the first restrictions on the reorientational motion of the dynamical $R3m$ polar nanoregions. Now, they can reorient only amongst four neighboring <111> directions, forming, on average, tetragonal-like distortions. These processes are accompanied by the appearance of large (of the order of wavelength of light) dynamic fluctuations that cause worsening of the optical quality of the sample (which is also reflected on the whole spectrum [36]).

With further decrease of the temperature, the optical quality of the crystal improves again. Below ~ 550 K, the four-site reorientational motion of the PNR's slows down, which is marked by the narrowing of the VV component of the CP and the increase of its VV and VH intensities. Simultaneously, the VH component broadens and reaches a maximum at ~ 450 K. This indicates rearrangements in the crystalline structure leading to the appearance of new restrictions on the ion motion. The analogy with KTN suggests that, below ~ 450 K, the motion of the $R3m$ clusters becomes restricted to two

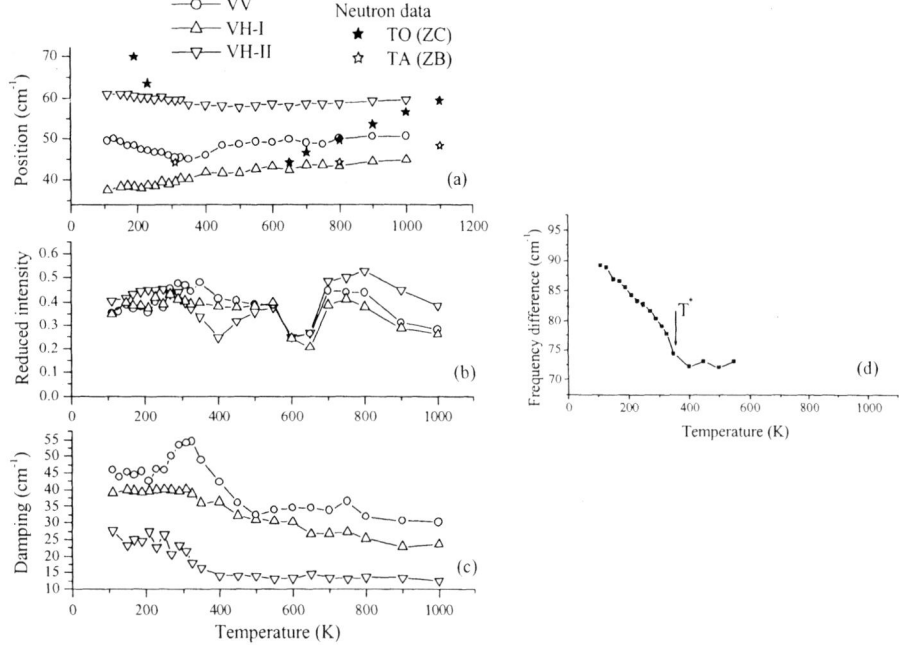

FIGURE 4. (a-c): Temperature dependences of the fitting parameters: position (a), reduced intensity (b) and damping constant (c) for the triplet line located at 45 cm^{-1}. Phonon frequencies, measured by neutron spectroscopy[26, 27, 28, 45], are shown for comparizon.
(d): Magnitude of splitting between components of the broad band located at 500-600 cm^{-1}.

neigboring <111> orientations, giving an average monoclinic-like distortion. Such a rearrangement causes some decrease in intensity of the VV component (with minimum at ~ 400 K), while the VH intensity keeps growing. Below ~ 400 K, the slowing down of the two-site relaxational motion causes a narrowing of the CP and an increase in the intensities of both components. Below $T_f \approx 350$ K, these effects become especially dramatic. Further decreasing the temperature below ~ 300 K, leads to the complete prohibition of intersite reorientational motion of the $R3m$ clusters, i.e. to appearance of static $R3m$ clusters. This is marked by a sharp decrease in intensity of both components of the CP. At $T_{do} \approx 210$ K (which is the temperature of an electric field-induced phase transition), the freezing process of the PNR's from dynamic to static is complete. Below this temperature, the central peak is narrow and its intensity is small in both scattering geometries.

Figure 4(a-c) presents the temperature evolution of the fitting parameters for the peak located at 45 cm^{-1}, showing its position (a), reduced intensity (b) and damping constant (c). This peak has a triplet structure, containing one component in VV (circles) and two components in VH (up and down triangles) geometries. The fitting parameters for this peak exhibit changes at the above mentioned temperatures T_d, T^* and T_{do}, confirming their significance in the structural evolution of the crystal. However, the origin of this peak (see Table I in Ref. [36]) requires clarification. From a comparison with the frequencies of the phonon modes measured by neutron spectroscopy, it is clear that this peak cannot be due to the zone center soft TO_1 mode (black stars in Fig. 4). On the other hand, the lower frequency VH component, and possibly the VV component could be due to disorder-induced scattering TA phonon from the zone boundary (white stars in Fig. 4). However, the higher frequency VH component would still not be accounted for.

In an attempt to account for both VH components simultaneously, we have tried to make use of a coupled oscillator model [36]. Results of the fit confirmed the importance of coupling processes in the formation of this line. These processes, however, occur without a significant contribution from the zone center TO_1 phonons, but more likely under the influence of the zone boundary TA phonons, with different polarizations propagating in different directions [30]. Interaction between such phonons is possible if mediated by relaxational motion of the polar nanoregions.

The importance of the temperature T^* is emphasized by the temperature dependence of the splitting between components of the band located at 500-600 cm^{-1} (Fig.4(d)), which is analogous to the one observed in KTN. The work to explain this phenomenon is currently in progress.

Conclusions

By means of neutron and light scattering spectroscopy, we have carried out an investigation of the development of the low temperature phase in relaxor ferroelectric materials (PZN-xPT and PMN). Our results show that the formation of this phase is preceded by appearance of the precursor clusters. These clusters nucleate at very high (several hundred degrees higher than the maximum of the dielectric constant) temperatures as highly dynamic objects. With lowering of the temperature, their motion becomes progressively more restricted, starting from Burns temperature T_d. The appearance of the static polar regions is marked by the temperature T^*. Finally, the reorientational motion freezes out. The process of slowing down is strikingly similar to the one in KTN. However, in the case of lead relaxors, due to the presence of frustrating fields, it does not result in the establishment of the long-range order. Phonon-related peaks appear in the light scattering spectrum, due to the coupling between phonons of different polarizations, mediated by the relaxational motion of the PNR's.

Acknowledgements

This research has been supported by DOE under Contract No. DE-FG02-00ER45842 and by ONR under Grant No. N00014-99-1-0738 (Z.-G. Ye).

REFERENCES

1. J. Kuwata, K. Uchino, and S. Nomura, Ferroelectrics **37**, 579 (1981).
2. T. Shrout, Z.P. Chang, N. Kim, and S. Markgraf, Ferroelectric Letters, **12**, 63 (1990).
3. S.-E. Park and T.R. Shrout, J. Appl. Phys. **82**, 1804 (1997).
4. S-F. Liu, S-E. Park, T.R. Shrout, and L.E. Cross, J. Appl. Phys. **85**, 2810 (1999).
5. D. Viehland, A. Amin, and J. F. Li, Appl. Phys. Lett. **79**, 1006 (2001).
6. Y. Yokomizo, T. Takahashi, and S. Nomura, J. Phys. Soc. Jpn. **28**, 1278 (1970).
7. C.A. Randall and A.S. Bhalla, Jpn. J. Appl. Phys. Part 1 **29**, 327 (1990).
8. D. Viehland, S.J. Jang, L.E. Cross, and M. Wuttig, Phys. Rev. B **46**, 8003 (1992).
9. G. Burns and F.H. Dacol, Solid State Commun. **48**, 853 (1983) and Phys. Rev. B **28**, 2527 (1983).
10. O. Svitelskiy and J. Toulouse, J. Phys. Chem. Solids **64**, 665 (2003).
11. P. Bonneau, P. Garnier, G. Calvarin, E. Husson, J. R. Gavarri, A. W. Hewat, and A. Morell, J. Solid State Chem. **91**, 350 (1991).
12. N. de Mathan, E. Husson, G. Calvarin, J.R. Gavarri, A.W. Hewat, and A. Morell, J. Phys.: Condens. Matter **3**, 8159 (1991).
13. A. Verbaere, Y. Piffard, Z.-G. Ye, and E. Husson, Mat. Res. Bull. **27**, 1227 (1992).
14. Y. Uesu, H. Tazawa, K. Fujishiro, and Y. Yamada, J. Korean Phys. Soc. **29**, S703 (1996).
15. S. Vakhrushev, S. Zhukov, G. Fetisov, and V. Chernyshov, J. Phys.: Condens. Matter **6**, 4021 (1994).
16. T. Iwase, H. Tazawa, K. Fujishiro, Y. Uesu, and Y. Yamada, J. Phys. Chem. Solids **60**, 1419 (1999).
17. Y. Matsushima, N. Ishizawa, N. Wakiya, and N. Mizutani, J. Ceram. Soc. Jpn. **108**, 617 (2000).
18. E. Husson, L. Abello, and A. Morell, Mat. Res. Bull. **25**, 539 (1990).
19. T. Egami, W. Dmowski, M. Akbas, and P.K. Davies, First Principles Calculations for Ferroelectrics, ed. by R.E. Cohen, Fifth Williamsburg Workshop, Williamsburg, VA February 1998, AIP Conf. Proc. **436**, p.1 (1998).
20. I.W. Chen, J. Phys. Chem. Solids **61**, 197 (2000).
21. R. Blinc, A. Gregorovic, B. Zalar, R. Pirc, V.V. Laguta, M.D. Glinchuk, Phys. Rev. B **63**, 024104 (2001).
22. N. N. Krainik, L. S. Gokhberg, and I. E. Myl'nikova, Sov. Phys. -Solid State **12**, 1885 (1971).
23. Z.-G. Ye and H. Schmid, Ferroelectrics **145**, 83 (1993).
24. M. L. Mulvihill, L. E. Cross, W. Cao, and K. Uchino, J. Am. Ceram. Soc. **80**, 1462 (1997).
25. I.G.Siny, S.G.Lushnikov, R.S.Katiar, V.H.Schmidt, Ferroelectrics, **226**, 191 (1999).
26. P. M. Gehring, S. Wakimoto, Z.-G. Ye, and G. Shirane, Phys. Rev. Lett. **87**, 277601 (2001).
27. S. Wakimoto, C. Stock, Z.-G. Ye, W. Chen, P.M.Gehring, and G.Shirane, Phys. Rev. B **66**, 224102 (2002).
28. A.Naberezhnov, S.Vakhrushev, B.Dorner, D.Strauch, and H.Moudden, Eur.Phys.J. B **11**, 13 (1999).
29. T.-Y. Koo, P.M. Gehring, G. Shirane, V. Kiryukhin, G. Lee, and S.-W. Cheong, Phys. Rev. B **65**, 144113 (2002).
30. S.A. Prosandeev, Eric Cockayne and B.P.Burton, First principles calculations of lattice dynamics in some PMN supercells, This Proceedings.
31. L. Zhang, M.Dong and Z.-G. Ye, Mater. Sci. Eng. B **78**, 96 (2000).
32. W. Chen and Z.-G. Ye, J. Mater. Sci. **36**, 4393 (2001).
33. Z.-G. Ye, P. Tissot, and H. Schmid, Mater. Res. Bull., **25**, 739 (1990).
34. D. La-Orauttapong, J. Toulouse, J.L. Robertson, and Z.-G. Ye, Phys. Rev. B. **64**, 212101 (2001).
35. D. La-Orauttapong *et al.*, e-print: cond-mat/0209420.
36. O. Svitelskiy *et al.*, e-print: cond-mat/0301501.
37. A.D. Bruce and R.A. Cowley, Structural Phase Transitions (Taylor and Francis, London, 1981).
38. K.H.Michel, J.Naudts, B.De Raedt. Phys. Rev. B, **18**, 648 (1978).
39. J. M. Rowe, J. J. Rush, D. G. Hinks, and S. Susman, Phys. Rev. Lett. **43**, 1158 (1979).
40. E.Husson, L.Abello, A.Morell, Mat. Res. Bull., **25**, 539 (1990).
41. H.Ohwa, M.Iwata, N.Yasuda, Y.Ishibashi, Ferroelectrics, **229**, 147 (1999); **218**, 53 (1998).
42. G.Yong, J.Toulouse, R.Erwin, S.M.Shapiro, B.Hennion, Phys.Rev. B 62 (2000) 14736.
43. J.Chen, H.M.Chan, M.P.Harmer, J. Am. Ceram. Soc., **72**, 593 (1989).
44. C.Boulesteix, V.Varnier, A.Llebaria, E.Husson, J. Sol. State Chem., **108**, 141 (1994).
45. P.M.Gehring, S.B.Vakhrushev, G.Shirane, AIP Conf. Proc. Fundamental Physics of Ferroelectrics 2000. Aspen. Winter Workshop, 314 (2000).

Correlations between the Structure and Dielectric Properties of $Pb(Sc_{2/3}W_{1/3})O_3 - Pb(Ti/Zr)O_3$ Relaxors

Pavol Juhás[*], Wojtek Dmowski[*], Ilya Grinberg[†], Takeshi Egami[*], Andrew M. Rappe[†] and Peter K. Davies[*]

[*]*Dept. of Materials Science and Engineering, University of Pennsylvania, Philadelphia, PA 19104*
[†]*Dept. of Chemistry, University of Pennsylvania, Philadelphia, PA 19104*

Abstract. The effects of Ti and Zr on the structure and ordering in the $(1-x)Pb(Sc_{2/3}W_{1/3})O_3 - (x)PbTiO_3$ (PSW–PT) and $(1-x)Pb(Sc_{2/3}W_{1/3})O_3 - (x)PbZrO_3$ (PSW–PZ) systems were studied using synchrotron x-ray and neutron diffraction. Rietveld refinement was carried out to determine the average long-range crystallographic structure and pair distribution function (PDF) analysis to probe the local displacements of the atoms. For $x < 0.25$ the B-cations form a 1:1 ordered doubled perovskite structure (space group $Fm\bar{3}m$). The refined occupancies were consistent with the "random site model", where the ordered structure consists of one B-sublattice occupied by Sc and the other by a random mixture of the remaining cations. The B-site order is reduced by incorporation of Zr, but highly stabilized by Ti with the degree of order in excess of 95% for $x \leq 0.25$. The results of PDF analysis show that on the local scale the Pb and O atoms are significantly displaced from their average lattice positions. The PDF curves were simulated by several models of simple Pb and O shifts. The short-range PDF of PSW could be approximated by allowing Pb shifts along [100] and rotations of BO_6 octahedra around $[10\bar{1}]$. This model was inadequate for a longer distances ($r > 4.25$ Å) suggesting the real cation displacements are more complicated. Distortions of the local structure in the PSW–PT system were modeled also by density functional theory calculations. The obtained magnitudes of local Pb displacements coincide with the temperature of paraelectric transition $T_{\varepsilon,max}$.

1. INTRODUCTION

The dielectric and piezoelectric properties of the lead based $PbBO_3$ perovskites show great variance with the composition and structural order of the B-site cations. Although the B-site chemistry is critical for the overall dielectric response of these systems, significant contributions come from the shifts of the A-site Pb cations. The magnitude and direction of these displacements is determined by the Pb environment, in particular by the oxygen neighbors. The Pb–O bonds are influenced by bonding of oxygen to its nearest B-site neighbors, and as a result oxygens mediate the interaction between Pb displacements and B-site chemistry. The exact relationship between structure and dielectric response is complicated and driven by many coupled interactions. We have attempted to investigate these relations in the two solid solutions of $Pb(Sc_{2/3}W_{1/3})O_3$–$PbTiO_3$ (PSW–PT) and $Pb(Sc_{2/3}W_{1/3})O_3$–$PbZrO_3$ (PSW–PZ). These systems appeared to be convenient for such study, because in spite of similar chemistries they show a completely different response to the B-site substitution.

PSW is similar in properties to the Pb(Mg$_{1/3}$Nb$_{2/3}$)O$_3$ (PMN) family of compounds, since it displays a relaxor type response and forms a B-site ordered structure with 1:1 rock salt periodicity. The effects of cation substitutions and thermal treatments on the B-site order and dielectric properties in $(1-x)$Pb(Sc$_{2/3}$W$_{1/3}$)O$_3-(x)$PbTiO$_3$ and $(1-x)$Pb(Sc$_{2/3}$W$_{1/3}$)O$_3-(x)$PbZrO$_3$ have been previously reported [1]. The results indicated that the 1:1 Pb[$\beta'_{1/2}\beta''_{1/2}$]O$_3$ order in PSW can be represented by a modified "random site" structure. According to this model, the structure of the fully ordered PSW end-member can be represented as Pb[Sc]$_{1/2}$[Sc$_{1/3}$W$_{2/3}$]$_{1/2}$O$_3$ and the ordered solid solutions by Pb[Sc]$_{1/2}$[Sc$_{(1-4x)/3}$W$_{(2-2x)/3}$M$_{2x}$]$_{1/2}$O$_3$, where M^{4+} is either Zr or Ti. Because of the $(1-4x)/3$ term, this substitution pattern for the solid solutions is only possible for $x \leq 0.25$, for $x > 0.25$ the M^{4+} cation must substitute on both lattice sites, to give stoichiometries with Pb[Sc$_{(4-4x)/3}$M$_{(4x-1)/3}$]$_{1/2}$[W$_{(2-2x)/3}$M$_{(2x+1)/3}$]$_{1/2}$O$_3$. The order parameter of PSW–PT and PSW–PZ samples has been evaluated from the conventional powder x-ray diffraction (XRD) by comparing the measured intensity of the (111) reflection to its calculated value, see Fig. 1. For PSW–PT the cation order was found to be much more stable and to extend to higher substitution levels compared to its PSW–PZ counterpart. This observation could be interpreted in terms of the radii difference of the ordered sites predicted by the random site model - which increase with x for the Ti system, but decrease for the substitution of Zr. All of the investigated compositions showed relaxor ferroelectric behavior, however PSW–PT displayed unusual trends in the temperature of the permittivity maximum $T_{\varepsilon,max}$, which decreased for $x \leq 0.2$ and then increased for $x > 0.25$. In contrast, for the PZ system $T_{\varepsilon,max}$ varies linearly with the Zr content, as displayed in Fig. 2. These observations can be rationalized in terms of the B-site occupancies predicted by the random site model. For the substitution of PT complete order occurs for $x \leq 0.25$ (Fig. 1), therefore one lattice site (β') is occupied exclusively by Sc and the other (β'') by a mixture of Sc, W and Ti. However for $x > 0.25$ Ti must substitute on both ordered sites and the presence of Ti–O–Ti or Ti–O–W bonds becomes likely. It is precisely at this composition where the sharp growth in $T_{\varepsilon,max}$ is observed - Fig. 2. The appearance of Ti–O–Ti and Ti–O–W neighbors can induce longer range coupling of the ferroelectrically active Ti^{4+} and W^{6+} cations. It is also possible that over-bonding of oxygens in Ti–O–W and Ti–O–Ti may alter the directions and magnitudes of lead displacements. For PSW–PZ the degree of order is less than 100% even for low substitution rates, implying that both ordered sites contain a mixture of several cations. Consequently there is no abrupt change in the pattern of B-site occupation and the linear behavior of $T_{\varepsilon,max}$ versus x is not unexpected.

Our previous data strongly supported the random site model as a correct description of the ordered PSW–PT and PSW–PZ compounds. Nevertheless more detailed structure analysis was necessary to reliably determine the B-site occupancies and to probe the alterations in the atomic displacements that accompany the changes in composition, order and dielectric response. Therefore we performed a detailed Rietveld and pair distribution function (PDF) studies of the PSW–PZ and PSW–PT systems using synchrotron x-ray and neutron diffraction (ND).

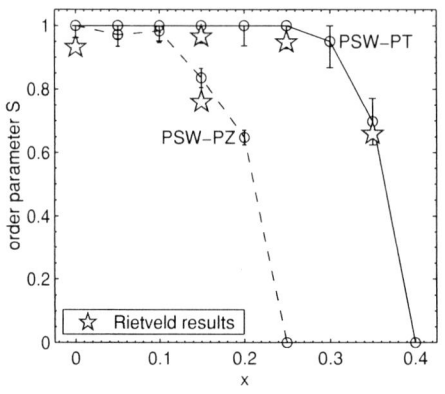
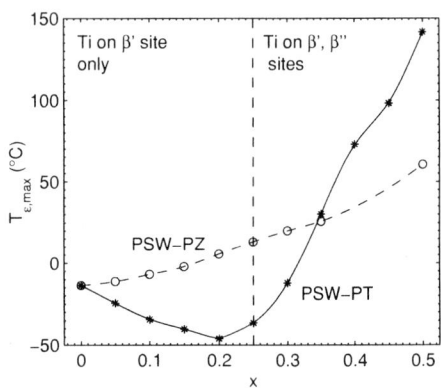

FIGURE 1. Degree of order S vs. composition x for PSW–PZ and PSW–PT.

FIGURE 2. Variation of $T_{\varepsilon,max}$ (1MHz) with x for PSW–PT and PSW–PZ.

2. EXPERIMENTAL PROCEDURES

Samples of $(1-x)$PSW $-$ (x)PT, $x = 0$, 0.15, 0.25, 0.35 and $(1-x)$PSW $-$ (x)PZ, $x = 0.15$, 0.35 were prepared by solid-state methods from high-purity oxides ($> 99.9\%$) via the "columbite route". The annealing treatment was conducted in a way known to maximize the B-site order. The details of the sample preparation are reported elsewhere [1]. X-ray diffraction data were measured using the beamline X7A at the National Synchrotron Light Source (NSLS), Brookhaven National Laboratory. Powder samples were placed in a 0.1 mm glass capillaries, which were rotated during the measurement to improve orientational averaging. A constant wavelength of $\lambda = 0.50095$ Å was selected using a Ge (111) monochromator. Diffraction intensities were measured by a position sensitive detector in the Q range of 1.1 Å^{-1} - 10 Å^{-1} ($d = 0.62$ Å - 5.8 Å). The data were analyzed by Rietveld method using the GSAS software package [2]. The Bragg reflections were fitted by an asymmetric peak profile function #3. The wider profiles of ordering reflections were modeled using the GSAS procedure for stacking fault broadening, which introduced additional broadening terms for the reflections outside of $\{200\}$ sub-lattice. The scale factor, background function, lattice constant, profile coefficients, temperature factors, oxygen position and B-cation occupancies were all refined together.

Neutron diffraction time of flight experiments were performed at the SEPD beamline at the Intense Pulse Neutron Source (IPNS), Argonne National Laboratory. The diffraction data were measured for 5 hours at 20 K and 290 K. The incident spectrum and scattering effects from the specimen chamber were calibrated by runs using vanadium rod and empty sample container. The diffraction data were analyzed by Rietveld and pair distribution function methods. The Rietveld analysis was carried out using data from the 144.85° detector bank in the Q range of 1.88 Å^{-1} - 15.7 Å^{-1} (d range 0.4 Å - 3.35 Å). The additional broadening of the ordering reflections was facilitated by the GSAS stacking fault model for the $\{200\}$ sub-lattice. The lattice parameter, background function, profile coefficients, B-site occupancies, temperature factors and oxygen po-

TABLE 1. Results of Rietveld analysis of XRD (^{x}) and ND (^{n}) patterns from PSW–PT and PSW–PZ.

x	a (Å)	S	Δz_O (a)	U_{Pb}	$U_{O\perp}$	$U_{O\parallel}$	U_B (Å^2)	R_{F^2} (%)	R_{wp} (%)
0 ^{x}	8.1349	0.933	0.0054	0.055	0.034	–	0.0039	6.7	4.0
0 ^{n}	8.1360	0.847	0.0082	0.049	0.029	0.006	0.0054	40	4.5
0.15Ti ^{x}	8.1011	0.969	0.0075	0.043	0.017	–	0.0033	9.7	3.7
0.15Ti ^{n}	8.0994	0.802	0.0080	0.048	0.023	0.005	0.0056	36	4.4
0.25Ti ^{x}	8.0761	0.948	0.0073	0.048	0.023	–	0.0081	6.8	4.0
0.25Ti ^{n}	8.0785	0.644	0.0073	0.047	0.020	0.006	0.0056	36	4.3
0.35Ti ^{x}	8.0593	0.659	0.0069	0.050	0.039	–	0.0110	8.0	3.6
0.35Ti ^{n}	8.0619	0.516	0.0038	0.047	0.022	0.009	0.0043	42	4.4
0.15Zr ^{x}	8.1615	0.760	0.0041	0.059	0.044	–	0.0120	7.1	3.4
0.15Zr ^{n}	8.1644	0.589	0.0055	0.049	0.035	0.008	0.0055	42	4.7
0.35Zr ^{x}	8.1953	0	–	0.051	0.045	–	0.0090	6.9	2.5
0.35Zr ^{n}	8.2050	0	–	0.053	0.043	0.009	0.0056	37	5.4

sitions were all refined together. For the PDF analysis, the scattering intensities were re-grouped to 5 detector banks at 21.8°, 44.0°, 90.0°, 139.7° and 150.0°, which allowed good intensity in the Q range of 1.2 Å^{-1} - 30 Å^{-1}. The experimental PDF curves were calculated from raw intensities using the PDFGetN software, and the PDF simulations were carried out using the PDFFit program [3, 4].

3. RESULTS

3.1. Rietveld Refinements

Results of the Rietveld refinement for the XRD and ND spectra of PSW are displayed in Fig. 3. The ordering reflections were observed in all samples (PSW–PT, x = 0.0, 0.15, 0.25, 0.35 and PSW–PZ x = 0.15) with the exception of PSW–PZ, x = 0.35. There was no sign of peak splitting in any of the measured x-ray or neutron spectra at 20 K or 290 K and all diffraction patterns were consistent with the cubic perovskite structure. The samples displaying B-site order were refined using the doubled perovskite structure, space group $Fm\bar{3}m$. A simple perovskite lattice, ($Pm\bar{3}m$) was used for disordered (0.65)PSW–(0.35)PZ. The total occupancies of all lattice sites were fixed to 1, and the occupancies of the B-cations were required to satisfy the overall stoichiometry. Such conditions permit 2 free parameters for the B-chemistry. However, only one can be refined, since all structure factors of ordering reflections are proportional to the same value ($F_{ord} \approx f_{\beta'} - f_{\beta''}$). Therefore the B-occupancies were set to linearly change with a single parameter S from a completely random structure ($S = 0$) to the one with maximum order at $S = 1$. The oxygen positions were refined in accordance to the $Fn\bar{3}m$ symmetry, which allows O shifts along β'–O–β'' bonds. XRD data were simulated using isotropic temperature coefficients with a common value for all B-cations. In the case of neutron diffraction, which can access wider Q range, separate temperature coefficients were used for the β' and β'' sites, and the O factors were refined as anisotropic with components perpendicular ($U_{O\perp}$) and parallel ($U_{O\parallel}$) to the β'–O–β'' bond. The refined lattice con-

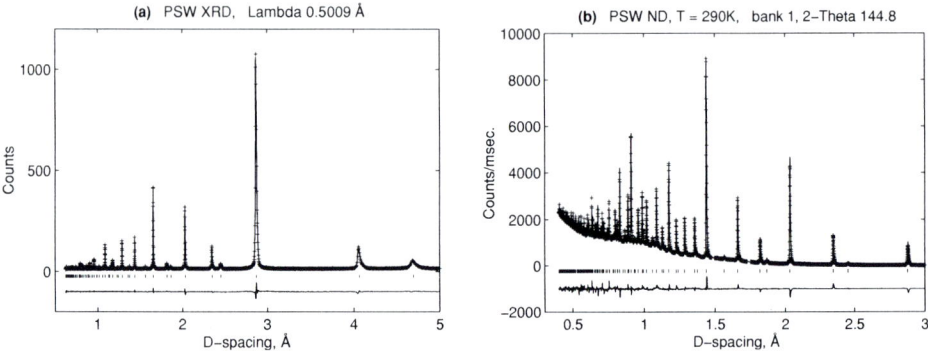

FIGURE 3. Rietveld refinement of (**a**) x-ray and (**b**) neutron diffraction patterns of PSW.

stants, order parameters S, the oxygen shifts Δz_O from β' to β'' and temperature factors U for XRD and ND spectra at 290 K are listed in Table 1.

All of the model structures could be refined to an excellent agreement with the experimental data, with the final residua $R_{wp} \approx 4\%$. The values of order parameter S obtained from neutron refinements were considerably smaller than their XRD counterparts. Because the final residua of structure factors R_{F^2} for XRD were an order of magnitude smaller than for ND (Table 1), the order parameters S from the XRD data should be considered reliable, and they also agree very well with the previous results, Fig. 1. The Rietveld analysis confirmed essentially complete order in PSW–PT for $x \leq 0.25$ and proved that the random site model is a correct description of the B-chemistry in PSW–PT and PSW–PZ systems. This model is also the arrangement that maximizes the x-ray intensities of the super-reflections, because Sc has the lowest atomic number of all B-cations ($F_{ord} \approx f_{\beta'} - f_{\beta''}$). Thus it is very unlikely that any other B-site structure could equally well simulate the observed strong intensities of the ordering reflections.

The large values of R_{F^2} for the ND refinements probably arise from the many overlapping peaks for high Q. While the wide Q range appears unfavorable for the occupancies, it allows better accuracy for the temperature coefficients. The R_{wp} factors for the neutron data were considerably improved by using anisotropic temperature factors for O and attributing separate values of U to the ordered β-sites. The refined temperature factors of Pb and O were very high and they correspond to an unrealistically large magnitudes of vibrations of ~ 0.22 Å for Pb and ~ 0.17 Å for O. Contrary to the expectations, the temperature factor of Pb in (0.75)PSW–(0.25)PT decreased from ~ 0.052 Å^2 to ~ 0.047 Å^2 after heating from 20 K to 290 K. This indicates that large Debye-Waller factors were not due to temperature vibrations, but they were rather caused by a local displacements of Pb and O atoms from their average lattice positions. These shifts are not correlated over larger distances, therefore the long range structure remains cubic, however they show up as increased temperature factors. The increase in U_{Pb} at low temperatures can be explained by the Pb shifts becoming more correlated and less likely to flip. For the O atoms, the temperature factor $U_{O\parallel}$ along the B–O bond is 2 - 4 times smaller than the perpendicular component $U_{O\perp}$. This suggests that local O shifts occur mainly in the directions transversal to B–O, which could correspond to rotations of the BO$_6$ octahe-

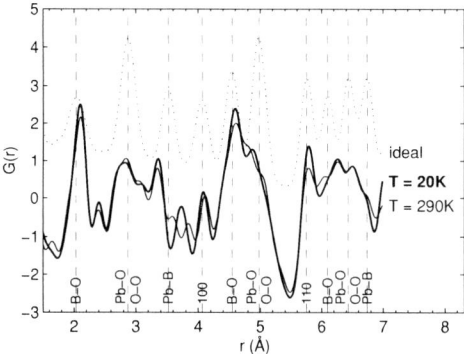

FIGURE 4. PDF curves for PSW at 20 K – thick line, and at 290 K – thin line. Dotted line denotes PDF calculated for the Rietveld - refined structure.

dra. The temperature coefficients for B-cations have reasonable values and they were found to increase with temperature. All of the refined Δz_0 were positive and indicated off-centering of oxygen away from the β' site, which is rich in the larger Sc cations.

3.2. Pair Distribution Function Analysis

The Pair Distribution Function (PDF) provides information about inter-atomic distances in the material and it is obtained by Fourier transformation of the entire diffraction spectrum. The usage of PDF for a crystalline materials is discussed elsewhere [5, 6, 7]. In this paper the PDF is expressed through the function $G(r)$ defined as

$$G(r) = \frac{1}{Nr}\sum_{i,j}\left[\frac{b_i b_j}{\langle b \rangle^2}\delta(r-r_{ij})\right] - 4\pi r \rho_0$$

where b is the neutron scattering length and r_{ij} is a distance of i and j atoms. Fig. 4 shows the experimental PDF curves of PSW obtained by neutron scattering at 20 K and 290 K. The dotted line on top is the PDF calculated for the average Rietveld structure and the vertical grid marks the expected bond lengths. The experimental PDF curves are considerably different from the calculated one and confirm a significant distortion of the local structure from the average lattice. The first peak of $G(r)$ at 2.1 Å is due to the B–O nearest neighbors, and has a full width at half maximum of ~0.21Å. This width is only slightly higher than the difference of the ionic radii of Sc^{3+} and W^{6+} ($\Delta R_\beta = 0.145$, $r_{Sc^{3+}} = 0.745$, $r_{W^{6+}} = 0.60$), which indicates that there is no significant off-centering of the B-cations. This conclusion is supported by the Rietveld analysis, which yielded small temperature coefficients for the B-cations. Therefore it can be assumed that the B-cations are fixed at their average positions. While the first PDF peak showed fair agreement with the calculated curve, there were considerable differences for the second and third peaks. The second nearest distance of Pb–O and O–O pairs is split to at least 3 overlapping peaks at 2.4 Å, 2.8 Å and 3.1 Å. This splitting is unlikely due to the O–O bonds, because distortions of BO_6 octahedra would display also in the first B–O peak. As a result there

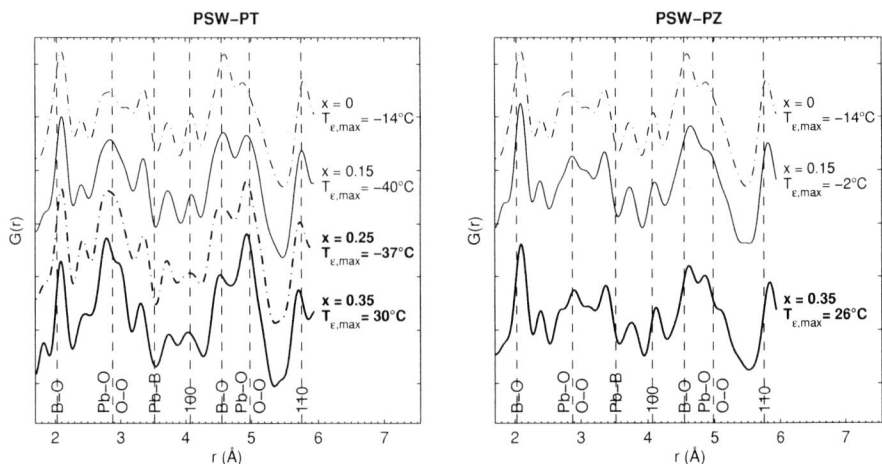

FIGURE 5. PDF curves for PSW–PT and PSW–PZ at 20 K. Composition and $T_{\varepsilon,max}$ values are noted on the right side of plots.

must be a large variance in the Pb–O bond lengths, which may arise from the shifts of Pb cations or from the rotations of the BO_6 octahedra. Lead displacements are confirmed by the third nearest distance of Pb–B, which is divided between 2 peaks at ∼3.4 Å and ∼3.7 Å. The $G(r)$ curves measured at 20 K and 290 K were quite similar, however they had a noticeable difference at $r \approx 3.5$ Å, corresponding to the Pb–B neighbors. The two peaks of the Pb–B lengths at 20 K become more spread with the heating to 290 K, and eventually create 3 maxima in the PDF. Increased temperature thus seems to add more options and larger randomness to the Pb displacements.

Effect of composition on the PDF curves of $(1-x)$PSW $-(x)$PT, $x = 0, 0.15, 0.25, 0.35$ and $(1-x)$PSW $-(x)$PZ, $x = 0.15, 0.35$ is displayed in Fig. 5. The most apparent change in $G(r)$ of PSW–PT is the alternation of amplitudes for B–O and Pb–B peaks at $r \approx 4.6$ Å and $r \approx 5.0$ Å. However, this is only a compositional effect due to the negative scattering length of Ti. The impact of Ti is noticeable also on the first B–O peak, which becomes narrower with x and develops a split at its left side foot. Because of the negative value of b_{Ti}, the Ti–O distance in the (0.65)PSW–(0.35)PT is represented by a local minimum at $r \approx 1.9$ Å. This minimum is offset from the main B–O peak by ∼0.2 Å, which is close to the difference of the radii of Sc and Ti ($\Delta R_\beta = 0.14$, $r_{Sc^{3+}} = 0.745$ Å, $r_{Ti^{4+}} = 0.605$ Å). Thus it is not possible to conclude without PDF modeling if there is any appreciable shift of the Ti atoms. Perhaps the most important structural feature in the compared curves is the alternation of the second peak at $r \approx 2.8$ Å, which is formed by Pb–O and O–O lengths. For PSW it has a right-side shoulder, then shows very broad, diffuse profile at the Ti compositions of $x = 0.15, 0.25$ and finally re-develops a shoulder at $x = 0.35$. The observed "ruggedness" of the Pb–O peak coincides with the temperature of permittivity maximum $T_{\varepsilon,max}$. The broad, diffuse Pb–O peak at $x = 0.15, 0.25$ suggests more randomness and shorter correlation length of the Pb displacements, which appears consistent with the drop in $T_{\varepsilon,max}$. A similar trend can be observed in the PSW–PZ

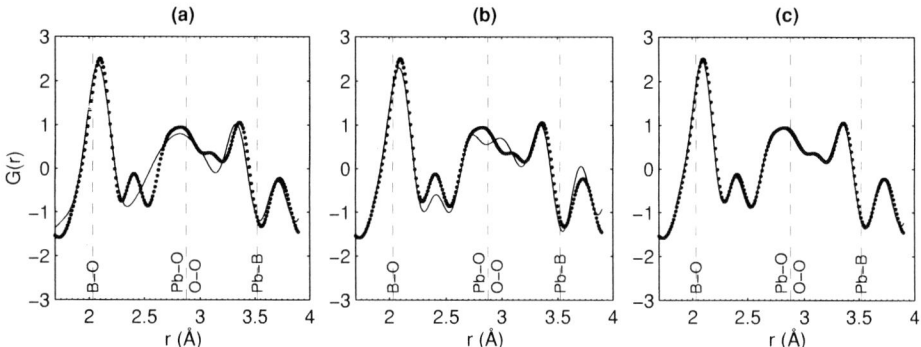

FIGURE 6. Simulated (solid line) and measured (dots) PDF for PSW. (a) [100] Pb shifts, (b) [100] Pb shifts and BO$_6$ rotations around [10$\bar{1}$], (c) [100] Pb displacements with arbitrary shifts of oxygens.

system, where the "ruggedness" of the main Pb–O peak grows with the substitution of Zr, accompanied by the increase in $T_{\varepsilon,max}$.

4. MODELING AND DISCUSSION

The models of local structure used a $2 \times 2 \times 2$ periodic cell based on the average structure. All temperature coefficients were refined as isotropic, with a separate factor for the Sc-rich and mixed-cation β-sites. Several simple models of Pb and O displacements had been examined, and their magnitudes, PDF scale factor and temperature coefficients were fitted to the experimental data.

In the first model, the Pb cations were allowed to shift along [100], [110], [111] and ±[111], in an anti-parallel pattern from β' to β'' sites. The oxygens could move in the B–O direction by changing the size of the BO$_6$ octahedra. The best fit for 20 K data was always obtained by the model using [100] Pb displacements, which nicely reproduced the doubled peak of Pb–B distances, Fig. 6(a). Since the average position of Pb is in the center of the B$_8$ cube, its shift in the [100] direction creates 4 short and 4 long Pb–B lengths, as observed in the experiment. However, this model performs poorly for the Pb–O distances, where it could not replicate the shortest Pb–O distance of 2.4 Å. Therefore additional O displacements were introduced by allowing rotations of the BO$_6$ octahedra. General rotations of 8 connected octahedra in a $2 \times 2 \times 2$ cell, are described by 6 parameters, e. g. by two axes r, s at [0 0 0] and [0.5 0.5 0.5], and their rotation angles. However, the refinement of all 6 parameters was numerically unstable, and the rotation axes had to be fixed in some special directions. These were chosen as (i) $r = s = [10\bar{1}]$, (ii) $r = s = [\bar{2}11]$ and (iii) $r = [10\bar{1}]$, $s = [\bar{2}11]$. The type (i) rotation shifts 4 oxygens of the O$_{12}$ cage directly to the Pb in the center, while the remaining 8 oxygens move away. The rotation (ii) creates the largest difference between the shortest and longest Pb–O lengths, and (iii) is a combination of the previous types. The best results were obtained using type (i) rotation, especially in the shorter range of PDF for 1.7Å $< r <$ 3.9 Å, as presented in Fig. 6(b). Although the BO$_6$ rotations were able to reproduce the

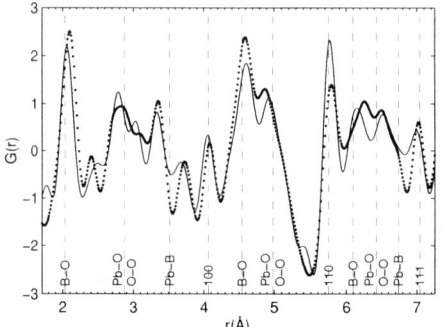

TABLE 2. Average cation displacements from DFT calculations

x	d_{Pb} (Å)	$d_{W,Ti}$ (Å)	$T_{\varepsilon,max}$ (°C)
0.0	0.4253	0.1764	-14
0.25	0.3947	0.1764	-40
0.625	0.3260	0.1895	244
1.0	0.4150	0.2851	490

FIGURE 7. PDF for DFT simulation of PSW, dots mark the experimental data.

shortest Pb–O distance, the overall agreement with the Pb–O peak was far from perfect. In addition, when the refinement range was expanded to 8 Å, the rotation angles were considerably diminished and the agreement at short distances was lost. Displacements of the B cations were attempted, but they failed to induce any appreciable change in the PDF. It appeared that the real oxygen displacements were more random than allowed by the constraints of BO_6 rotations, with weak correlations even within the basic perovskite cell. Such structure was simulated by allowing all oxygens to shift arbitrarily, and the PDF of this model could be refined to a perfect agreement with the experimental data, see Fig. 6(c). The components of O shifts longitudinal to B–O were found to be about 2 times shorter than the transversal ones. The reliability of the simulated structures was verified by calculating the bond valence sums [8]. For the final structures the bond valence sums deviated from their ideal values by $\sim0.2 - \sim0.6$, which is similar or greater than for the initial Rietveld structure. This suggests that models for the local displacement can be improved and this work is currently in progress.

The experimental PDF data were also used to test the results of ab-initio density functional theory (DFT) simulations of the PSW–PT system. Calculations were performed using a $2 \times 2 \times 2$ or $3 \times 2 \times 2$ cell with periodic boundary conditions. The system energy was evaluated using a local density approximation exchange-correlation functional and minimized with respect to the atomic coordinates. The DFT calculations were carried out for 4 compositions with $x = 0, 0.25, 0.625$ and 1.0 and respective chemical formulas of $Pb_{12}[Sc_8W_4]O_{36}$, $Pb_8[Sc_4W_2Ti_2]O_{24}$, $Pb_8[Sc_2W_1Ti_5]O_{24}$ and $Pb_8Ti_8O_{24}$. Comparison of the DFT simulation of PSW with the experimental PDF is shown in Fig. 7. The calculated curve agrees well with the experimental data and essentially all of the peak positions are reproduced. The model structure of PSW had just one arrangement of the B cations, which was periodically repeated. Many more arrangements are possible in the real structure and their absence may account for the differences between the calculated and experimental curves. The average magnitudes of Pb and B-cation displacements with respect to their oxygen cage are listed in Table 2. The calculated cation shifts hint at a possible mechanism for the changes in $T_{\varepsilon,max}$. For $x < 0.25$ the displacements of the active W and Ti cations remain constant, possibly due to the absence of Ti–O–Ti or W–

O–Ti clusters. At the same time the Pb shifts decrease together with the unit cell volume and the value of $T_{\varepsilon,max}$ drops. For a larger Ti content, the W and Ti displacements show significant growth and their contribution to the dielectric response may become dominant, canceling the impact of decrease in Pb shifts. For a compositions close to PT long range ferroelectric domains with uniform Ti shifts are established, which may couple with the Pb cations and allow their shifts to grow back.

5. CONCLUSIONS

By carrying out synchrotron x-ray and neutron diffraction it was concluded that the average structure of PSW–PT and PSW–PZ, $x \leq 0.35$ systems is cubic with a "random site" arrangement of the B-cations. The Rietveld analysis confirmed almost complete B-site order in the solid solution of PSW–PT for $x \leq 0.25$. On the local scale the crystal structure is considerably distorted from the average lattice. The distortion is realized through displacements of Pb and O atoms, while the B-cations remain at their average positions. The short-range PDF can be approximated by uniform Pb shifts in [100] direction and bound rotations of BO_6 octahedra around $10\bar{1}$ axis. For a longer range the O shifts need to be more random, however their major components are in the plane perpendicular to the B–O bond. The DFT simulated structure of PSW displayed a good agreement with the experimental PDF. The DFT results indicate that a possible reason for the decrease in $T_{\varepsilon,max}$ in the PSW–PT at $x < 0.25$ is a reduction of Pb displacements, before the contributions from the off-centering of Ti take over the dielectric response.

ACKNOWLEDGMENTS

The authors thank Dr. Beatriz Noheda from the NSLS, Brookhaven National Lab and Dr. Simine Short from the IPNS, Argonne National Lab for help with the experiments. The NSLS and IPNS are supported by the Department of Energy, Division of Materials Sciences and of Chemical Sciences. This work was funded by the Office of Naval Research through grant N00014-01-1-0860.

REFERENCES

1. Juhás, P., Davies, P. K., and Akbas, M. A., "Chemical Order and Dielectric Properties of Lead Scandium Tungstate Relaxors," in *AIP Conference Proceedings*, American Institute of Physics, 2002, vol. 626, pp. 108–116.
2. Larson, A. C., and von Dreele, R. B., *GSAS*, Los Alamos National Laboratory, Report LAUR 86-748 (2000).
3. Peterson, P. F., Gutmann, M., Proffen, T., and Billinge, S. J. L., *J. Appl. Crystallogr.*, **33**, 1192 (2000).
4. Proffen, T., and Billinge, S. J., *J. Appl. Crystallogr.*, **32**, 572–575 (1999).
5. Egami, T., *Mater. T. JIM*, **31**, 163–176 (1990).
6. Toby, B. H., and Egami, T., *Acta Crystallogr. A*, **A48**, 336–46 (1992).
7. Billinge, S. J. L., *Local Atomic Structure and Superconductivity of $Nd_{2-x}Ce_xCuO_4$: A Pair-Distribution-Function Study*, Ph.D. thesis, University of Pennsylvania (1992).
8. Brese, N. E., and Okeeffe, M., *Acta Crystallogr. B*, **47**, 192–197 (1991).

Cation Ordering in Single Crystals of 1:1 and 1:2 Complex Perovskite Solid Solutions

I. P. Raevski, S. A. Prosandeev, S. M. Emelyanov, V. G. Smotrakov, V. V. Eremkin, F. I. Savenko, I. N. Zakharchenko, E. S. Gagarina, O. A. Bunina, and E. V.Sahkar

Rostov State University, Stachki Ave. 194, Rostov-on-Don 344090, Russia

Abstract. The results of X-ray and dielectric studies of ordering effects in as-grown (1-x) $PbSc_{1/2}Ta_{1/2}O_3$ –(x)$PbSc_{1/2}Nb_{1/2}O_3$(PST-PSN) and (1-x)$PbMg_{1/3}Nb_{2/3}O_3$-(x)$PbSc_{1/2}Nb_{1/2}O_3$(PMN-PSN) single crystals are reported. Regularities of cation ordering during the crystal growth seem to be the same for solid solution crystals of both 1:1 and 1:2 ternary perovskites

Compositional ordering effects are usually studied via prolonged heat treatment of sintered ceramic samples [1-7]. As the crystal growth takes much more time than sintering of ceramics, it inevitably includes annealing of the inner parts of the crystal, which were formed at higher temperatures. In the present study, we compare the effect of ordering on the properties of as-grown solid solution crystals of 1:1 and 1:2 ternary perovskites. Both PST-PSN and PMN-PSN solid solution *crystals* were grown by the flux method. For all compositions of each system we used the same flux ($PbO-B_2O_3$), crystallization temperature interval: (1150 to 1060)^{0}C for PST-PSN and (1150 to 980)^{0}C for PMN-PSN, and the cooling rate: 5 K/h for PST-PSN; 10K/h at $T > 1100^0$C and 1K/h in the (1100 to 980)^{0}C range for PMN-PSN. X-ray supercell reflections due to B-cation ordering were observed for as-grown crystals from the $0 \leq x \leq 0.8$ (PST-PSN) and $0.1 \leq x \leq 0.65$ (PMN-PSN) compositional ranges. The concentration dependence of ordered domain sizes estimated from the widths of the superstructure reflections compared to the fundamental ones is shown in Fig. 1.

Though the maximal ordered domain sizes are similar (~50 nm) in both systems, PSN-PST crystals with $x \leq 0.4$ display a sharp permittivity peak while a strongly diffused $\varepsilon(T)$ maxima are observed for all PMN-PSN compositions, especially for those with $x < 0.6$ (Fig. 2). In PMN-PSN crystals, the diffusion of the dielectric permittivity maximum is the lowest for $x \approx 0.6$ and increases towards the end members of the solid solution. For compositions with $x > 0.6$ a sharp step appears

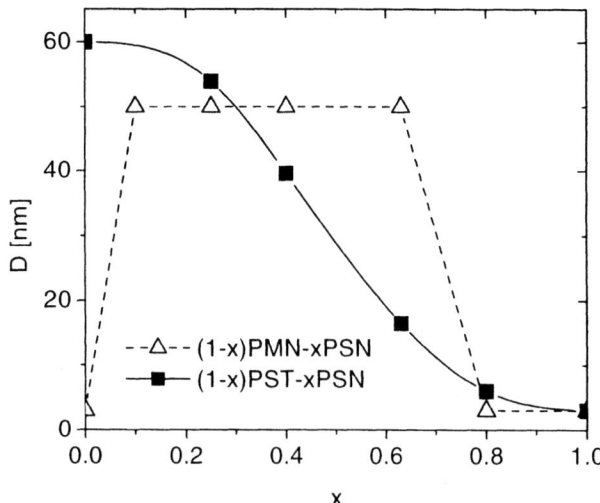

FIGURE 1. The concentration dependences of ordered domain sizes for PST-PSN and PMN-PSN crystals

on the $\varepsilon(T)$ curve at temperature T_s somewhat below T_m (Fig.2). These results are similar to those obtained earlier for annealed PMN-PSN ceramics [7]. This step may be attributed to a spontaneous transition from the relaxor to normal ferroelectric state [8]. A drop of Vogel-Fulcher excitation energy, E_0, is observed at $x \approx 0.6$ (Fig.3). We relate this drop in E_0 to the change of the local order: we think that, at $x < 0.6$ Mg ions control random fields on Pb, but, at $x > 0.6$ Sc ions are responsible for these fields.

Previously it was found out that the mean ordering degree s of the as-grown crystals of PSN, PST and some other ternary 1:1 perovskites depend on the relation between the crystallization temperature and the temperature T_t of the compositional order-disorder phase transition, the time of crystallization and the diffusion rate for B-site cations in the temperature interval of crystallization [5,9-11]. The temperature T_t of the compositional order-disorder phase transition for PST-PSN was found to be an approximately linear function of composition [9]. For the PMN-PSN system T_t was reported to be a parabolic function of x [6,7].

Below we will derive an analytical expression for the ordering temperature in PMN-PSN in the framework of the Bragg-Williams approximation [12]. Consider two sublattices with equal number of sites $N/2$ and three sorts of ions occupying these sites with the probabilities:

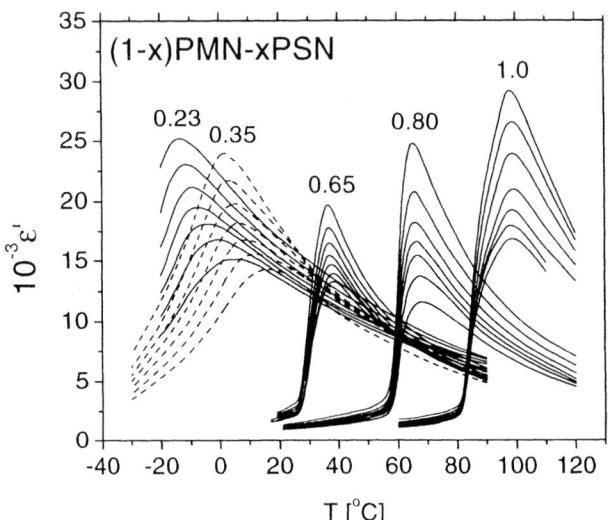

FIGURE 2. $\varepsilon'(T)$ dependencies in PMN-PSN crystals, measured at different frequencies: 10^{-2}, 10^{-1}, 10^0, 10^1, 10^2, 10^3, 10^4 Hz (from top to bottom). Numbers correspond to x values.

$$\begin{array}{ll} p_1^{(1)} = \frac{1}{3}(1-s)(1-x) & p_1^{(2)} = \frac{1}{3}(1+s)(1-x) \\ p_2^{(1)} = \frac{1}{6}(1-s)(1-x) & p_2^{(2)} = \frac{1}{6}(1+s)(1-x) \\ p_3^{(1)} = \frac{1}{2}(1-s)x & p_3^{(2)} = \frac{1}{2}(1+s)x \\ p_4^{(1)} = \frac{1}{2}(1+s) & p_4^{(2)} = \frac{1}{2}(1-s) \end{array} \qquad (1)$$

Here subscript shows the sort of the ions: 1: $Mg_{2(1-x)/3}$, 2: $Nb_{(1-x)/3}$, 3: Sc_x, 4: $Nb_{1/2}$, and the superscript is the sublattice's number. We consider ordering in $Pb[(1-x)(Mg_{2/3}Nb_{1/3})-(x)Sc]_{1/2}[Nb]_{1/2}O_3$ between the ions in the first and second brackets [6,7,13]. The ions in the first bracket are assumed fully disordered (the random layer model). The free energy in the form suggested by Bragg and Williams is

$$F = \sum N_{ij}w_{ij} + k_B NzT \sum p_i (\ln p_i - 1) \qquad (2)$$

where $N_{ij} = zN\left(p_i^{(1)}p_j^{(2)} + p_j^{(1)}p_i^{(2)}\right)$, z is the number of nearest neighbors, s is a degree of the order in the considered structure, w_{ij} are pair energies. After substitution (1) to (2) and finding the equilibrium degree of ordering one can obtain

$$T \ln \frac{1-s}{1+s} = s\left(a + bx - cx^2\right) \qquad (3)$$

where the coefficients a, b and c can be expressed in terms of the pair energies w_{ij}. The equation obtained allows one to find the dependence of the degree of the

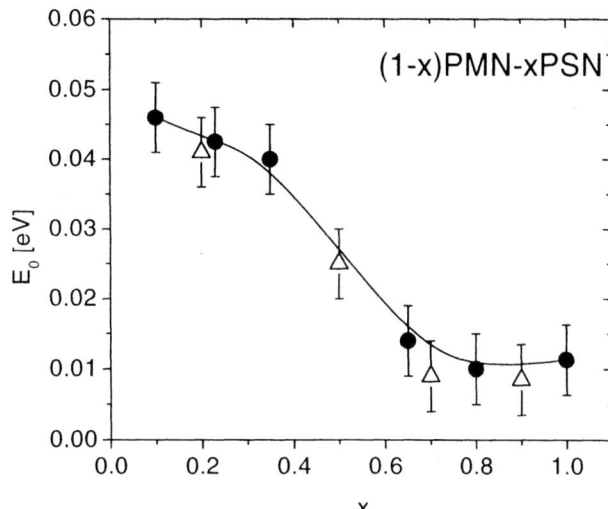

FIGURE 3. Concentration dependence of the Vogel-Fulcher excitation energy E_0 for PMN-PSN crystals (filled circles) in comparison with the same data for ordered ceramics [7] (open triangles).

order on temperature and concentration. There is an (ordering) temperature, T_t, above which the crystal shows disorder. This temperature can be easily found by expanding the logarithm in the series with respect to s and leaving only the first power of s. The bifurcation point (where a new solution appears distinct from $s = 0$) is at:

$$T_t = \tfrac{1}{2}\left(a + bx - cx^2\right) \tag{4}$$

The ordering temperature can be expressed over the corresponding temperatures at $x = 0$ (T_1) and $x = 1$ (T_2):

$$T_t = T_1 + (T_2 - T_1 + c)\,x - cx^2 \tag{5}$$

At $c = 0$ the concentration dependence is linear. At $c > 0$ the ordering temperature for intermediate concentrations can be higher than for the boundaries.

Calculations using expression (5) and experimentally estimated values of $T_t \approx 1270$ K for PMN [7] and $T_t \approx 1470$K for PSN [2] give a $T_t(x)$ dependence (Fig. 4) which is close enough to experimental one for PMN-PSN ceramics [5], especially if one takes into account that the boundary obtained in [7] corresponds to some residual order and, hence, the true T_t values are somewhat higher.

During the crystal growth of PSN and PST, the disordered structure is formed first, even when the crystallization temperature is much lower than T_t and the

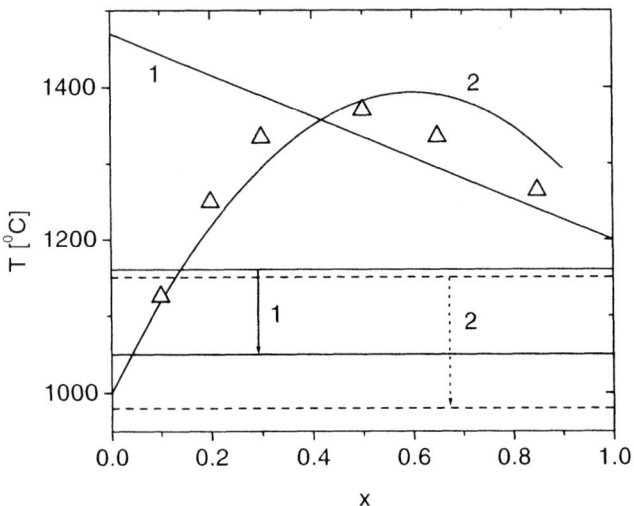

FIGURE 4. Relation between the interval of crystallization (arrows) and $T_t(x)$ dependences calculated using expression (6) for PST-PSN (1) and PMN-PSN (2) crystals. Triangles show experimental T_t values for PMN-PSN ceramics [7].

disordered phase is non-equilibrium [5,10,11]. The ordering then starts via a diffusion mechanism. The actual degree of ordering for as-grown crystals depends upon t/τ, where t is the time of crystallization, and τ the time necessary for achieving equilibrium degree of ordering s_e via diffusion; τ depends on T while s_e on T/T_t.

Our results of X-ray and dielectric studies of the as-grown PST-PSN crystals showing the monotonous decrease in size of the ordered regions with x, i.e. with increase of the T/T_t ratio, are in qualitative agreement with this picture. The same is true for the PMN-PSN system. Superstructure reflections due to B-site ordering are absent in pure PMN having T_t close to the lower boundary of the temperature interval of crystallization, but appear at $x \geq 0.1$, for which T_t exceeds this boundary. Besides the decrease of ordering in both systems at large x due to the increase of T/T_t ratio (Fig. 4), the absence of superstructure reflections for $x \geq 0.8$ is likely to be caused by general lowering of the intensity of these reflections with x due to smaller difference in atomic scattering factors of Sc and Nb as compared to Sc and Ta or Mg and Nb (in PSN intensity of superstructure reflections is known to be very low even for high degree of ordering [2]).

A relaxor-like dielectric behavior in PMN-PSN crystals with rather large (~50 nm) ordered domains is consistent with the random-site model of 1:1 B-site cation ordering implying a very short correlation length for ferroelectric coupling due to large random fields in one of the B-sublattices [6,7].

Basing on the Bragg-Williams approximation in the alloy theory we explicitly (in the framework of this theory) obtained the analytical expression for T_t corresponding to all the observed types of the $T_t(x)$ dependences.

According to model [10], a monotonous diffusion of $\varepsilon'(T)$ maximum for PST-PSN crystals is naturally explained by a decrease of T_t with x. A parabolic shape of the $T_t(x)$ dependence for the PMN-PSN system implies a higher ordering degree of the crystals at intermediate concentrations in comparison with the boundary ones, in agreement with our experimental findings. Thus the same regularities of cation ordering during the crystal growth are valid for solid solution crystals of both 1:1 and 1:2 ternary perovskites.

The study was partially supported by RFBR (grants # 01-03-33119 and 01-02-16029).

REFERENCES

1. Park, S.E. and Hackenberger, W., *Current Opinion in Solid State and Materials Science*, **6**, 11-18 (2002).
2. Stenger, C.G.F., and Burggraaf, A.J., *Phys. Stat. Sol.*, **A61**, 653-664 (1980).
3. Setter, N., and Cross, L.E., *J. Appl. Phys.*, **51**, 4356-4360 (1980).
4. Bokov, A.A., and Raevski, I.P., *Ferroelectrics*, **90**, 125-133 (1989).
5. Bokov, A.A., and Raevski, I.P., *Ferroelectrics*, **144**, 147-156 (1993).
6. Davies, P.K., and Akbas, M.A., *J. Phys. Chem. Solids*, **61**, P.159-166 (2000).
7. Davies, P.K., Farber, L., Valant, M., and Akbas, M.A., *Fundamental Physics of Ferroelectrics 2000*: Aspen Center for Physics Winter Workshop, ed. by R.E.Cohen. AIP Conf. Proc. Melville, N.-Y., **535**, 38-46 (2000).
8. Chu, F., Reaney, I.M., and Setter, N., *Ferroelectrics*, **151**, 343-348 (1994).
9. Raevski, I.P., Malitskaya, M.A., Gagarina, E.S., Smotrakov, V.G., and Eremkin, V.V., *Ferroelectrics*, **235**, 221-230 (1999).
10. Bokov, A.A., Raevski, I.P., Smotrakov, V.G., and Zaitsev S.M., *Sov.Phys.-Crystallogr.*, **32**, 769-771 (1987).
11. Bokov, A.A., Raevski, I.P., Smotrakov, V.G., and Prokopalo, O.I., *Phys. Stat. Sol.*, **A93**, 411-417 (1986).
12. Krivoglaz, M.A. and Smirnov, A.A., *The theory of order-disorder in alloys*, American Elsevier Publishing Company, INC., New York, 1965.
13. Burton, B., *Phys.Rev.*, **B59**, 6087-6091 (1999).

First Principles Investigation of Novel Ferroelectric Perovskite Alloys Based on A-site Substitution

S.V. Halilov*†, M. Fornari** and D.J. Singh*

Center for Computational Materials Science, Naval Research Laboratory Washington, DC 20375
†*Department of Materials Science, University of Pennsylvania, Philadelphia, PA 19104*
**Department of Physics, Central Michigan University, Mt. Pleasant, MI 48859*

Abstract.
We report first principles studies of the lattice instabilities of as yet unsynthesized perovskites based on alloys in the $(Th,Pb,Ba,Bi,Y)ScO_3$ system. These systems are characterized by strong structural instabilities of the cubic perovskite lattice and A-site driven ferroelectricity. Tetragonal ferroelectric structures in these systems are found to show rather large values of the c/a ratio, which when present near a morphotropic phase boundary are thought to be associated with large piezoelectric actuation. The relationship between octahedral rotational and ferroelectric instabilities is discussed as are the conditions for obtaining a morphotropic phase boundary.

INTRODUCTION

$PbZr_xTi_{1-x}O_3$ (PZT) ceramic alloys with compositions near the morphotropic phase boundary (MPB) around $x = 0.5$ form the basis of most piezoelectric transducer devices [1, 2, 3]. However, the last five years have seen renewed interest in novel piezoelectric materials based on perovskite ferroelectrics. This is stimulated in part by the finding of new higher performance single crystal relaxor materials, [4] and in part by better fundamental understanding of the physics underlying the piezoelectric behavior of these materials. In particular, there is now a more clear understanding of the microscopic basis of piezoelectricity in perovskite ferroelectrics in terms of electric field induced polarization rotation from a rhombohedral ferroelectric phase near a MPB [5, 6]. This understanding enables us to identify perovskite based materials that are potentially good piezoelectrics based on their lattice instabilities. However, compared to Pb based materials with various B-site substitutions, relatively little work has been done to date investigating alloys based on A-site substitution and as a result our understanding of the chemical trends in the lattice instabilities of these materials is limited.

Here we report investigation of perovskites in the $(Th,Bi,Pb,Ba)ScO_3$ system by first principles calculations. This is an extension of our recent study of $(Cd,Pb)TiO_3$, [7] were we performed calculations that suggested that this could be a technologically interesting MPB system if it can be synthesized in an appropriate concentration range. In any case, the results suggest that Cd would be a potentially useful dopant for improving the properties of Pb based MPB systems. As in that recent study of the $(Pb,Cd)TiO_3$ system [7], we focus here on the ferroelectric modes, the balance between rhombohedral and tetragonal ferroelectric states, the lattice strain on the tetragonal side and competing instabilities associated with octahedral rotation [8, 9].

METHOD

Calculations for ordered supercells were carried out using density functional theory within the local density approximation using the general potential linearized augmented planewave method [10] with local orbitals to relax linearization errors and to treat high lying semicore states [11]. The calculations were done with convergence parameters like those in our recent work on the (Cd,Pb)TiO$_3$ system [9]. Spin-orbit effects are important in the electronic structure of actinides and structural properties of Th metal. [12] However, based on test calculations, they are not important for the structural energetics of tetravalent Th on the perovskite A-site. The calculations reported here were done in a scalar relativistic approximation. Except where otherwise noted, the local density approximation (LDA) equilibrium volumes are used.

BiScO$_3$ AND YScO$_3$, (Th,Pb)ScO$_3$

Th is a particularly interesting element for exploring chemical trends in the lattice instabilities of perovskites. As was emphasized early on, covalency plays an important role in the balance between various lattice instabilities in ferroelectric perovskites [13, 14] and in particular covalency with the A-site ion is important in stabilizing the tetragonal ferroelectric state of PbTiO$_3$. Without this there would be no morphotropic phase boundary (MPB) and therefore little piezoelectric response in solid solutions like PZT. Th is the only tetravalent ion likely to enter the perovskite A-site with transition metal B-site ions. The high valence of Th and the relativistic contraction of the actinides (which brings down the s and p states), favors hybridization between unoccupied p states and neighboring O p states. Furthermore, in light actinides $5f$ states may also participate in bond formation. However, Th perovskites would seem more likely with trivalent than divalent B-site ions; here we choose Sc and consider perovskite oxides where charge balance is retained by an admixture of a divalent A-site ion, D, i.e. Th$_{0.5}$D$_{0.5}$ScO$_3$. (Note that perovskites with high valent A-site ions combined with low valent B-sites are unusual.) The equilibrium perovskite lattice parameter calculated for an fcc (111) ordered supercell, with D=Pb and the ideal cubic perovskite structure was 7.55 a_0, which is very similar to BiScO$_3$. The tolerance factor, $t = (r_A + r_O)/\sqrt{2}(r_B + r_O)$, where r_A, r_B and r_O are the ionic radii of A, B and O plays a primary role in determining the lattice instabilities of simple perovskites. YScO$_3$ and BiScO$_3$ would seem to be chemically reasonable compounds to alloy into Th$_{0.5}$D$_{0.5}$ScO$_3$. These have t=0.8 and t=0.9, respectively, implying a tendency towards A-site driven ferroelectricity in competition with octahedral rotation. The average t for (Th,Pb)ScO$_3$ is similar to BiScO$_3$. At this lattice parameter, the fundamental LDA band gap is between the O $2p$ derived valence band and Th $5f$ derived conduction bands.

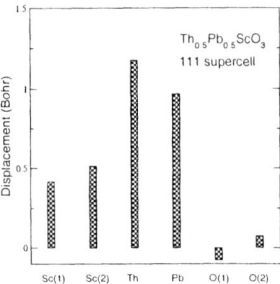 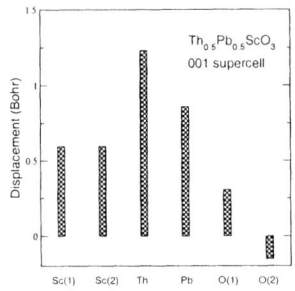

FIGURE 1. Displacement pattern with repect to the average O position for the rhombohedral ferroelectric state in an fcc (111) ordered ThPbSc$_2$O$_6$ supercell (left) and the tetragonal ferroelectric state of (001) ordered supercell (right). Note the A-site character.

The energetics of rhombohedral ferroelectric (R), tetragonal ferroelectric (T) and octahedrally rotated (ROT) states were calculated for this 10 atom supercell (for R and ROT) and a 10 atom (001) ordered supercell (for T). For the ferroelectric states full relaxations of the internal atomic coordinates were used with the symmetry constrained to correspond to the ferroelectric state (octahedral rotation was not allowed). For the R and ROT structures the lattice parameters were held fixed at the pseudocubic values, while for the T structure, the c/a ratio was varied, but the volume was held fixed.

We obtained strongly A-site driven (see Fig. 1) ferroelectricity in Th$_{0.5}$Pb$_{0.5}$ScO$_3$, with comparable off-centerings of both Pb and Th ions. Significantly, the ferroelectric instabilities are strong and the tetragonal (T) state is preferred over the rhombohedral (R). The T and R are 49 mRy and 41 mRy, respectively, lower in energy than the ideal cubic structure on a per Sc basis. The Th is more ferroelectrically active than the Pb as measured by the magnitude of its off-centering. Furthermore a large tetragonal strain is obtained for the T structure: $c/a \approx 1.08$ (see Fig. 2).

This would suggest alloying with a perovskite having a strong ferroelectric instability and a rhombohedral ground state in order to obtain a potentially piezoelectrically active MPB. However, there is a further complication. As in Pb based perovskites with strong A-site driven ferroelectricity [7, 8, 9, 15, 16], there is a competing ROT instability (see below). However, before turning to the ROT instability, we consider possible alloys having Th$_{0.5}$Pb$_{0.5}$ScO$_3$ as one end-point. We calculated the R and T ferroelectric instabilities of perovskite BiScO$_3$ and YScO$_3$ as well as their ROT instabilities. As expected, BiScO$_3$ strongly favors the R state, while YScO$_3$, with its very small A-site ion has energetically close R and T states. Interestingly, in YScO$_3$ the tetragonal strain is both large and soft. In both BiScO$_3$ and YScO$_3$ the ferroelectricity is strongly associated with the A-site. For example, in R BiScO$_3$ the calculated shift of the Bi ions with respect to the average O position is more than twice as large as the Sc shift. Based on the calculated energies of the R and T states (see Table 1), the occurance of an MPB would be much more likely in Th$_{0.5}$Pb$_{0.5}$ScO$_3$ – BiScO$_3$ than in Th$_{0.5}$Pb$_{0.5}$ScO$_3$ – YScO$_3$, and would occur on the Th$_{0.5}$Pb$_{0.5}$ScO$_3$ rich side of the phase diagram.

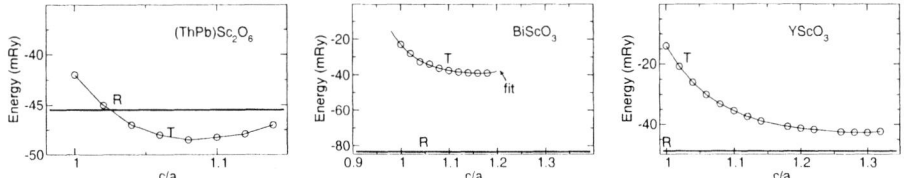

FIGURE 2. Calculated energetics of the R and T ferroelectric states of (Th,Pb)ScO$_3$ supercells (left) BiScO$_3$ and (center) YScO$_3$ (right), relative to the undistorted cubic perovskite on a per Sc basis. The horizontal lines are the R energy while the data points are the T energy as a function of the tetragonality, c/a. Note the softness with respect to c/a. Note also the different scales.

TABLE 1. Calculated energetics of ferroelectric and octahedral rotational instabilities (ROT) relative to the ideal perovskite. Energies are given for rotations about (111) and (001) and are in mRy on a per Sc basis.

Compound	T	R	ROT(111)	ROT(001)
(Th,Pb)ScO$_3$	-49	-41	-65	-58
BiScO$_3$	-39	-82	-87	-63
YScO$_3$	-43	-49	-140	-109

OCTAHEDRAL ROTATION

However, all of this neglects the competing ROT instabilities. We calculated the energetics of these for the Th$_{0.5}$Pb$_{0.5}$ScO$_3$ supercells and for BiScO$_3$ and YScO$_3$. These are given in Table 1.

As expected from the small A-site ionic radius of YScO$_3$, the rotational instabilities of the perovskite structure are extremely strong and should suppress any ferroelectric tendency. In BiScO$_3$, the ROT and ferroelectric R states are very close in energy, with the ROT state slightly favored. Th$_{0.5}$Pb$_{0.5}$ScO$_3$ also has a strong ROT instability, which is somewhat stronger than the ferroelectric T instability. Therefore, we expect that competition with octahedral rotation would suppress ferroelectricity in the part of the Th$_{0.5}$Pb$_{0.5}$ScO$_3$ – BiScO$_3$ phase diagram where a piezoelectrically active MPB might otherwise occur (n.b. rotational and ferroelectric instabilities have different symmetries and therefore do not interact at lowest order so coexistence would seem reasonable, but they are driven by similar physics – the large volume available to the A site ion – so for the large displacements present here they are expected to be antagonistic). We calculated the rotational energetics in BiScO$_3$ with the volume expanded by 8.5 %. Unlike the case of the titanates [9], we found almost no volume dependence, presumably reflecting a relative softness of the ScO$_6$ octahedra relative to TiO$_6$. Therefore, it would seem that small volume changes by chemical substitution are unlikely to sufficiently shift the balance between ferroelectric and octahedrally rotated states in Th$_{0.5}$Pb$_{0.5}$ScO$_3$ – BiScO$_3$ to yield a ferroelectric MPB.

Another possibility would be to partially substitute a large ion on the A-site. The rational for this is that, because of the shared O ions connecting the octahedra, rotations of neighboring octahedra must be at least short range correlated, while off-centerings of the *A*-site cations on different sites can be relatively independent. This is reflected for example in the relatively weak dispersion of the ferroelectric mode away from the Γ point in the Pb based ferroelectrics and the much stronger dispersion of rotational (*e.g.* R_{25}) modes away from the zone boundary [17]. Replacing the Pb in our ThPbSc$_2$O$_6$ fcc supercell with Ba, we find a significant reduction in the tendency towards rotated states (-44 mRy per Sc *vs.* -65 mRy). Therefore, depending on where in the Th$_{0.5}$Pb$_{0.5}$ScO$_3$ – BiScO$_3$ phase diagram the MPB would occur without rotations, partial substition of Pb by Ba might be sufficient to make the MPB appear. Another possibility would be partial replacement of Bi with La.

SUMMARY AND CONCLUSIONS

LDA calculations of the lattice instabilities of hypothetical perovskite Th$_{0.5}$Pb$_{0.5}$ScO$_3$, BiScO$_3$ and YScO$_3$ have been done. Considering only ferroelectric instabilities, a piezoelectrically active MPB is expected in the Th$_{0.5}$Pb$_{0.5}$ScO$_3$ – BiScO$_3$, probably on the Th$_{0.5}$Pb$_{0.5}$ScO$_3$ rich side. However, the calculations show that instabilities associated with rotation of the ScO$_6$ octahedra would suppress ferroelectricity in this region. We suggest that partial substitution of a larger *A*-site ion, such as Ba for Pb and/or La for Bi might be sufficient to make the ferroelectric state appear in the region where the MPB is expected. Based on the tolerance factors alloys in the Th$_{0.5}$Pb$_{0.5}$ScO$_3$ – BiScO$_3$ pseudobinary system are expected to be on the borderline of perovskite stability, *i.e.* where synthesis, while perhaps difficult, may well be possible. It would be interesting to attempt their synthesis and measure their structural and dielectric properties. Furthermore, as in the case of Cd, the present results suggest that Th may be an interesting additive for modifying the properties of MPB piezoelectic systems, although the toxicity of Th may make this an unattractive route for practical applications.

ACKNOWLEDGMENTS

We are grateful for helpful discussions with L. Bellaiche, R. Cohen, P. Davies, T. Egami, J. Iniguez, I. Mazin and A. Rappe. This work was supported by the Office of Naval Reseach and the Center for Piezoelectrics by Design. Some calculations were done at the ASC HPCMO computer facility. Some calculations were done using the CHSSI supported DoD-AE code.

REFERENCES

1. K. Uchino, *Piezoelectric Actuators and Ultrasonic Motors* (Kluwer Academic, Boston, 1996).
2. B. Jaffe, W.R.J. Cook, and H. Jaffe, *Piezoelectric Ceramics* (Academic Press, New York, 1971).
3. M.E. Lines and A.M. Glass, *Principles and Applications of Ferroelectrics and Related Materials* (Clarendon, Oxford, 1977).
4. S.E. Park and T.R. Shrout, J. Appl. Phys. **82**, 1804 (1997).
5. H. Fu and R.E. Cohen, Nature **281**, 403 (2000).
6. L. Bellaiche, A. Garcia, and D. Vanderbilt, Phys. Rev. Lett. **84**, 5427 (2000).

7. S.V. Halilov, M. Fornari and D.J. Singh, Appl. Phys. Lett. **81**, 3443 (2002).
8. D.J. Singh, Phys. Rev. B **52**, 12559 (1995).
9. M. Fornari and D.J. Singh, Phys. Rev. B **63**, 092101 (2001).
10. D.J. Singh, *Planewaves, Pseudopotentials and the LAPW Method* (Kluwer Academic, Boston, 1994).
11. D. Singh, Phys. Rev. B **43**, 6388 (1991).
12. M.D. Jones, J.C. Boettger, R.C. Albers and D.J. Singh, Phys. Rev. B **61**, 4644 (2000).
13. R.E. Cohen and H. Krakauer, Phys. Rev. B **42**, 6416 (1990).
14. R.E. Cohen, Nature **358**, 137 (1992).
15. R. Ranjan, Ragini, S.K. Mishra, D. Pandey and B.J. Kennedy, Phys. Rev. B **65**, 060102 (2002).
16. B. Noheda, L. Wu and Y. Zhu, Phys. Rev. B **66**, 060103 (2002).
17. P. Ghosez, E. Cockayne, U.V. Waghmare and K.M. Rabe, Phys. Rev. B **60**, 836 (1999).

Ab initio study of silver niobate

Ilya Grinberg* and Andrew M. Rappe*

Department of Chemistry and the Laboratory for Research on the Structure of Matter, University of Pennsylvania, Philadelphia, PA 19104-6323

Abstract. Using DFT calculations, we investigate the local structure and the distortion patterns of the perovskite $AgNbO_3$ to examine the feasibility of using silver on the perovskite A-site in lead-free piezoelectrics. Our calculations show that in a 5-atom $AgNbO_3$ unit cell, silver atoms can off-center by about 0.5 Å, forming short covalent bonds similar to the short Pb-O bonds in Pb-based ferrroelectrics. In the more realistic 40-atom supercell, Ag behaves similar to Pb in $PbZrO_3$, forming short covalent Ag-O bonds through a combination of large octahedral rotations and small Ag off-center distortions. Unlike $PbZrO_3$, $AgNbO_3$ is not antiferroelectric due to the presence of large Nb off-center distortions. In the $AgNbO_3$-$PbTiO_3$ solid solution, both A-cations off-center significantly and display a preference for distortion away from Nb atoms and toward Ti atoms. The interplay between this preference and maximization of local dipole alignment should lead to an MPB in the $AgNbO_3$-$PbTiO_3$ phase diagram.

INTRODUCTION

Ferroelectrics are a technologically important class of materials and are used in a wide variety of applications. The study of ferroelectric solid solutions is also interesting from a scientific standpoint, as the coupling of long-range electrostatic interactions with short-range chemical bonding presents unique challenges for theoretical and experimental investigations. Currently, lead-based perovskites such as $Pb(Zr,Ti)O_3$ (PZT) are used in high-performance piezoelectric applications. The presence of Pb is crucial for high performance as Pb displays large off-centering which gives rise to large internal polarization and better coupling to electric field. Since lead compounds are toxic, lead-free oxides have been investigated for possible use in environmentally friendly piezoelectrics [1].

To preserve large internal polarization of Pb-based oxides, the proposed lead-free solid solutions must display off-centering behavior on both the A and B perovskite sites. For the B-cation, there are many possible candidates, as transition metals in the $3d$, $4d$ and $5d$ rows of the periodic table all display off-centering behavior, when placed in a large enough O_6 cage. For the A-site, the options are more limited. Of the cations found at the perovskite A-site at ambient pressure and temperature, three have been shown experimentally to exhibit off-centering behavior: Pb, Bi and Li. In a pioneering theoretical work, Halilov, Fornari and Singh have recently demonstrated off-centering behavior for Cd as well [2]. However, Bi and Cd are both toxic, and therefore are not good candidates for environmentally-friendly piezoelectrics. Bi-based perovskites are also difficult to synthesize under ambient conditions. Lithium is not toxic and displays a very large off-centering of 1.0 Å [3], but its small size gives rise to limited solubility in the perovskite phase.

In this study, we examine the behavior of silver atoms on the perovskite A-site for possible use in ferroelectric applications. Our reasoning is as follows. Experimentally $CdTiO_3$ is known not to be a ferroelectric due to the small tolerance factor. However, Halilov *et al.* have shown that in a 5-atom unit cell Cd does exhibit large off-centering and that the Cd distortions are enhanced when Cd site volume is expanded by alloying with $PbTiO_3$ [2]. The off-centering is due to the mixing of Cd $5p$ and O $2p$ states. As Ag is a neighbor of Cd in the periodic table, it is likely that it will display similar behavior. Since Ag is non-toxic, Ag off-centering behavior may open the possibility of truly environmentally-friendly high performance piezoelectrics.

The most widely studied perovskite with Ag on the A-site is $AgNbO_3$ [4, 5, 6]. Assuming the +1 Ag ionic radius of 1.38 Å in the twelve-coordinated A-site, the tolerance factor for $AgNbO_3$ is 0.956. $AgNbO_3$ can be easily made using conventional solid-state synthesis methods. Dielectric properties of $AgNbO_3$ and solutions of $AgNbO_3$ with $AgTaO_3$ as well as Li, Na, and K subsitutions were recently studied [4]. $AgNbO_3$ assumes six structural phases with increasing temperature. At T<340 K it is found in either a weak ferroelectric or ferrielectric orthorhombic phase; at 340 K it undergoes a transition to an antiferroelectric phase, a second antiferroelectric phase appears at 540 K and at 626 K $AgNbO_3$ becomes paraelectric. Two more paralectric phases appear at higher temperatures. The lattice parameters of the low-temperature phase have been determined by X-ray diffraction to be a=3.91 Å and c=3.94 Å. The structure is characterized by octahedral rotations typical for perovskites with low tolerance factor. While typically perovskites with low tolerance factor are antiferroelectric at low temperature, $AgNbO_3$ is not. The origin of this anomalous behavior is unclear and will be investigated using DFT calculations.

METHODOLOGY

We use density functional theory [7, 8] (DFT) calculations with the local density approximation [9] (LDA) for the exchange-correlation functional for all calculations. The calculations are done with our in-house plane wave code. For Ag and Nb, we use optimized designed non-local pseudopotentials [10, 11], created with the OPIUM package [12] and tuned to achieve optimal eigenvalue and tail norm transferability [13]. We use the same oxygen pseudopotential as in previous studies [14]. The calculations were done in small $1\times1\times1$, 5-atom unit cells and large $2\times2\times2$ 40-atom supercells. Ionic minimizations are started from randomized perfect perovskite positions, and no symmetry is imposed during structural relaxation as the energy is minimized with respect to all ionic coordinates. Due to the well-known inability of DFT methods to predict the correct volume for ferroelectric perovskites, all calculations are done at the experimental volume.

TABLE 1. Effects of volume expansion on tetragonality and cation displacements in a 5-atom unit cell of $AgNbO_3$.

	$V^{1/3}$=3.93 Å	$V^{1/3}$=4.00 Å	$V^{1/3}$=4.1 Å
c/a	1.03	1.03	1.26
Ag disp.	0.50	0.63	0.91
Nb disp.	0.22	0.23	0.53

RESULTS

5-atom unit cell

For 5-atom calculations, we find that at the experimental volume the optimized c/a ratio is 1.03. Both silver and niobium off-center significantly; silver distorts by 0.5 Å, similar to the Pb off-centering in $PbTiO_3$ and to the Cd off-centering in $(Pb,Cd)TiO_3$. In $PbTiO_3$, Pb atoms distort along the (100) direction, splitting Pb-O bonds into three equal groups of four short (2.4-2.5 Å), medium (2.7-2.9 Å) and long (3.1-3.3 Å) bonds, By contrast, the silver atom in 5-atom $AgNbO_3$ distorts along the (521) direction, creating a distribution of Ag-O bonds that can be roughly split into four short (2.3-2.6 Å), five medium (2.66-3.00 Å) and three long (3.11-3.31 Å) Ag-O bonds. Niobium off-centers by 0.25 Å, also along the (521) direction, forming three short bonds of 1.83 Å, 1.89 Å and 1.93 Å and three long bonds of 2.01 Å, 2.05 Å and 2.15 Å. Brown's rules of valence [15] analysis of the Nb-O distances shows that Nb displacement significantly changes the bond order of the Nb-O bonds, creating a difference of about 0.5 between the valence of the short and long Nb-O bonds. Significant changes in bond order suggest that in $AgNbO_3$, Nb should display an anomalous Born effective charge larger than the formal +5 charge.

Halilov et al. have found that volume expansion has a dramatic effect on the preferred c/a ratio of $CdTiO_3$. For a 5.4% volume expansion of $AgNbO_3$, we find no increase in tetragonality, with optimized c/a ratio of 1.03. As expected, the volume expansion increases the magnitudes of the Ag and Nb off-center distortions. However, a much larger 13.5% volume expansion has a dramatic effect on tetragonality with optimized c/a=1.26. The large increase in tetragonality is accompanied by a doubling of the Ag and Nb distortion magnitudes. The results at three different volumes are summarized in Table I. The nonlinear response of $AgNbO_3$ to volume expansion is similar to the one found by Tinte, Rabe and Vanderbilt in their study of the effects of negative pressure on the structure of $PbTiO_3$ [16].

40-atom supercell

Calculations in a small 5-atom unit cell are unphysical for perovskites with low tolerance factors, since a 5-atom unit cell does not allow octahedral rotation, which accommodates the mismatch between the A-O and B-O sublattices. We therefore carried

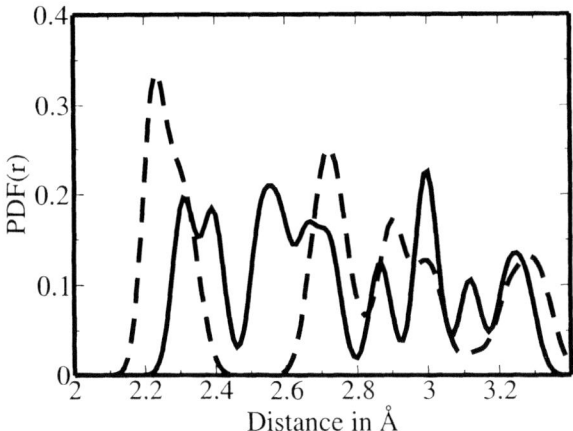

FIGURE 1. Ag-O partial PDFs obtained from 5-atom (solid) and 40-atom (dashed) $AgNbO_3$ relaxed structures.

out calculations in a 2×2×2 40-atom $AgNbO_3$ supercell, which is a more realistic representation of the material.

Optimization of the lattice parameters at the experimental volume showed that an orthorhombic phase (c=7.88 Å, a=7.82 Å) is preferred, in excellent agreement with experimental results [6]. As expected, the relaxed 40-atom supercell displays large octahedral rotations, which stabilize the structure by 0.3 eV per 5-atom unit cell. The octahedra tilt by about 14.5°; this is similar to the magnitude of tilts found in $PbZrO_3$. We find that the nature of Ag-O bonding is unchanged in the larger supercell. Ag-O bonds are split into groups of three short (2.2-2.3 Å), six medium (2.7-3.0 Å) and three long (3.1-3.3 Å) bonds. However, despite the presence of short covalent Ag-O bonds, Ag distortions are extremely small (0-0.1 Å). This is because the O_6 tilting brings the oxygen atoms closer to Ag, eliminating the need for a large distortion to make short Ag-O covalent bonds. A comparison of the Ag-O PDFs for 5-atom and 40-atom calculations (Figure 1) shows that both PDFs have similar splits of Ag-O distances into short, medium and long Ag-O bonds. Despite the similar nature of bonding, the motifs by which the desired bonding is achieved are different in the 5-atom and 40-atom unit cells.

Nb distortions in the 40-atom supercell are 0.25 Å, slightly larger than in the 5-atom unit cell. Unlike the $Zr-O_6$ polarization in PZ, the $Nb-O_6$ polarization in AN is of mostly covalent character, displaying a clear split between three short Nb-O distances of 1.9 Å and three long Nb-O distances of 2.17 Å (Figure 2). A comparison of the Nb-O partial PDFs for the 5-atom and the 40-atom unit cells show that the splitting in the Nb-O peak is larger for the 40-atom cell. This is due to the larger volume of the rotated O_6 cage, which creates additional space for the Nb off-center distortions. We find that Nb ions displace in (221), (2$\bar{2}$1), (22$\bar{1}$) and (2$\bar{2}\bar{1}$) directions, averaging out to a ferroelectric polarization in the (100) direction and anti-ferroelectric distortions in the (010) and (001) directions.

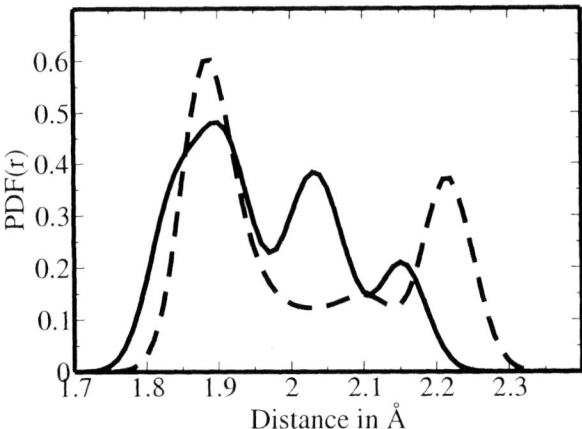

FIGURE 2. Nb-O partial PDFs obtained from 5-atom (solid) and 40-atom (dashed) $AgNbO_3$ relaxed structures.

$AgNbO_3$-$PbTiO_3$ solid solution

Our study of $AgNbO_3$ showed that Ag off-centering is eliminated by the large O_6 rotations in pure $AgNbO_3$ due to the low tolerance factor. To increase the tolerance factor, thereby reducing the O_6 rotation, we can alloy $AgNbO_3$ with a perovskite of larger volume. The increase in the size of the Ag site should also contribute to the unlocking of the Ag off-center behavior. Since our primary interest is in high-performance morphotropic phase boundary piezoelectrics, we choose to alloy $AgNbO_3$ with $PbTiO_3$. We expect that the large Pb and Ti displacements in $PbTiO_3$ will give an additional impetus for Ag off-centering. While the alloying introduces the toxic Pb atoms on the A-site, it is nevertheless a good first step in investigation of Ag behavior in solid solution, as $PbTiO_3$ and its solid solutions have been extensively studied [14, 17, 18, 19, 20] and are now well-understood. For a preliminary study, we choose a composition of 87.5% $PbTiO_3$ and 12.5% $AgNbO_3$. We perform our calculations at the interpolated experimental volume, which is 4.4% larger than the experimental volume of pure $AgNbO_3$.

The results for AN-PT calculations are presented in Table II along with the results of a 5-atom DFT calculations for $PbTiO_3$ and $AgNbO_3$ for comparison. As expected, we find that alloying with $PbTiO_3$ dramatically increases the Ag displacement to 0.7Å, larger than the 0.5 Å found for the 5-atom $AgNbO_3$ unit cell. However, $AgNbO_3$ alloying decreases the magnitude of the Pb distortions, from 0.5 Å in pure $PbTiO_3$ to 0.3 Å in $AgNbO_3$-$PbTiO_3$. The decrease in Pb distortion is due to the decrease in the volume of the Pb site, and due to the significant 6° octahedral rotations induced by $AgNbO_3$ alloying. For the B-site behavior, the relaxed structure shows Nb and Ti distortions that are slightly smaller than those in end member compounds, most likely due to smaller Pb distortions.

Our research on PZT showed that the difference in the strength of Pb-Zr and Pb-Ti repulsive interactions leads to the local preference of Pb atoms to move away from

TABLE 2. Comparison of the 5-atom $AgNbO_3$, 40-atom $AgNbO_3$, $PbTiO_3$ and $AgNbO_3$-$PbTiO_3$ local structure data. Displacement magnitudes are in Å, and O_6 cage tilt angles in degrees.

	c/a	Ag disp.	Nb disp.	Pb disp.	Ti disp.	O_6 tilt angle
5-atom $AgNbO_3$	1.03	0.5	0.22			0
40 atom $AgNbO_3$	1.00	0.1	0.25			14.5
$PbTiO_3$	1.05			0.5	0.28	0
$(AgNbO_3)_{1/8}$-$(PbTiO_3)_{7/8}$	1.03	0.7	0.22	0.3	0.25	6

Zr and toward Ti [20]. The competition between the local preference to move toward Ti and the preference to obtain maximum local dipole alignment then gives rise to compositional phase transitions. In PZT, the stronger Pb-Zr repulsion is due to the significant size difference between Zr and Ti (0.74 Å for Zr, 0.60 Å for Ti). To examine the A-B interactions in $AgNbO_3$-$PbTiO_3$, we calculated the (Pb,Ag)-Nb and (Pb,Ag)-Ti partial PDFs from our relaxed structure (Figure 4). A clear separation between Pb-Nb and Pb-Ti peaks can be observed, similar to the separation between the Pb-Zr and Pb-Ti peaks in PZT. This indicates that Pb-Nb repulsion is greater than Pb-Ti repulsion. The preference to avoid Nb and displace toward Ti, together with the preference for dipole alignment should lead to a MPB between rhombohedral and tetragonal phases in the $AgNbO_3$-$PbTiO_3$ phase diagram.

Since the size difference between Nb and Ti is much smaller (0.64 Å for Nb and

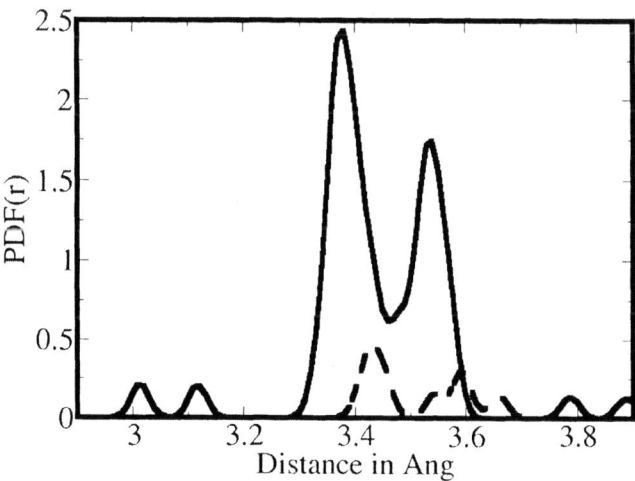

FIGURE 3. (Pb,Ag)-Ti (solid) and (Pb,Ag)-Nb (dashed) partial PDFs obtained from relaxed $(AgNbO_3)_{1/8}$-$(PbTiO_3)_{7/8}$ relaxed structure. Ag-Ti peaks are at 3.0 Å and 3.1 Å, Pb-Ti peaks are at 3.3 Å and 3.5 Å.

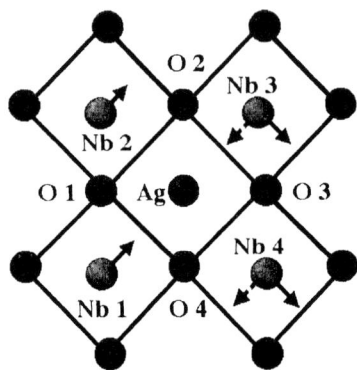

FIGURE 4. Nb distortion patterns in the 40-atom AgNbO$_3$ supercell. Solid arrows represent actual distortions found by DFT calculations. In the proposed antiferroelectric distortion pattern, shown by dashed arrows for Nb atoms 3 and 4, oxygen atoms 2 and 4 would be overbonded.

0.60 Å for Ti) than the Zr/Ti size difference, while the separation between the A-B PDF peak positions is the same, it is unlikely that the greater strength of the Pb-Nb repulsion is due to the size effect. Rather, it is due to the difference in the valence of the two cations. An oxygen atom with Nb and Ti nearest neighbors is overbonded [21] relative to an oxygen which has two Ti nearest neighbors, and is therefore less able to create a strong short covalent bond with a Pb atom. Thus, the difference in the Pb-Nb and Pb-Ti repulsion strength is due to an indirect, through-oxygen effect.

DISCUSSION

Despite the tolerance factor of 0.956 (similar to PbZrO$_3$), our DFT calculations confirm experimental findings that AgNbO$_3$ at low temperature is not antiferroelectric. We believe that this difference is a direct consequence of the different character of the B-cation behavior in the two materials. As illustrated by the Nb-O PDFs in Figure 2, Nb atoms displace by a significant 0.25 Å, creating three short Nb-O bonds of high bond order and three long Nb-O bonds of low bond order. As shown in Figure 4, an anti-ferroelectric distortion pattern coupled with creation of short, high bond order Nb-O bonds would produce over- and under-bonded oxygen atoms. A ferroelectric distortion pattern, which does not lead to oxygen overbonding is therefore more favorable. In PbZrO$_3$, the Zr off-centering is in large part due to O-Zr-O bond angle bending which does not appreciably change the bond order of the Zr-O bonds. Therefore, an antiferroelectric distortion pattern does not lead to oxygen atom under- or over-bonding and is preferred.

Our results show that Ag behavior on the perovskite A-site is similar to that of Pb. Short covalent bonds are a well-known requirement of the Pb-O bonding at the perovskite A-site [22]. However, the way these bonds are achieved depends on the environment of the Pb atoms. In oxides with a high tolerance factor such as $PbTiO_3$, a strong off-center distortion creates short covalent Pb-O bonds. In oxides with low tolerance factor such as $PbZrO_3$, large octahedral rotations bring oxygen atoms closer to Pb, eliminating the need for a large Pb displacement. Similarly, short covalent Ag-O bonds are a requirement of the Ag-O bonding on the A-site. In our 40-atom $AgNbO_3$ calculations, the short Ag-O bonds are created through octahedral rotations; in the 5-atom calculations the octahedral rotations are no longer possible and the short Ag-O bonds are created through off-center diplacements. This gives rise to similar Ag-O partial PDFs for the 5-atom and the 40-atom calculations. Without octahedral rotations, $AgNbO_3$ is similar to $PbTiO_3$; with the rotations it is similar to $PbZrO_3$.

Our results, as well as the previous work of Halilov *et al.*, show that absence of ferroelectricity in simple perovskites with low tolerance factor does not necessarily imply an inability of the A-cation to off-center. The lack of ferroelectric behavior can be caused by octahedral rotations creating the short covalent bonds and not by an inability to create covalent bonds. Thus, both $CdTiO_3$ as examined by Halilov *et al.* and $AgNbO_3$ as presented in this paper display ferroelectricity when the rotations are turned off by constraining the material to a 5-atom unit cell or by alloying with $PbTiO_3$. Other A-site cations found in simple perovksites with low tolerance factor and thought to be ferroelectrically inactive, may similarly display ferroelectric activity in solid solutions.

CONCLUSION

Using DFT calculations, we investigated the chemistry of $AgNbO_3$ and the $AgNbO_3$-$PbTiO_3$ solid solution. We find that silver ions create short covalent Ag-O bonds and are therefore ferroelectrically active in the absence of large $B\text{-}O_6$ octahedral rotations. Calculations on 40-atom $AgNbO_3$ unit cells confirm the presence of ferroelectricity in the low temperature phase of pure $AgNbO_3$. The origin of preference of $AgNbO_3$ for a ferroelectric phase over the antiferroelectric phase is due to the strong Nb distortions which would create over- and under-bonded oxygen atoms in the antiferroelectric distortion pattern. Our calculations on $AgNbO_3$-$PbTiO_3$ show that alloying $PbTiO_3$ with $AgNbO_3$ increases Ag displacements but decreases Pb displacements and the c/a ratio. Despite the similar size of the Nb and Ti cations, Pb-Nb repulsion is stronger than Pb-Ti repulsion, and Pb atoms show a preference for distortions away from Nb ions and toward Ti ions. This difference in the Pb-Nb and Pb-Ti repulsion strength is due to indirect, through-oxygen overbonding effect. The combination of Pb preference for distortion toward Ti atoms and the desire for maximum local dipole alignment should lead to a MPB with good piezoelectric properties in the $AgNbO_3$-$PbTiO_3$ phase diagram.

ACKNOWLEDGMENTS

This work was supported by the office of Naval Research grant number N-000014-00-1-0372 and through the Center for Piezoelectrics by Design. AMR acknowledges the support of the Camille and Henry Dreyfus Foundation. Computational support was provided by the Center for Piezoelectric Design and the DOD HPCMO at the ERDC SGI O3800. We would like to thank Gerhard Theurich and Eric Walter for help with pseudopotentials, the Center for Piezoelectric Design for support of the OPIUM project and Peter K. Davies for stimulating discussions.

REFERENCES

1. Chiang, Y. M., Farrey, G. W., and Soukhojak, A. N., *Appl. Phys. Lett.*, **73**, 3683–3685 (1998).
2. Halilov, S. V., Fornari, M., and Singh, D. J., *Appl. Phys. Lett.*, **81**, 3443–3445 (2002).
3. Prosandeev, S., Cockayne, E., and Burton, B. P., "First Principles Studies of $KNbO_3$, $KTaO_3$ and $LiTaO_3$ Solid Solutions," in *Fundamental Physics of Ferroelectrics-2002*, edited by R. E. Cohen, AIP Conference Proceedings 626, American Institute of Physics, New York, 2002, pp. 64–73.
4. Kania, A., *J. Phys. D Appl. Phys.*, **34**, 1447–1455 (2001).
5. Koh, J. H., Khartsev, S. I., and Grishin, A., *Appl. Phys. Lett.*, **77**, 4416–4418 (2000).
6. Kania, A., and Kwapulinski, J., *J. Phys. Cond. Matt.*, **11**, 8933–8946 (1999).
7. Hohenberg, P., and Kohn, W., *Phys. Rev.*, **136**, B864–71 (1964).
8. Kohn, W., and Sham, L. J., *Phys. Rev.*, **140**, A1133–8 (1965).
9. Perdew, J. P., and Zunger, A., *Phys. Rev. B*, **23**, 5048–79 (1981).
10. Rappe, A. M., Rabe, K. M., Kaxiras, E., and Joannopoulos, J. D., *Phys. Rev. B Rapid Comm.*, **41**, 1227–30 (1990).
11. Ramer, N. J., and Rappe, A. M., *Phys. Rev. B*, **59**, 12471–8 (1999).
12. Theurich, G. J., Walter, E. J., and Rappe, A. M., *http://opium.sourceforge.net* (2003).
13. Grinberg, I., Ramer, N. J., and Rappe, A. M., *Phys. Rev. B Rapid Comm.*, **63**, 201102 1–4 (2001).
14. Ramer, N. J., Mele, E. J., and Rappe, A. M., *Ferroelectrics*, **206**, 31–46 (1999).
15. Brown, I. D., "The Bond-Valence Method," in *Structure and Bonding in Crystals II*, edited by M. O'Keeffe and A. Navrotsky, Academic Press, New York, 1981, pp. 1–30.
16. Tinte, S., Rabe, K. M., and Vanderbilt, D., *Ferroelectrics Workshop* (2003).
17. Garcia, A., and Vanderbilt, D., *Phys. Rev. B*, **54**, 3817–24 (1996).
18. Dmowski, W., Egami, T., Farber, L., and Davies, P. K., "Structure of $Pb(Zr,Ti)O_3$ Near the Morphotropic Phase Boundary," in *Fundamental Physics of Ferroelectrics-2001*, edited by R. E. Cohen, AIP Conference Proceedings 582, American Institute of Physics, New York, 2001, pp. 33–44.
19. Cohen, R. E., *Nature*, **358**, 136–8 (1992).
20. Grinberg, I., Cooper, V. R., and Rappe, A. M., *Nature*, **419**, 909–11 (2002).
21. Burton, B. P., and Cockayne, E., *Phys. Rev. B Rapid Comm.*, **60**, 12542–5 (1999).
22. Egami, T., Dmowski, W., Akbas, M., and Davies, P. K., "Local Structure and Polarization in Pb-Containing Ferroelectric Oxides," in *First-Principles Calculations for Ferroelectrics–Fifth Williamsburg Workshop*, edited by R. E. Cohen, AIP Conference Proceedings 436, American Institute of Physics, New York, 1998, pp. 1–10.

Off-center atomic displacements in BaTiO$_3$ quantum dots

Huaxiang Fu and Laurent Bellaiche

Physics Department, University of Arkansas, Fayetteville, Arkansas 72701

Abstract. Structural properties of BaTiO$_3$ quantum dots are studied using a first-principles derived effective-Hamiltonian approach. Truncated long-range dipole-dipole interaction and modification of the short-range interaction due to the existence of vacuum are taken into account. Our calculations show that there are significant off-center atomic displacements in these dots; the amplitudes of such displacements are comparable with those occuring in bulk BaTiO$_3$. However, unlike in the bulk system, the net polarization in dots is found to be zero. Our results also show that the local displacements in the dots tend to flip across the distance of the entire dot, resulting in an unusual and complex pattern.

INTRODUCTION

As temperature decreases, bulk BaTiO$_3$ undergoes sequences of structural transitions[1]: from the paraelectric cubic phase to a tetragonal phase at 403K, then to an orthorhombic phase at 278K, and finally to a rhombohedral phase at 183K. Some of these phase transitions have an order-disorder character, as indicated by the existences of substantial atomic displacements at cubic phases and the broad Fourier spectrum over the Brillouin zone.[2] The microscopic mechanism that are responsible for the existence of ferroelectric phases are, in particular, the long-range (LR) dipole-dipole interaction and the short-range (SR) interatomic covalent coupling. The ferroelectric phases of bulk BaTiO$_3$ have now been understood, at a microscopic level, from both first-principles calculations[3] and effective Hamiltonian simulations.[2]

Compared to bulk materials, quantum dots (i.e., finite-size nanocrystals) do not have the imposed periodic boundary condition. As a result, the microscopic forces that determine the material phases are altered, and the delicated balance between the LR and SR interactions in bulk can be strongly modified with respect to the bulk case. It is therefore interesting to examine whether there are significant off-center displacements in dots, and whether these displacements will align themselves to form ferroelectric phases. Specifically, the following three differences exist between the bulk and dots: (1) The LR dipole interaction that is crucial to yield ferroelectric phases is truncated in dots. Unlike in bulk, the electrostatic potentials at different cells are thus not equal. (2) The SR interaction among the atoms near the boundary of dots are altered. The atomic displacements at the surfaces will thus be different from those inside the dots. (3) Depending on the sizes of dots, dielectric screening may be different from the bulk value, which will affect the LR dipole energy.

Here we study the off-center atomic displacements in BaTiO$_3$ dots using an effective

Hamiltonian approach.[2, 4, 5] The Hamiltonian is modified with respect to those developed in Ref.4 to take into account the differences between bulk and dots. We found that in dots there are significant off-center atomic displacements with amplitudes comparable to those occuring in bulk. Our calculations further show that unlike in bulk, local modes in dots prefer to form domains with opposite displacements, such that the dipole energy can be maximally reduced. Our calculations also indicate that the conclusion of forming domains will not be significantly changed if the short-range interaction at surface varies . We also found that atomic displacements of neighboring cells in dots rotate to different degrees, depending on the sizes of dots. The results can be tentatively explained by the competition between the dipole energy and inhomogeneous strain energy.

METHOD

The energy variations due to atomic displacements are described by the effective Hamiltonian, which includes the LR dipole-dipole interaction, the site-site SR interaction between local modes up to the third neighbors, homogeneous and inhomogeneous elastic strain energies, and the atomic displacement-strain coupling. The parameters corresponding to these interactions are those used in Ref.6.

Supercell methods are generally employed to handle systems without periodic boundary condition. This approach will, however, introduce an artificial LR Coulomb interaction unless the size of cell is very large. Here, we choose not to use the periodic supercell approach. We handle the dipole-dipole interaction in real space, and there is no artificial interaction coming from the periodic dipole images. The shapes of dots are assumed to be cubic or rectangular, and the surfaces are assumed to be BaO-terminated. Truncation of bulk into finite-size nanocrystals will cause the amplitudes of local modes at surface to be different from those near the center of dot. A site-site interaction between the local mode and the vacuum is thus added. The parameter of this additional interaction is determined from forces acting on surface local modes, and these forces are obtained from local-density-functional calculations. In addition, an interaction between inhomogeneous strains of the cells at surfaces and the vacuum is included in the Hamiltonian for dots. These additional SR interactions are found to be important for the local displacements at surface. For instance, the added site-site interaction will cause displacements at surface directed towards the vacuum.

Monte Carlo simulation is used to obtain atomic displacements for a given temperature. Monte Carlo (MC) simulations are carried out starting from high temperature (300K); 40000 MC steps are performed for each temperature and each step sweeps all sites available. Temperature is then annealed by 50K each time down to 10K. Results presented in this paper are obtained at 10K, and no essential difference are found between 10K and 50K.

RESULTS AND DISCUSSIONS

Different sizes of $BaTiO_3$ dots are considered in our calculations. The size of dot is measured by the number of bulk cubic cells along three directions, for instance, the size of a dot in which there are n_1, n_2, and n_3 bulk five-atom cells along the x, y and z directions, respectively, is denoted as $n_1 \times n_2 \times n_3$. For the convenience of demonstration, calculation results will be presented mostly at a cross-section plane, specified by its normal direction and its order index among those planes with the same normal direction. For example, the y=7 plane indicates the 7th plane with its normal direction along the y axis. Results for a 12x12x12 dot at 10K is given in Fig.1, where the projected amplitudes and directions of local displacements are shown for the y=6 plane. We choose the y=6 plane, because it is one of the two central planes along the y direction and is thus the plane that subjects to the least surface effect. However, the surface effect still occurs at the edge of this cross section. It is evident in Fig.1 that the local displacement of each 5-atom cell has a nonzero amplitude, even for those cells near the surface. This is also true at higher temperature. In this dot, the typical amplitude of the local mode u along a given axis-direction is 0.022 (in unit of bulk lattice constant), which is comparable to that in bulk. Therefore, each cell in dots has a rather large off-center displacement. However, as shown in Fig.1, the polarizations of different cells are *not* aligned along the same direction, i.e., the dot is not ferroelectric as a whole. Instead, the displacements at the considered cross section tend to rotate from one cell to the other. The local displacements at two opposite surfaces are nearly anti-parallel. As a result, the overall polarization is close to zero, and the dot is paraelectric.

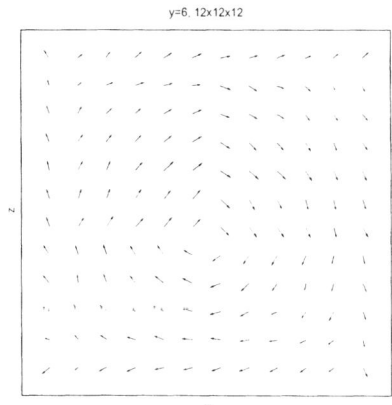

FIGURE 1. Amplitudes and directions of the local modes at the y=6 cross section of a 12x12x12 dot.

To examine what causes the very unusual mode pattern observed in Fig.1, we first investigate the surface effect. We turn off the site-site interaction between surface and vacuum, and the simulation results for the y=6 cross section of the 12x12x12 dot is given in Fig.2. The local displacements at surface in Fig.2 show less component perpendicular to the surface, compared to Fig.1. This is because the vacuum-surface interaction pulls Ti atoms towards the vacuum.[7, 8] However, the mode pattern in Fig.2 is similar to Fig.1,

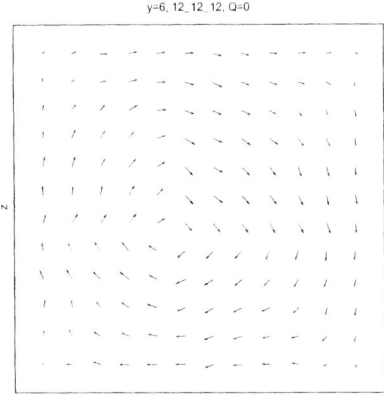

FIGURE 2. Amplitudes and directions of the local modes at the y=6 cross section of the 12x12x12 dot. The surface-vacuum interaction is turned off in this calculation.

and the tendency for local modes to rotate remains. Surface effects can thus be excluded as the major reason for the observed rotation wave of local displacements. Again, the local displacements at two opposite surfaces are nearly anti-parallel, similar to what in Fig.1.

We tentatively interpret the local-mode pattern as a result of the competition of two driving forces: the LR dipole-dipole energy and the inhomogeneous strain energy. First, let's consider the dipole interaction alone, and examine how two dipoles with *fixed* center positions will align their dipole directions. Here we refer the line connecting the centers of two dipoles as the dipole axis. The lowest-energy alignment is that both dipoles point along the dipole axis (Fig.3a). Indeed, if we examine local dipoles at the same row (or the same column) in Fig.1, the components along the dipole axis are mostly pointing the same direction. The second-lowest-energy alignment is that two dipoles point along opposite directions, both perpendicular to the dipole aixs (Fig.3b). For the local dipoles at the same row (or column) in Fig.1, we now examine their components *perpendicular* to the dipole axis. Indeed, the perpendicular components like to point at opposite directions, consistent with Fig.3b. However, the flip of perpendicular components does *not* occur within the nearest-neighbor distance. Instead, the flip occurs over the distance across the entire dot, i.e., from one side of surface to the opposite side of surface. This may be because an immediate flip would cause a large local strain and thus cost more strain energy.

Now, we investigate a rather large dot, and local-mode amplitudes and directions for a 24x24x24 dot are shown in Fig.4, where the center plane y=12 is plotted. Interestingly, local modes in the this large dots are much more ordered than those in Fig.1. The cross section in Fig.4 can be roughly divided into four domains, polarization of each domain pointing along the diagonal direction. The local modes at the center of each domain are nearly parallel. The interfaces between different domains are found to be sharp. No sizable net polarization is observed because the polarizations of different domains point along different directions. The domain structure can be seen more clearly in Fig.5,

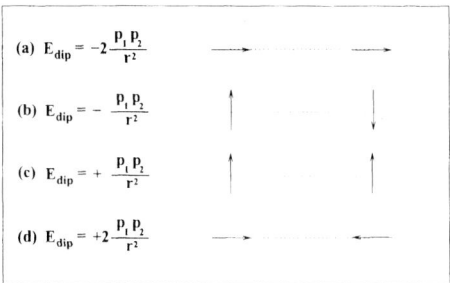

FIGURE 3. Schematic illustration of possible alignments of two dipoles in the increasing order of energy. Dash line in the figure is the dipole axis.

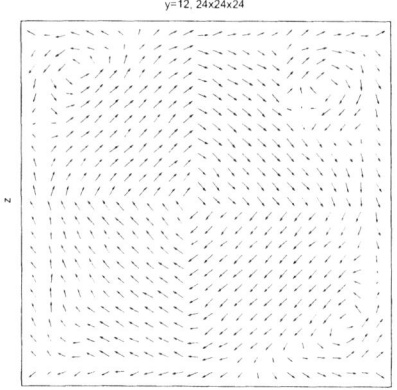

FIGURE 4. Amplitudes and directions of the local modes at the y=12 cross section of a 24x24x24 dot.

where describes, for a fixed (y,z) site, the x-axis polarization averaged over all possible sites with different x coordinates. Two separate domains are obvious, and no alloy-like mixing of positive and negative x-polarizations occur (i.e., no alloy-like mixing of "+" and "x" signs occur in Fig.5).

By contrasting the results of two different-sized dots, we may conclude that increasing the size further, which is computationally time-consuming, would form larger and more uniform domain. To verify this statement, we performed simulations for a quantum wire, mimicked by a 50x4x4 dot with a very large length along the x dimension. The result for such a wire is given in Fig.6. It can be seen that along the wire direction the local-mode directions are indeed more uniformly aligned with less rotation. Also, there exist domains in which the local-mode components perpendicular to the dipole axis change abruptly. For sufficiently large dots, we expect that polarization of each domain will point along one of the eight directions equivalent to the [111], as in the bulk rhombohedral phase.

In summary, we studied the local modes in $BaTiO_3$ quantum dots using an atomistic

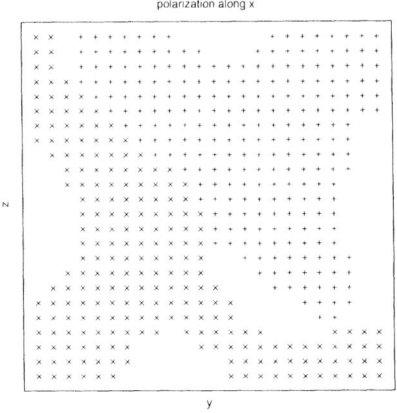

FIGURE 5. Signs of averaged x-polarizations for the 24x24x24 dot. For a given (y,z) site, the x-polarizations along a line parallel to x-axis are averaged. "+" indicates a positive x-polarization with amplitude larger than 0.015 (in unit of bulk lattice constant), and "x" indicates a negative x-polarization with amplitude larger than 0.015. No sign indicates no significant average x-polarization.

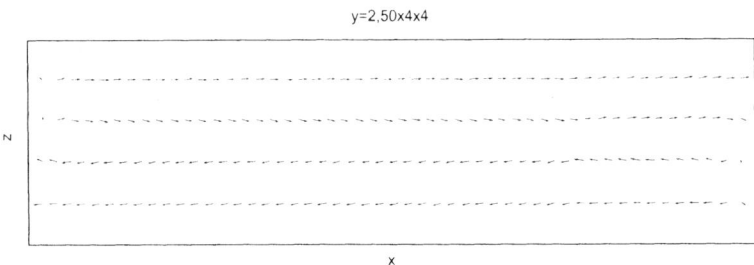

FIGURE 6. Amplitudes and directions of the local mode at the y=2 cross section of a 50x4x4 wire.

approach. The truncated long-range dipole interaction and the existence of surface are taken into account. We found that the local-mode components perpendicular to the dipole axis prefer to flip directions. However, this flip does not occur within a short distance, and instead it occur over the distance across the boundary of dots. We believe that the underlying mechanism responsible for these effects is the competition between the LR dipole-dipole interaction and the SR strain energy. More calculations and more analysis are needed to fully assess the validity of this mechanism. We also found that, as the size of dot increases, ferroelectric domain forms. In each domain, the local modes are uniformly aligned, and the polarization points along the [111]-equivalent directions.

ACKNOWLEDGMENTS

This work was supported by the Office of Naval Research Grants (N00014-01-1-0366 and N00014-01-1-0600) and by the ONR-sponsored center for piezoelectrics design (N00014-01-1-0365). The computing facilities were supported by the NSF (DMR-0116315 and DMR-9983678). We thank J. Iniguez for sending us the parameters of bulk $BaTiO_3$. We also thank I.A. Kornev for helpful discussion.

REFERENCES

1. M.E. Lines and A.M. Glass, Principles and applications of ferroelectrics and related materials (Clarendon, Oxford, 1979).
2. W. Zhong, D. Vanderbilt, and K.M. Rabe, Phys. Rev. Lett. **73**, 1861 (1994).
3. R. E. Cohen, Nature (London) **358**, 136 (1992); R. E. Cohen and H. Krakauer, Ferroelectrics **136**, 65 (1992).
4. W. Zhong, D. Vanderbilt, and K.M. Rabe, Phys. Rev. B **52**, 6301 (1995).
5. L. Bellaiche, A. Garcia, and D. Vanderbilt, Phys. Rev. Lett. **84**, 5427 (2000).
6. J. Iniguez and D. Vanderbilt, Phys. Rev. Lett. **89**, 115503 (2002).
7. J. Padilla and D. Vanderbilt, Phys. Rev. B **56**, 1625 (1997).
8. R.E. Cohen, J. Phys. Chem. Solids **57**, 1393 (1996).

First Principles Calculations of Ionic Vibrational Frequencies in $PbMg_{1/3}Nb_{2/3}O_3$

S. A. Prosandeev*[†], E. Cockayne* and B. P. Burton*

*Ceramics Division, Materials Science and Engineering Laboratory, National Institute of Standards and Technology, Gaithersburg, Maryland 20899-8520
[†]Physics Department, Rostov State University, 5 Zorge St., 344090 Rostov on Don, Russia

Abstract. Lattice dynamics for several ordered supercells with composition $PbMg_{1/3}Nb_{2/3}O_3$ (PMN) were calculated with first-principles frozen phonon methods. Nominal symmetries of the supercells studied are reduced by lattice instabilities. Lattice modes corresponding to these instabilities, equilibrium ionic positions, and simulated infrared (IR) reflectance spectra are reported.

There is ample evidence of 1:1 (NaCl-type; the "random layer model" [1]) short-range order (SRO) in PMN [2], but first principles (FP) calculations with sufficiently large supercells to realistically approximate a disordered PMN crystal with SRO are prohibitively time consuming. Relatively small supercells that might reasonably approximate the case of 1:1 SRO include the $[001]_{NCC'}$ structure [3] which was predicted to be the PMN cation-ordering ground state (CGS) [4].

One objective of this study is to fully relax different small PMN supercells consisting of 15 and 30 ions in order to determine their "displacive ground states" (DGS), and to compare their energies, dynamical charges, vibrational frequencies, and infrared (IR) reflectance spectra. These results are important to understand possible effects of local ordering on IR reflectance and Raman spectra.

A second objective to understand the nature of the soft vibrational modes in the ordered structures of PMN. It has been reported that relaxors exhibit both ferroelectric (FE) and antiferroelectric (AFE) characteristics, [5] and that competition between FE and AFE fluctuations is the cause of glass-type properties in PMN.

All FP calculations were done with the Vienna *ab initio* simulation package (VASP) [6]. Several supercells of PMN composition were considered (Fig. 1). Our FP computations show that all these structures are dynamically unstable when the ions are placed on ideal perovskite positions, and full relaxation often leads to surprisingly low symmetry. For example, the FP DGS of the $[001]_{NNM}$ structure (a $[001]_{2:1}$ superlattice) is monoclinic (type M_C [7]), with those Pb close to a Mg plane displaced in the (0.18 Å, 0.05 Å, -0.05 Å) and (0.18 Å, 0.05 Å, 0.05 Å) directions. Those Pb ions which are between the Nb planes are mostly displaced in the x direction by 0.27 Å.

Tetragonal $[001]_{NCC'}$ PMN is also dynamically unstable, and has has a wide spectrum of instabilities that are associated with FE, AFE, octahedral tilting and other modes. We relaxed the $[001]_{NCC'}$ structure after a random initial perturbation of the ions.

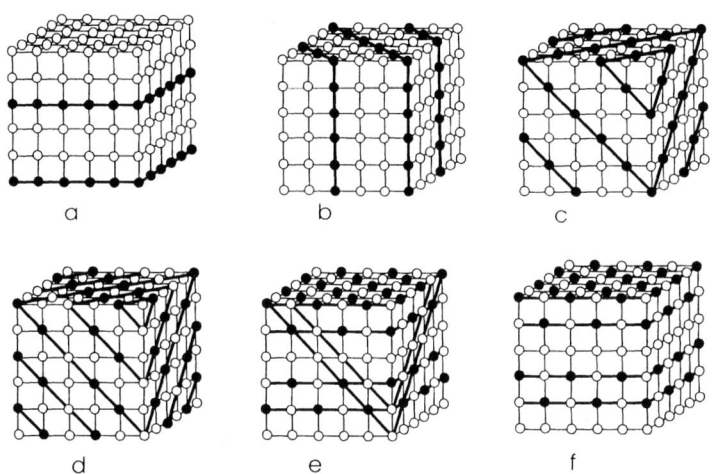

FIGURE 1. PMN structures considered in the present study: $[001]_{NNM}$ (a), $[110]_{NNM}$ (b), $[111]_{NNM}$ (c), $[111]_{NHNH'NH''}$ (d), $[001]_{NCC'NC'C}$ (e), and $[001]_{NCC'}$ (f)

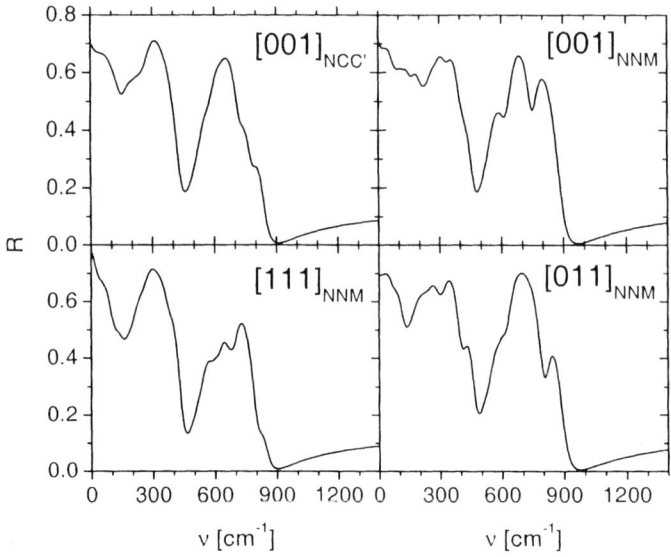

FIGURE 2. Computed IR reflectance spectra for different ordered structures of PMN

TABLE 1. Ionic coordinates (in Å) in the $[001]_{NCC'}$ structure

type	x	y	z	type	x	y	z	type	x	y	z
Mg_1	7.98	-0.02	8.16	Mg_2	4.00	-0.00	4.08	Nb_1	4.01	-0.09	8.11
Nb_2	3.95	-0.04	0.08	Nb_3	7.93	-0.02	0.00	Nb_4	3.99	-3.98	4.12
O_1	2.04	0.40	8.10	O_2	5.96	-0.31	8.04	O_3	4.40	1.98	8.12
O_4	3.66	-1.98	8.10	O_5	2.05	0.17	0.00	O_6	6.03	-0.03	12.02
O_7	4.18	2.01	12.12	O_8	4.00	-1.98	12.07	O_9	1.98	0.00	3.86
O_{10}	6.06	0.06	3.96	O_{11}	4.02	2.06	3.87	O_{12}	4.06	-2.02	3.92
O_{13}	0.27	-0.01	10.21	O_{14}	4.00	-3.92	1.94	O_{15}	0.11	-0.04	6.09
O_{16}	3.97	0.14	10.11	O_{17}	4.12	0.02	1.95	O_{18}	3.95	0.20	6.13
Pb_1	5.75	1.99	10.07	Pb_2	5.96	1.98	2.41	Pb_3	5.84	1.87	6.30
Pb_4	1.89	1.74	10.10	Pb_5	5.95	-2.04	2.40	Pb_6	2.06	1.83	6.27

Relaxed coordinates are listed in Table 1. The basis vectors (in Å) are: (4.00, -4.00, 0.00), (4.01, 4.01, 0.01), and (0.00, -0.01, 3×4.07). Relaxations of ionic coordinates are a complex mixture of octahedral deformations around Mg-ions (with rather large frozen angles) plus AFE, and FE, Pb-displacements in opposition to neighboring O-ions. The x-direction displacements differ from y-direction displacements. Pb-ions have either two, or four, Mg nearest neighbors (nn): Pb-ions with two Mg nn are displaced, by 0.25 Å, towards the centroid of the Mg-Mg nn pair. Pb-ions with four Mg nn are not displaced. We have not proven that the structure in Table 1 is the PMN $[001]_{NCC'}$ DGS (it is possible that relaxation from a different initial perturbation would lead to a state with lower energy). The reported structure is dynamically stable, however, and has lower energy than PMN $[001]_{NCC'}$ that is relaxed with only FE displacements, or only octahedral tilting.

Dynamical charges $Z^\star$ for the $[001]_{NCC'}$ structure are listed in Table 2.[8] For Nb, there is a strong correlation between the number of nn Nb ions and the dynamical charge. Nb ions that have Nb nn in both the $\pm\alpha$ directions have particularly large $Z^\star_{\alpha\alpha}$ (9.11). The two symmetry-independent cases of Nb ions that have one Nb nn in the $\pm\alpha$ directions have $Z^\star_{\alpha\alpha}$ of 6.48 and 7.87, respectively. The Nb ions that have no Nb nn in the $\pm\alpha$ directions have $Z^\star_{\alpha\alpha} = 6.00$. Dynamical charges for Pb also exhibit significant anisotropy, and environment dependence.

TABLE 2. Ionic dynamical charges (in |e|) for PMN $[100]_{NCC'}$ (tetragonal symmetry imposed).

ion i	$Z^\star_{izz}$	$Z^\star_{ixx}$	$Z^\star_{iyy}$	ion i	$Z^\star_{izz}$	$Z^\star_{ixx}$	$Z^\star_{iyy}$	ion	$Z^\star_{izz}$	$Z^\star_{ixx}$	$Z^\star_{iyy}$
Mg_1	2.74	2.65	2.65	Mg_2	2.74	2.65	2.65	Nb_1	6.48	6.00	6.00
Nb_2	7.87	9.11	9.11	Nb_3	7.87	9.11	9.11	Nb_4	6.48	6.00	6.00
O_1	-2.61	-3.62	-2.82	O_2	-2.61	-3.62	-2.82	O_3	-2.61	-2.82	-3.62
O_4	-2.61	-2.82	3.62	O_5	-1.99	-7.04	-2.09	O_6	-1.99	-7.04	-2.09
O_7	-1.99	-2.09	-7.04	O_8	-1.99	-2.09	-7.04	O_9	-2.61	-3.62	-2.82
O_{10}	-2.61	-3.62	-2.82	O_{11}	-2.61	-2.82	-3.62	O_{12}	-2.61	-3.62	-2.82
O_{13}	-4.15	-2.57	-2.57	O_{14}	-5.71	-2.42	-2.42	O_{15}	-3.93	-2.44	-2.44
O_{16}	-5.71	-2.42	-2.42	O_{17}	-4.15	-2.57	-2.57	O_{18}	-3.93	-2.44	-2.44
Pb_1	3.52	4.41	4.41	Pb_2	3.52	4.41	4.41	Pb_3	4.17	3.93	3.93
Pb_4	3.52	4.41	4.41	Pb_5	3.52	4.41	4.41	Pb_6	4.17	3.93	3.93

For the 30-ion $[001]_{NCC'}$ structure, computed diagonal frequencies of the dynamical matrix are listed in Table 3. The frequencies of Nb-, Mg- and bending O-vibrations are

TABLE 3. The diagonal frequencies of the dynamical matrix for the 30-ion $[001]_{NCC'}$ supercell of PMN (in cm^{-1}).

ion	z	x	y	ion	z	x	y	ion	z	x	y
Mg_1	346	328	340	Mg_2	276	340	344	Nb_1	279	271	290
Nb_2	322	276	270	Nb_3	318	286	251	Nb_4	301	288	283
O_1	385	537	333	O_2	262	608	251	O_3	399	331	521
O_4	256	267	648	O_5	300	640	248	O_6	263	657	257
O_7	297	272	625	O_8	281	246	633	O_9	292	577	302
O_{10}	281	612	278	O_{11}	285	278	600	O_{12}	279	299	613
O_{13}	555	264	294	O_{14}	661	267	261	O_{15}	735	214	207
O_{16}	557	241	250	O_{17}	733	233	253	O_{18}	596	244	245
Pb_1	108	88	76	Pb_2	81	78	81	Pb_3	93	81	68
Pb_4	105	76	89	Pb_5	88	79	70	Pb_6	92	67	76

all between 200 cm^{-1} and 350 cm^{-1}. The stretching O-vibrations along Mg-O and Nb-O bonds range from 520 cm^{-1} to 735 cm^{-1}. The Pb diagonal frequencies are in the range from 67 cm^{-1} to 108 cm^{-1}.

As in experiment [9], computed IR reflectance spectra (Fig. 2) consist of three main reststrahlen bands of the vibrational modes that are typical of perovskites: the first group is below 120 cm^{-1}; the second spreads from 150 cm^{-1} to 400 cm^{-1}; and the third is from 500 cm^{-1} to 800 cm^{-1}. The two lower bands split into two subbands each. Assignments of these bands can be made on the basis of the diagonal frequencies shown in Table 3. In experimental data [9], these groups of lines are rather compact as in the computed $[001]_{NCC'}$ and $[111]_{MNN}$ structures. However, the experimentally determined magnitude of the reflectivity in the interval from 500 cm^{-1} to 800 cm^{-1} is lower than in the computation. This could be connected with an overestimation of the oxygen dynamical charge and/or with large damping for some frequencies in this interval (an estimated damping constant of 60 cm^{-1} was used for all frequencies), or it could be that the systems studied here are not sufficiently representative of SRO-disordered PMN.

The lowest calculated optical frequency in the equilibrium 30-ion $[001]_{NCC'}$ structure is 24 cm^{-1}. It is lower than the lowest Pb diagonal frequency (60 cm^{-1}) shown in Table 3 because of the interaction among the ionic vibrations. This mode is basically a perovskite acoustic mode, but has some infrared oscillator strength due to the superlattice Mg-Nb arrangement, which folds certain non-zone center acoustic modes to zone-center IR-active modes. Pb-dominated modes are spread over the interval from 24 cm^{-1} to 129 cm^{-1}. Mostly ferroelectric Pb displacements are at 40 cm^{-1} and 60 cm^{-1} to 90 cm^{-1} although a significant FE contribution to the vibrations exists in the whole interval from 24 cm^{-1} to 129 cm^{-1}. Some modes in this interval are dominated by the displacement of a specific Pb ion (are "quasilocal"). Similarly, some modes in the frequency interval from 500 cm^{-1} to 800 cm^{-1} are dominated by oxygen displacement along a specific Nb-O-Mg bond.

Raman spectra show broad lines with gradual temperature dependence in a wide temperature interval [10, 11]. The presence of these lines would be forbidden if the ions were in symmetric environments. Ionic displacements due to disorder and symmetry breaking can explain the existence of these Raman lines. A possible measure of the intensities of these lines is the square of the projection of the ionic displacements, from symmetric positions, onto the vibrational modes: $S_i = |\langle \mathbf{v}_d | \mathbf{v}_i \rangle|^2$, where $\mathbf{v}_d$ is the vector of the frozen displacements, and $\mathbf{v}_i$ is the vector of the i-th vibration in the displacements' representation.

Low-frequency Raman lines (about 50 cm^{-1}) are almost certainly due to Pb-O stretching modes [11], and may be associated with the "quasilocal" vibrations described above. Note that, as is typical of Pb-based perovskites that also have large B-cations (e.g Mg), Pb-vibrational branches have relatively small dispersion and (for reference structures with Pb ions at ideal perovskite positions) are unstable across most of the Brillouin zone. The particular instabilities that freeze in should depend sensitively on the local electric fields produced by the Nb-Mg configuration. Freezing of lattice instabilities creates low-symmetry Pb-sites, and also allows AFE Pb-vibrations to couple with FE vibrations.

Structural instabilities (which exist in all the supercells studied) also imply that displacive relaxations will reduce their energies relative to the values reported in Burton and Cockayne [4] Column one in Table 4 gives the energies of each structures with symmetry restrictions imposed as in [4] (for example, tetragonal symmetry was imposed for $[001]_{NNM}$ and $[001]_{NCC'}$). Column two gives the relaxed (DGS) energy for each structure. Remarkably, the hierarchy of formation energies and predicted CGS ($[001]_{NCC'}$) remain the same.

There are many possible ordered derivatives of the random layer model [1] that have (111) Nb-layers which alternate with (111) ($Mg_{2/3}Nb_{1/3}$)-layers in which Nb and Mg are ordered; e.g. Fig. 1d. Depending on the projection axis, the mixed layers in this structure are ordered either in stripes, or in an Mg-honeycomb pattern with Nb's at hexagon centers. This structure has a high energy relative to the others in Fig. 1. Perhaps cations in the $Mg_{2/3}Nb_{1/3}$ layers would prefer to have more neighbors of the same species. This is consistent with the conclusion of Hoatson et al.[12] who obtained an improved inverse Monte Carlo fit to NMR data with the assumption of Mg-Mg and Nb-Nb clustering within mixed layers. The structure shown in Fig. 1e also has higher energy than $[001]_{NCC'}$.

TABLE 4. The supercell's energy (in eV per 30-ion supercell)

	ideal structure	relaxed structure
$[001]_{NNM}$	0.825	0.583
$[110]_{NNM}$	0.710	0.155
$[111]_{NNM}$	0.696	0.150
$[001]_{NCC'}$	0.523	0

In summary, our FP computations have shown that the relaxation of the ionic coordinates in the small supercells of PMN does not change the hierarchy of the energy of these structures: $[001]_{NCC'}$ remains the CGS among the small supercells considered. The lowest frequency vibrations in this structure are acoustic modes, but which have nonzero infrared oscillator strengths due to superlattice Mg-Nb ordering. The next-lowest modes are connected with mixed FE-AFE Pb vibrations and with Pb-O stretching modes. The computed IR reflectance spectrum, qualitatively, corresponds to experimental data [9], although there are some discrepancies in the line magnitudes at high frequencies.

S.A.P. appreciates discussions with J. Toulouse, O. Svitelskiy, J. Petzelt, S. Kamba and Yu. Yuzyuk.

REFERENCES

1. P. K. Davies and M. A. Akbas, J. Phys. Chem. Sol. **61**, 159 (2000).
2. H.B. Krause, J.M. Cowley and J. Wheatley, Acta. Cryst. **A35**, 1015 (1979).
3. All of the structures in this paper are layered structures and can thus be described using the notation $[xyz]_{AB...}$, where xyz is the stacking direction and $AB...$ the stacking sequence. In the stacking sequences, M refers to a pure Mg layer, N to a pure Nb layer, C to a $Mg_{1/2}Nb_{1/2}$ "checkerboard" layer, and H to a $Mg_{2/3}Nb_{1/3}$ "honeycomb" layer. Where more than one stacking of the same type of layer is possible, primes and double primes distinguish among these stackings.
4. B. P. Burton and E. Cockayne, Ferroelectrics **270**, 173 (2002).
5. T. Egami, Ferroelectrics **267**, 101 (2002).
6. G. Kresse and J. Hafner, Phys. Rev. **B47**, 558 (1993).
7. M_C refers to polarization in the $[ab0]$ direction. See, *e.g.*, figure 2 in L. Bellaiche, A. Garcia, and D. Vanderbilt, Phys. Rev. B **64**, 060103 (2002).
8. To reduce the computational cost associated with the very low symmetry of the DGS, the dynamical charges were calculated with respect to a PMN supercell in which tetragonal symmetry was maintained. The dynamical charges of the atoms will change slightly upon relaxation to the positions in Table 1, and some ions which are symmetry equivalent in Table 2 are not symmetry equivalent in the DGS (Table 1). We assume that the differences between the dynamical charges of the two cases will not significantly affect the calculated IR spectrum in this case.
9. I. M. Reaney, J. Petzelt, V. V. Voitsekhovskii, F. Chu, and N. Setter, J. Appl. Phys. **76**, 2086 (1994).
10. V. I. Torgashev, Yu. I. Yuzyuk, L. T. Latush, P. N. Timonin, and R. Farhi, Ferroelectrics **199**, 197 (1997).
11. E. Husson, L. Abello, and A. Morell, Mat. Res. Bull. **25**, 539 (1990).
12. G.L. Hoatson, D.H. Zhou, F. Fayon, D. Massiot, and R.L. Vold, Phys. Rev. **B66** 224103 (2002).

E-Field and Temperature Dependent Transformation in <102>-Cut PMN-PT Crystal

Chi-Shun Tu[1,*], L.-W. Huang[1], R. Chien[2], and V. Hugo Schmidt[2]

[1]*Department of Physics, Fu Jen University, Taipei, Taiwan 242, Republic of China*
[2]*Department of Physics, Montana State University, Bozeman, MT 59717, USA*

Abstract. Dielectric permittivity and domain structures have been measured as functions of temperature, frequency and E-field on a <102>-cut PMN-31%PT single crystal. By using relations of optical indicatrices and extinction, the unpoled PMN-31%PT crystal was evidenced to possess a rhombohedral phase at room temperature. As temperature increases, the extinction angles don't exhibit apparent change until ~350 K, indicating that the crystal keeps a mostly rhombohedral phase below ~350 K. Above ~350 K, the extinction angles begin to rotate through the monoclinic phase and then reach the tetragonal phase for most domains near 400 K. Near 410 K, the crystal transforms into the isotropic cubic phase. In addition, E-field (along the [102] direction) dependent domain structures have also been observed at room temperature. The optical indicatrices of the domain matrix begin to rotate near E= 3.25 kV/cm through monoclinic polarizations. However, the crystal doesn't exhibit total extinction even under an electric field of E= 37 kV/cm, indicating that a field-induced single domain has not been reached. The domain structures show a strong hysteresis behavior during processes of increasing and decreasing E-field.

INTRODUCTION

Monoclinic (M) and orthorhombic (O) phases have been evidenced in relaxor-based ferroelectric (FE) crystals $(PbMg_{1/3}Nb_{2/3}O_3)_{1-x}(PbTiO_3)_x$ (PMN-xPT) and $(PbZn_{1/3}Nb_{2/3}O_3)_{1-x}(PbTiO_3)_x$ (PZN-xPT), which depend on PT concentration, temperature range, history, strength of external E-field and crystallographic orientation [1-13]. In the PZN-PT system, the O phase was found in a narrow range 9%≤x≤10% [5]. PZN-9%PT even shows the O phase without a prior E-field poling [5].

By neutron analysis, a M_C (Z=1)-type (space group Pm) M phase was evidenced in unpoled PMN-35%PT powder samples at 80 K [10]. More recently, a M_C-type M phase has been found in unpoled ceramics of PMN-xPT for 37%≥ x ≥31% [11]. Various phase coexistences of R/M_C, T/M_C and $T/M_C/O$ at both 20 K and 300 K were proposed for 31%, 33% and 35% PT compositions respectively [11]. It is important to note that the effect of M_A-type polarization rotation (from <111> to <001>) will be largely cancelled in randomly oriented ceramics [12]. It was recently found that phase transformation from R to cubic phase is through polarization rotation of M domains (perhaps M_A-type) as temperature increases [13].

These phenomena open a new window in relaxor-based ferroelectrics. One may ask whether the intermediate phases (M or O) or various phase coexistences are intrinsic, or are merely due to spatial phase segregation. On the other hand, the earlier given morphotropic phase boundary (MPB) (between R and T phases) needs to be reconsidered [14]. In this report, both dielectric permittivity and domain structures were investigated on a <102>-cut PMN-31%PT, because the <102>-cut orientation would be most sensitive direction for observing movement of M domains.

EXPERIMENT

The lead magnesium niobate-lead titanate crystal PMN-31%PT was grown using a modified Bridgman method. The sample was cut perpendicular to the <102> direction. A JEOL6100 electron microscope was used to determine concentrations of local A- and B-site ions. A variable-frequency Wayne-Kerr Precision Analyzer PMA3260A with four-lead connections was used to obtain dielectric permittivity with gold electrodes deposited by R.F. sputtering. The domain structures were observed by using a Nikon E600POL polarizing microscope with a crossed polarizer-analyzer pair. Transparent conductive films of ITO (Indium Tin Oxide) were deposited on sample surfaces for E-field dependent measurements. The thickness of sample is near 40 μm. The experimental configuration for domain observation is illustrated in Fig. 1.

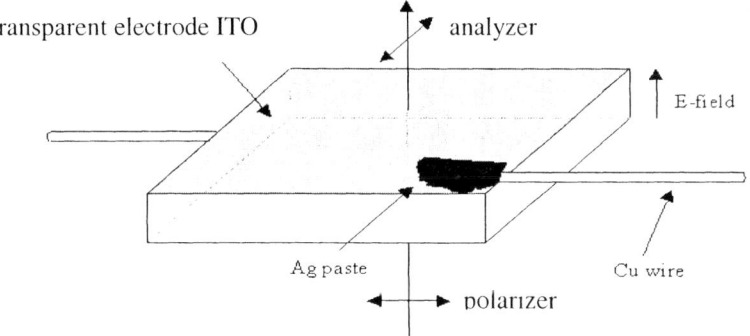

FIGURE 1. Experimental configuration for domain structures.

RESULTS AND DISCUSSIONS

Fig. 2 shows temperature- and frequency-dependent dielectric permittivity which exhibits a maximum near 410 K and an extra bump near 380 K superimposed on the broad background. A broad thermal hysteresis was seen in the region of 250-410 K. Temperature and E-field dependent domain structures are shown in Figs. 3 and 4 respectively. The "C" region in Figs. 3 and 4 indicates a crack caused by the polishing process. The accuracy of determining optical extinction angle is about ±3°.

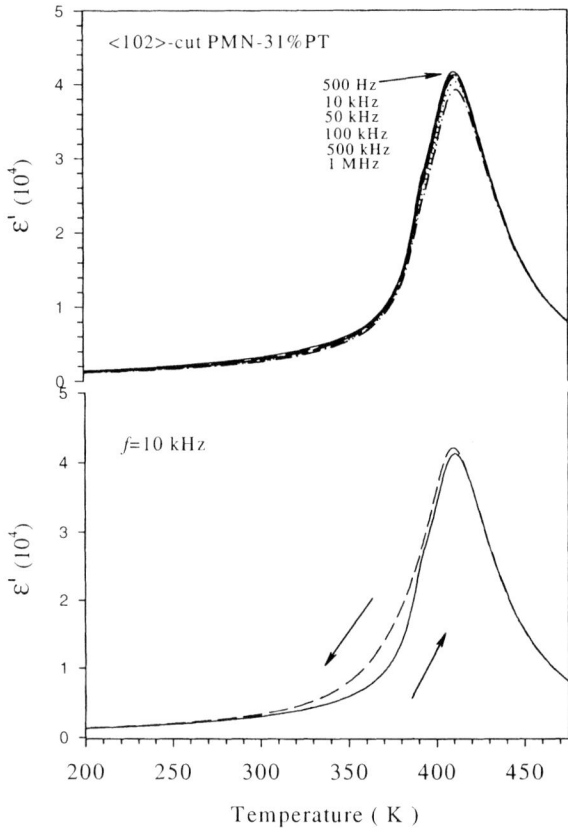

FIGURE 2. Temperature- and frequency-dependent dielectric permittivity.

Fig. 5 shows the relations among various phases and corresponding polarizations viewed along the [102] direction. Squares indicate the directions of tetragonal polarization vectors **P**. Triangles indicate directions for rhombohedral **P**'s. Circles indicate directions for orthorhombic **P**'s. Dot-dashed and dashed lines indicate directions that polarizations can take for monoclinic cells based on the double-size (Z=2) orthorhombic cell. Solid and dotted lines alternate between rectangles and circles, indicating directions that polarizations can take for monoclinic cells based on the simple (Z=1) cubic, tetragonal, or rhombohedral cells. Any polarization whose direction does not correspond to one of the 3 symbol types or 4 types of lines results from a triclinic-shaped cell. This indicates that most higher-symmetry phases (*O*, *R*, or *T*) with nearby polarization directions are related directly by monoclinic phases.

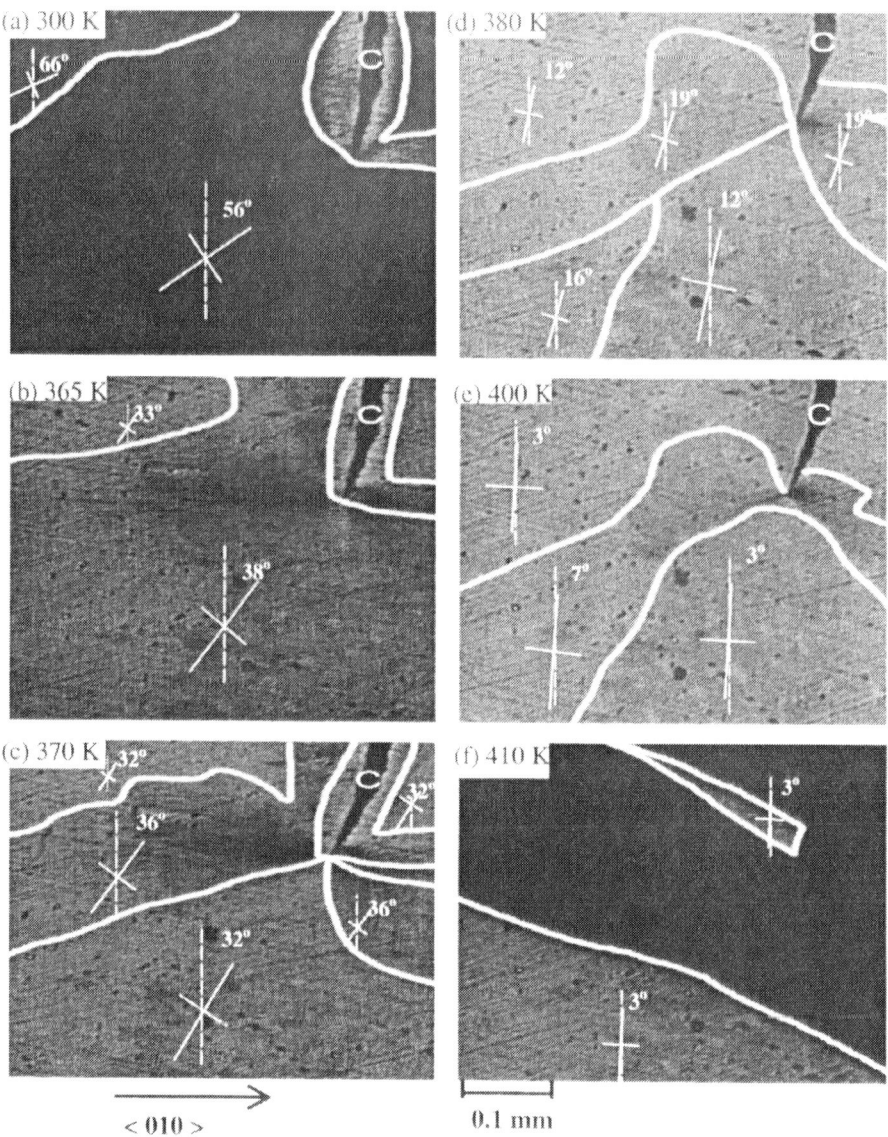

FIGURE 3. Temperature-dependent domain structures.

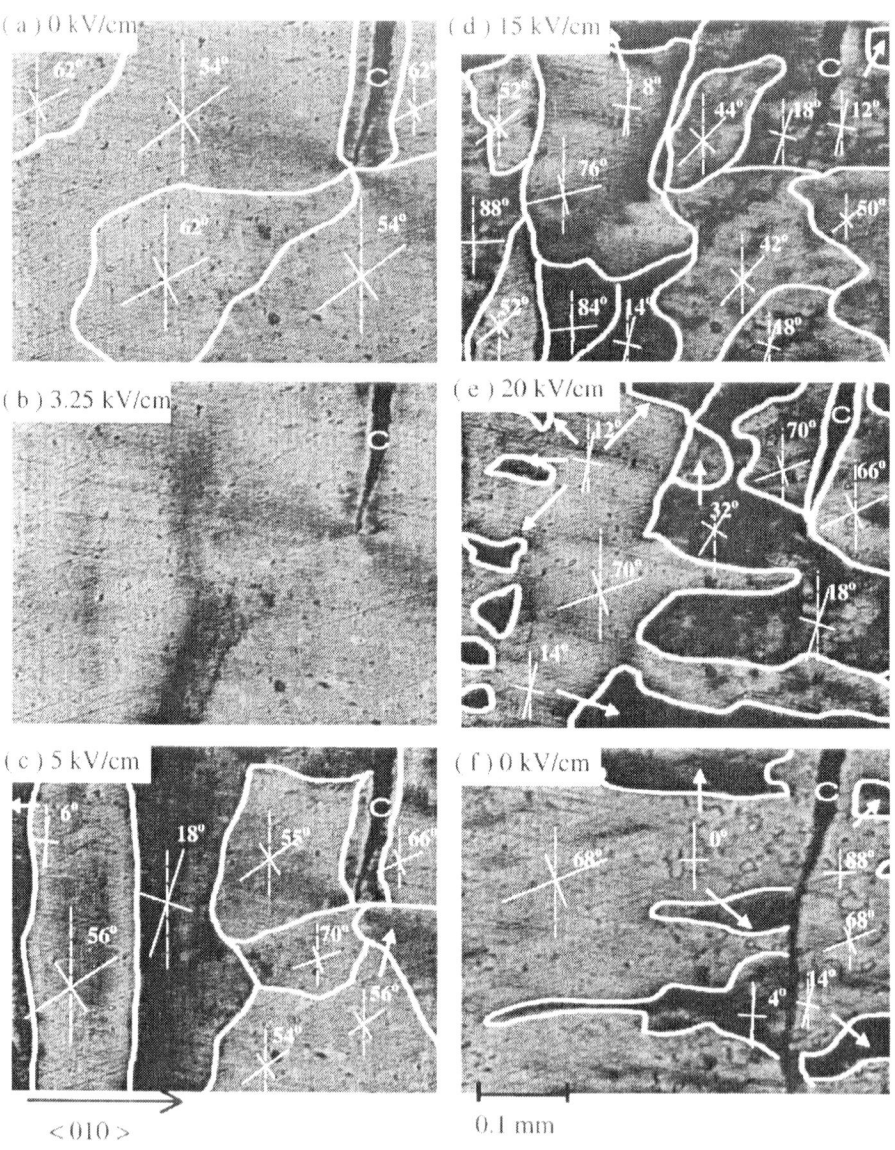

FIGURE 4. E-field dependent domain structures at room temperature.

The M_C cell **P** lies between two adjacent T and O **P** vectors (solid and dotted line). The M_A cell has **P** between two adjacent T and R **P** vectors (dashed lines), whereas the M_B cell has **P** between two adjacent O and R **P** vectors (dash-dotted lines). Solid lines between some symbols indicate no shift in extinction directions away from those in symbols connected by these lines, which represent some of the Z=1 Mc cell polarization directions.

Domains that are optically inactive will have extinction for optical electric field along the radial and circumferential axes indicated by solid crossed lines inside the symbols. *O* domains that have incomplete extinction due to some reasons, are given by open circles. Fig. 6 presents a projection along the optical propagation direction [102] with extinction angles between various *R* and *T* domains.

As shown in Fig. 3(a), the domain matrix at 300 K exhibits two main orientations of optical indicatrices with extinction angles (with regard to the <010> axis) at 56° and 66°, respectively. When observing the [102]-cut sample along [102] between a crossed polarizer-analyzer pair, as shown in Fig. 6, the allowed optical extinction angles for *R* domains are 24°, 37°, 53° or 66° measured from the <010> axis. For instance, the extinction angle between [010] and [111] (or [$\bar{1}$ 1 $\bar{1}$]) is 24° or 66°. Thus, domains are mostly rhombohedral at room temperature, perhaps mixed with monoclinic domains.

Orientations of optical indicatrices, as illustrated in Fig. 3(b-d), exhibit apparent changes in the region of 360-380 K where the dielectric permittivity ε' (Fig. 2) shows a plunge accompanied by a slight frequency dispersion. As temperature increases, orientations of optical indicatrices exhibit gradual evolution up to angles of 3° or 7° near 400 K. Then, as seen in Fig. 3(f), the cubic phase (which corresponds to the black region) begins to appear near 410 K and soon entirely occupies the entire crystal. It is consistent with T_m~410 K which corresponds to the maximum of ε' (see Fig. 2).

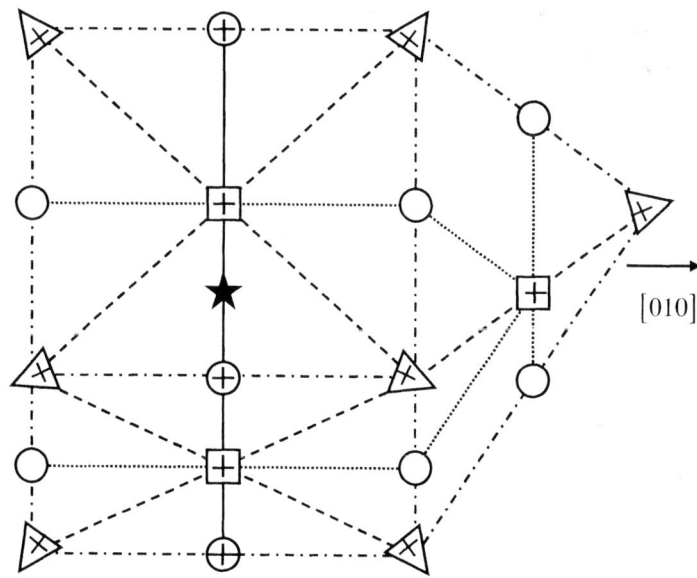

FIGURE 5. Relations among various phases and corresponding polarizations viewed along the [102] direction.

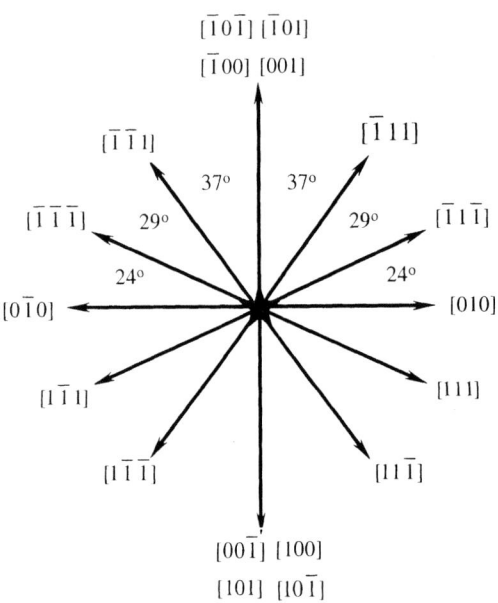

FIGURE 6. Projection of extinction angles along the optical propagation direction [102] for various domains.

What is the phase in the region of ~360-410 K? We see no evidence for O domains at 300 K. As we pointed out in the previous result [13], the O phase seen for prior E-field on a <011>-cut PMN-33%PT crystal possibly occurs only because that cut favors the O phase when E is applied along <011> [8]. Thus, the O phase was not expected for the <102>-cut crystal. However, as indicated by the blank circles in Fig. 5, we cannot determine if O domains exist as temperature increases.

Comparing extinction angles in Fig. 3 with Fig. 5, it is reasonable to believe that the crystal undergoes a distortion from R domains toward T domains through M phases as temperature increases above 360 K. The T phase seems to appear near 400 K [Fig. 3(e)], because most orientations of optical indicatrices of domains turn into 3°. In other words, optical extinction angles of domains near 400 K are almost parallel with <010> axes which are the polar axes of T domains, *i.e.* [010] or [100]. Ideally, the extinction should be zero degrees for T domains.

For E-field dependent domain structures, as shown in Fig. 4, orientations of optical indicatrices of domains begin with extinction angles 54° and 62° which correspond to R domains under E=0 kV/cm, and then exhibit apparent rotations (through M domains) in the region of 3.25-5 kV/cm which is close to the coercive field E_c~5 kV/cm of other PMN-PT crystals [2]. Above E=20 kV/cm, the domain matrix doesn't show much change with increasing E-field up to 37 kV/cm. The crystal doesn't exhibit total optical extinction even at E=37 kV/cm, possibly because [102] is not a preferred direction for R domain at room temperature. It is important to note that domain structures of the

<111>-cut PMN-33%PT showed total extinction under E=12 kV/cm applied along the <111> direction [13]. In addition, the <102>-cut PMN-31%PT crystal didn't reenter its previous domain pattern after E-field was removed, indicating a strong E-field hysteresis.

In conclusion, temperature dependent behavior shows that domains of the <102>-cut PMN-31%PT crystal develop from R (possibly mixed with M) domains toward T domains through rotation of polarizations of M phases. However, with increasing E-field up to 37 kV/cm, the domain structures didn't reach single domain and total optical extinction.

ACKNOWLEDGMENTS

This work was supported by NSC Grant 91-2112-M-030-006 and DoD EPSCoR Grant N00014-02-1-0657.

REFERENCES

1. Z.-G. Ye and M. Dong, *J. Appl. Phys.* **87**, 2312 (2000).
2. C.-S. Tu, C.-L. Tsai, J.-S. Chen, and V.H. Schmidt, *Phys. Rev.* B **65**, 104113 (2002).
3. M.K. Durbin, E.W. Jacobs, J.C. Hicks, and S.-E. Park, *Appl. Phys. Lett.* **74**, 2848 (1999).
4. D. Viehland, *J. Appl. Phys.* **88**, 4794 (2000).
5. D. La-Orauttapong, B. Noheda, Z.-G. Ye, P.M. Gehring, J. Toulouse, D.E. Cox, and G. Shirane, *Phys. Rev.* B **65**, 144101 (2002)
6. H. Fu and R.E. Cohen, *Nature* **403**, 281 (2000).
7. G. Xu, H. Luo, H. Xu, and Z. Yin, *Phys. Rev.* B **64**, 020102 (R) (2001).
8. Y. Lu, D.-Y. Jeong, Z.-Y. Cheng, and Q. M. Zhang, H. Luo, Z. Yin, and D. Viehland, *Appl. Phys. Lett.* **78**, 3109 (2001).
9. Z.-G. Ye, B. Noheda, M. Dong, D. Cox, and G. Shirane, *Phys. Rev.* B **64**, 184114 (2001).
10. J.-M. Kiat, Y. Uesu, B. Dkhil, M. Matsuda, C. Malibert, and G. Calvarin, *Phys. Rev.* B **65**, 064106 (2002).
11. B. Noheda, D.E. Cox, G. Shirane, J. Gao, and Z.-G. Ye, *Phys. Rev.* B **66**, 054104 (2002).
12. H. Fu and R.E. Cohen, *Nature* **403**, 281 (2000).
13. C.-S. Tu, V.H. Schmidt, I.-C. Shih, and R. Chien, *Phys. Rev.* B **67**, 020102 (R) (2003).
14. J. Kuwata, K. Uchino, and S. Nomura, *Ferroelectrics* **37**, 579 (1981).

* Email address: *phys1008@mails.fju.edu.tw*

Polarization Rotation and Monoclinic Phase in Relaxor Ferroelectric PMN-PT Crystal

V. Hugo Schmidt,[1*] R. Chien,[1] I.-C. Shih,[2] and Chi-Shun Tu[2]

[1]Department of Physics, Montana State Univ., Bozeman, MT 59717, USA
[2]Department of Physics, Fu Jen University, Taipei, Taiwan 242, R.O.C.

Abstract. A monoclinic phase is evidenced between rhombohedral and cubic phases in a <111>-cut single crystal PMN-33%PT from polarizing microscope observation of domain structures. Some rules for interpreting these observations for various cuts are reviewed, to illustrate how particular cuts are useful for distinguishing among particular types of phases and domains. Near 360 K the structure begins to distort from the rhombohedral toward the tetragonal phase through monoclinic domains (probably M_A-type but perhaps M_B-type). However, the present <111>-cut crystal seems to disfavor the tetragonal phase and persists in the monoclinic phase up to T~420 K, where the cubic phase begins to develop. Temperature-dependent orientations of optical indicatrices of domains indicate polarization rotations within the monoclinic planes. In addition, a previous electric-field-cooled process enhances long-range order in domain patterns observed upon heating. At room temperature, with increasing E-field along [111], the crystal undergoes successive phase transformations via polarization rotations (perhaps M_A-type), i.e. rhombohedral → tetragonal (associated with 90°-domain walls) → [111] rhombohedral. As E-field decreases, very different domain patterns appear (as compared with those with increasing field) which indicate a strong thermal hysteresis relation between E-field and strain.

Polarization rotation in high-strain relaxor crystals such as PMN-PT and PZN-PT occurs quite easily in response to temperature, pressure, and electric field changes, and probably polarization direction is influenced by crystal cut. This rotation can occur stepwise as the crystal transforms among the 4mm tetragonal (T), mm2 orthorhombic (O), and 3m rhombohedral (R) phases, or from one domain type to another within one of these phases. Continuous polarization rotation can occur between any pair of the above phases by means of monoclinic (M) phases. In principle, any other polarization directions could occur for a triclinic phase, though such a phase has not yet been reported in these crystals. Thus, including also the high-temperature parent cubic (C) phase, these crystals could exist in six of the seven crystal systems! We present here a way of displaying these possible structures and polarizations with respect to their experimental analysis by means of polarizing microscopy, and illustrate the method with experimental results for PMN-33%PT.

In polarizing microscopy the crystal can be rotated between two crossed linear polarizers. There are two key directions, the light propagation vector **k** (perpendicular to the crystal cut plane) and the optical electric field axis **E** of the incident light. The relation of **E** to the projection of the ferroelectric polarization vector **P** onto the plane perpendicular to **k** is important for interpreting the results. We display this relation by means of Figures that aid in visualizing the possible **P** vectors and their projections onto this plane. Each Figure consists of the projection along **k** of a cube aligned with the parent-phase cubic cell axes, with symbols representing where **P** crosses the visible faces of the cube. The origin of **P** is at the center of the cube. Some faces with sides parallel to **k** are shown folded out somewhat.

Simplest, and thus shown first, is the (001)-cut projection in Fig. 1, with all four sides folded out. The inner square outlines the front face of the cube. The triangles at its corners represent four R-phase **P** directions, and the circles at its edge midpoints represent four O-phase **P** directions. The square symbol at its center represents a T-phase domain whose **P** is parallel to **k**. This symbol is shown in solid black to indicate that extinction occurs for all **E** directions because **E** is perpendicular to the optic axis for this optically uniaxial domain.

The solid crosses in the other 24 symbols indicate the **E** axes for which extinction will occur for the corresponding domains. Extinction occurs if **E** is along a principal axis of the ellipse formed by the intersection of the plane perpendicular to **k** with the index ellipsoid. The index ellipsoid is a surface whose distance from the origin in a given direction is proportional to the index of refraction n for light polarized with **E** in that direction. The R and T domains are optically uniaxial, so the index ellipsoid is an ellipsoid of revolution.

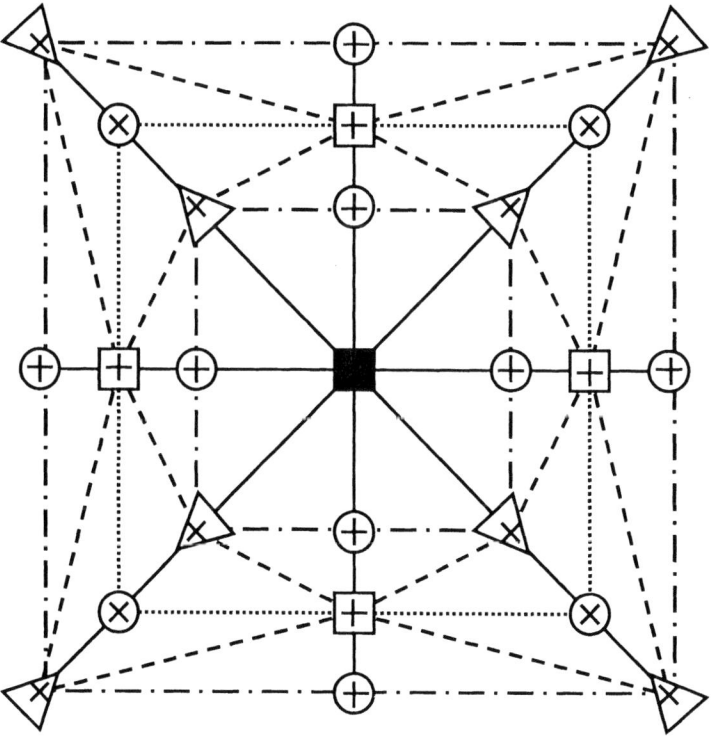

FIGURE 1. Relation between polarizations and various phases projected on (001) plane.

For the R and T domains, the optic axis is along **P**. For such domains in the (001)-cut crystal, the intersection of the (001) plane with the index ellipsoid forms an ellipse that has principal axes parallel and perpendicular to the projection of **P** onto the (001) plane. Extinction for these R and T domains occurs for **E** along the principal axis directions of these ellipses, as indicated by the directions of the lines forming the crosses inside the triangle and square symbols in Fig. 1. A concise and clear mathematical analysis for the general extinction problem appears in Sommerfeld [1]. A detailed treatment was published by Hartshorne and Stuart [2].

The O domains are optically biaxial. The optic axis directions are arbitrary in a certain plane and are of no practical use in polarizing microscopy for these crystals. Of importance are the principal axes of the index ellipsoid, which for O domains align with the axes of the double-size (Z=2) O unit cell. One of these axes aligns with **P** and makes a 45° angle with two cubic axes, another axis is the third cubic axis, and the third is perpendicular to the other two. For instance, for any of the four O domains shown in Fig. 1 having **P** represented by circles near the four corners, the unit cell projection is a rectangle (almost a square) with edges canted 45° relative to the page edges. In the third direction, the cell edges are perpendicular to the page and are about $\sqrt{2}$ shorter, that is, they coincide closely with the length of the original primitive cubic unit cell.

For such biaxial domains, under what conditions will the index ellipsoid intersection with the plane perpendicular to **k** be an ellipse with a principal axis along the projection of **P**, so that extinction will occur with **E** along this projection of **P**? There are two independent sufficient conditions. Case 1 is if **P** lies in this plane. Case 2 is if an index ellipsoid principal axis is perpendicular to **E** when **E** is aligned with the projection of **P**. For the (001) cut crystal represented by Fig. 1, all 12 O symbols satisfy Case 2. The 4 in the central plane (circles near the corners) satisfy Case 1 also, so the extinction directions for all 12 O domains are known to be along the axes shown in the O symbols in Fig. 1.

For the (001) cut we see that T domains will give extinctions at $0°+m90°$, where m is any integer. We will measure all angles in Figs. 1-3 from the vertical direction, and will henceforth omit m90° terms and only consider angles in the $0° \leq \phi < 90°$ range. We see that R domains only give extinctions at 45°, whereas O domains give extinctions at 0° and 45°. Thus the (001) cut is useful for distinguishing situations for which only R or only T domains exist, but a mixture of R and T domains could give the same extinction angles as a crystal with only O domains. We also see the importance of knowing which direction is which in the plane perpendicular to **k**.

All monoclinic phases for M domains that may exist in perovskite crystals belong to the m point group, but only the M_C phase has a Z=1 unit cell based on the primitive cubic unit cell. Any M_C cell **P** lies between two adjacent T and O **P** vectors (dotted lines in Fig. 1). The M_A and M_B "phases" in our opinion should be called a single phase whose cell is based on the Z=2 orthorhombic cell. The important distinction is that the M_A cell has **P** between two adjacent T and R **P** vectors (dashed lines), whereas the M_B cell has **P** between two adjacent R and O **P** vectors (dash-dot lines). If the plane of any set of three adjacent **P** vectors described above includes **k**, extinction occurs for the monoclinic phase for **E** in the same directions as for the adjacent T, R, or O phases, and the lines representing these monoclinic **P** directions are shown solid to indicate that the extinction directions are known. Otherwise, the extinction directions are not known, and complete extinction does not occur for any **E** direction because two of the three index ellipsoid principal axes change direction with wavelength. This situation occurs for the O domains shown as open circles in Fig. 2 for a (110) crystal cut.

Another origin of incomplete extinction in the m monoclinic and mm2 orthorhombic point groups is the optical activity that occurs unless **k** lies in a mirror plane. O domains that have incomplete extinction only because of optical activity are shown by circles with open crosses in Fig. 3 for a (111) crystal cut. These satisfy Case 1 above, whereas the O domains represented by the other three circles satisfy Case 2.

From Figs. 1-3 we see advantages for each of these crystal cuts. The (001) cut of Fig. 1 clearly distinguishes R from T phases by the 45° extinction angle difference. The (110) cut of Fig. 2 unambiguously distinguishes certain R domains by a special 35.3° extinction angle relative to T, some O, and other R domains. The (111) cut of Fig. 3 has the special feature that any extinctions at angles other than 0°, 30°, 60°, *etc.* must be from monoclinic or possibly triclinic domains. Our observation of extinctions at such other angles is interpreted by us as indicating monoclinic domains at intermediate temperatures in (111)-cut PMN-33%PT, as discussed below.

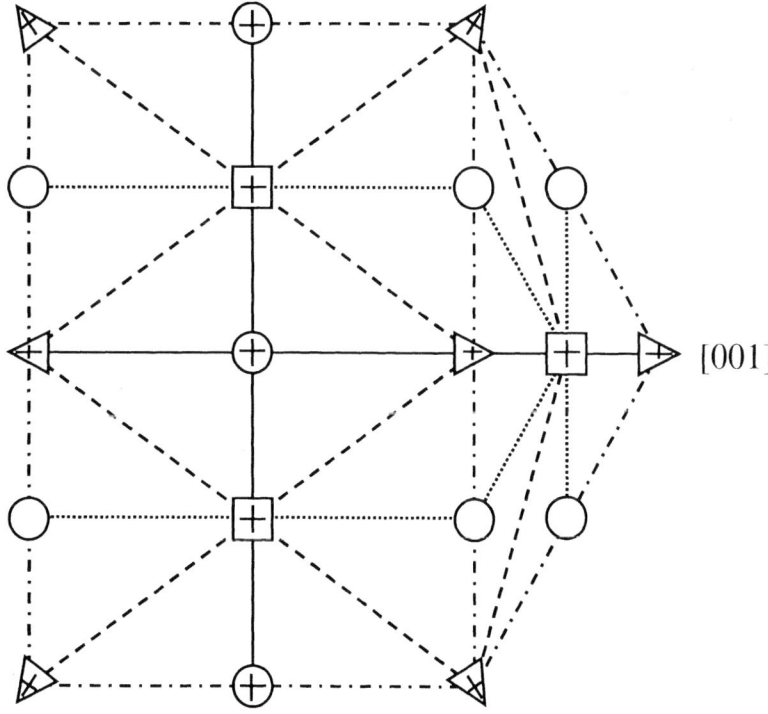

FIGURE 2. Relation between polarizations and various phases projected on (110) plane.

One indication of this is shown in Fig. 4, in which the extinctions at 0° and 30° seen at 275 K and attributed by us to R domains change to extinctions at 15° as shown at 420 K and attributed by us to monoclinic domains. Note that angles indicated in various domains in Figs. 4 and 5 are angles at which extinctions occur. These Figures are photographs taken at a polarizer angle of 0°, so only domains marked 0° in these Figures appear black in the photographs. From Fig. 3, the extinction angle change to 15° could occur for either M_A or M_B domain polarizations represented by lines leading away from R-domain triangles on the periphery of the Figure. At 424 K the crystal goes into the cubic m3m phase without having gone through the T phase, as evidenced by extinction for all angles above 424 K.

Another indication is from polarizing microscope pictures taken while dc fields of various strengths were applied along [111] as shown in Fig. 5. As field increases, the R domains represented by triangles on the periphery of Fig. 3 change to the R domain represented by the central triangle in Fig. 3, which gives extinction for all angles. Upon decreasing field, extinction at 15° occurs for some crystal regions, indicating a monoclinic phase. Details about the experimental method, and some results (but not results while field is applied), appear elsewhere [3].

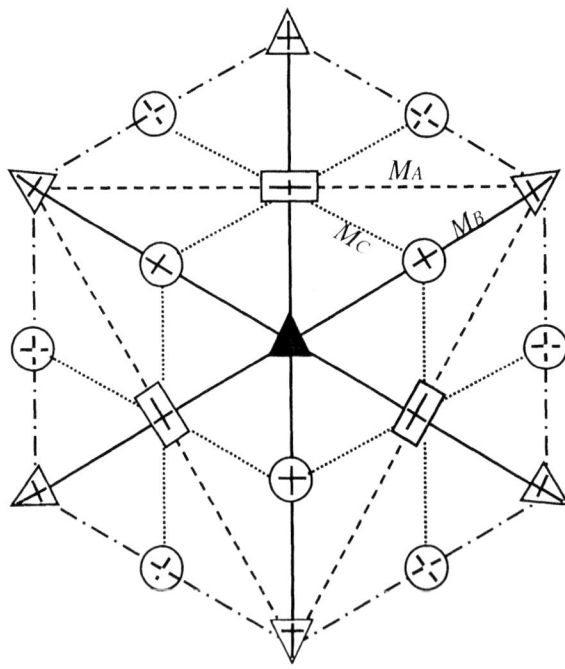

FIGURE 3. Relation between polarizations and various phases projected on (111) plane.

More work is needed to further explain this behavior. In particular, are the M domains single phases, or are they composed of microdomains of higher–symmetry phases, as suggested by Viehland [4], that on average have monoclinic symmetry? Such microdomains could be too small to observe with visible light. Another question is reconciliation of this result with the phase diagram by Noheda *et al.* [5], which for our composition suggests for increasing temperature the phase sequence M to T to C. Possible origins of this difference are differences in sample treatment (pulverization and oxygen loss, sample cut, thermal and electrical history) in the two experiments.

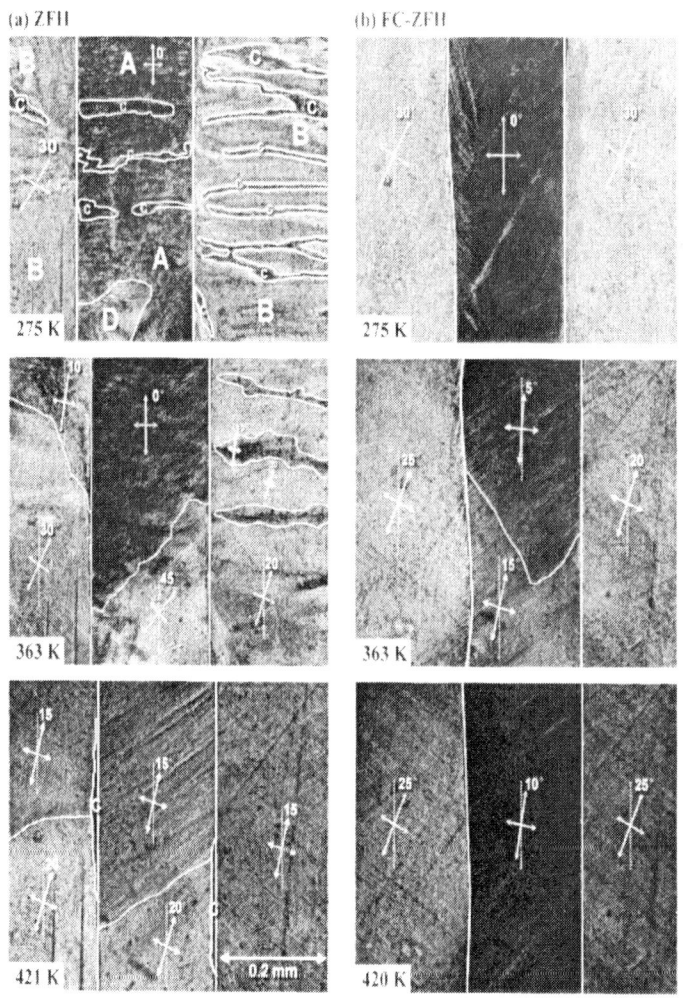

FIGURE 4. Polarizing microscope results of a (111)-cut PMN-33%PT crystal for various temperatures with and without prior field cooling

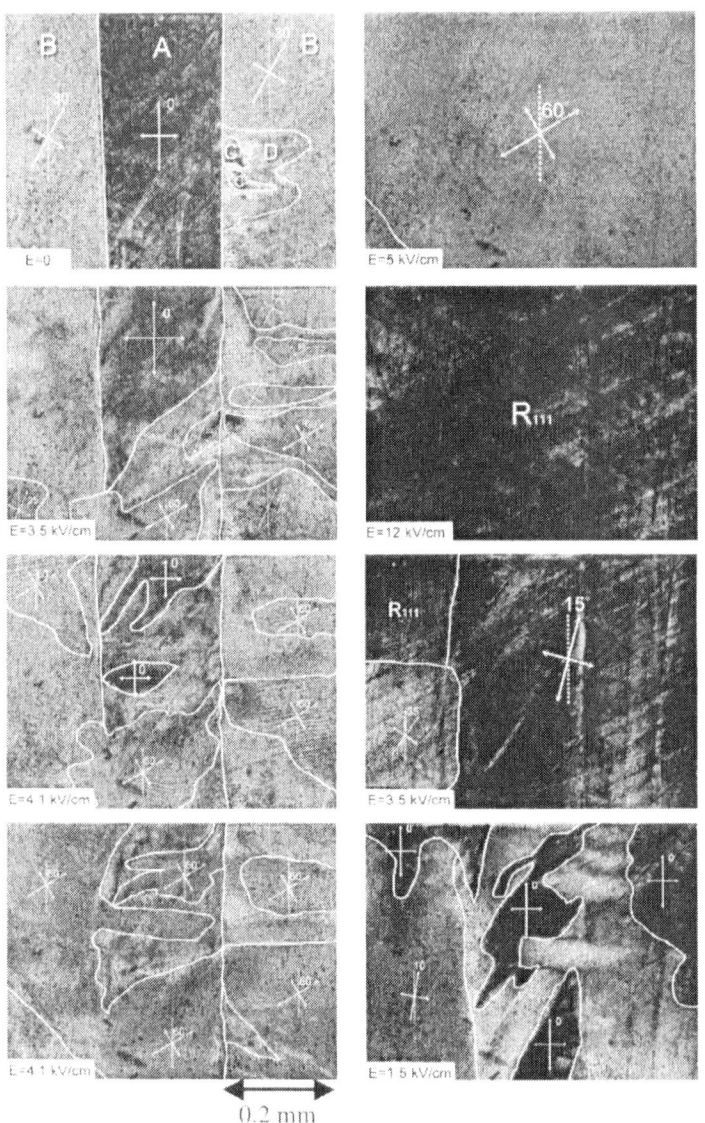

FIGURE 5. Polarizing microscope results at 300 K during applications of dc electric fields of various strengths along [111] for a (111)-cut PMN-33%PT crystal.

ACKNOWLEDGMENT

This work was supported by DoD EPSCoR Grants N00014-99-1-0523 and N00014-02-1-0657.

REFERENCES

1. A. Sommerfeld, *Optics*, Academic Press, New York, 1964, pp. 129-139.
2. N.H. Hartshorne and A. Stuart, *Crystals and the Polarising Microscope*, E. Arnold Ltd., London, 1970.
3. C.-S. Tu, V.H. Schmidt, I.-C. Shih, and R. Chien, *Phys. Rev. B* **67**, 020102(R) (2003).
4. D. Viehland, private communication.
5. B. Noheda, D.E. Cox, G. Shirane, J. Gao, and Z.-G. Ye, *Phys. Rev. B* **66**, 054104 (2002).

References contained in Refs. 3 and 5 provide background material on PMN-PT crystals.

*Email address: schmidt@physics.montana.edu

$PbTiO_3$ at Finite Temperature: An Ab-initio Molecular Dynamics Study

V. Srinivasan[1], R. Gebauer[1,2], R. Resta[1,3,4], and R. Car[1]

(1) Department of Chemistry and Princeton Materials Institute,
Princeton University, Princeton, NJ 08544

(2) Abdus Salam International Centre for Theoretical Physics (ICTP), I-34014 Trieste, Italy

(3) INFM DEMOCRITOS National Simulation Center, via Beirut 2, I-34014 Trieste, Italy

(4) Dipartimento di Fisica Teorica, Università di Trieste,
Strada Costiera 11, I-34014 Trieste, Italy

Abstract. $PbTiO_3$ is a prototypical ferroelectric material that exhibits a single structural phase transition (cubic to tetragonal): it is a soft mode driven, predominantly displacive, transition. In this paper, we study the behavior of $PbTiO_3$ at finite temperature by ab-initio molecular dynamics simulations. In this approach classical mechanics is used to describe nuclear dynamics, while the interatomic potential is generated on the fly from the ground state of the electrons within density functional theory. Fluctuations of volume and shape of the simulation cell are included by means of Parrinello-Rahman constant pressure scheme. Extensive convergence studies based on static calculations indicate that a 3×3×3 supercell containing 135 atoms, with a single k-point sampling, is sufficient to represent accurately the T = 0 energetics of this material. Although computationally demanding, ab-initio molecular dynamics simulations for $PbTiO_3$ using a 3×3×3 cell are feasible with current computational methodologies. Here we report preliminary results of simulations that are both below and above the phase-transition temperature. We discuss, in particular, how phonon softening occurs with temperature and how thermal expansion affects the results.

I INTRODUCTION

For many years the theory of structural phase transitions in ferroelectric (FE) materials has been the domain of phenomenological or empirical models, whose microscopic foundations are somewhat arbitrary, and whose basic mechanisms cannot be assessed. The first "modern" paper addressing FE phase transitions appeared less than a decade ago [1], followed by a few other papers in subsequent years [2–5].

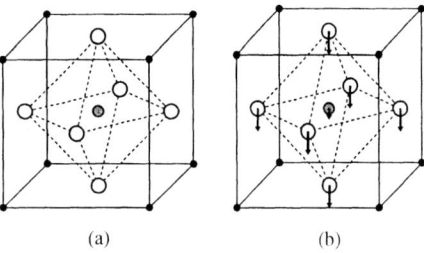

FIGURE 1. Crystal structure of PbTiO$_3$: solid, shaded, and empty circles represent Pb, Ti, and O atoms, respectively. (a): Cubic structure above T$_c$. (b): Tetragonal FE structure below T$_c$. The arrows indicate the *actual* magnitude of the atomic displacements, where the origin has been kept at the Pb site: at this scale, the Ti displacement and the macroscopic strain ($c/a = 1.06$) are barely visible.

Actually, all of these studies of FE phase transitions are *not* based on full-fledged ab-initio computations; instead, they are based on either model Hamiltonians or force fields, describing the nuclear system, and whose parameters are fitted to a set of frozen-nuclei ab-initio electronic structure computations. Once the nuclear model Hamiltonian obtained, this is used in a purely classical simulations at finite temperature: either classical Monte Carlo (MC) [1,2] or molecular dynamics [3-5]. It is remarkable that the experimental phase-transition sequence is reproduced in all the cases studied so far, although the calculated transition temperatures have large errors. The error sources are basically two: the simplifications intrinsic to the model Hamiltonian approach, and the shortcomings of the density functional used in the ab-initio computations. In most cases, the local-density approximation was adopted. To date, there is not a clear assessment of the relative importance of the two errors.

We are performing the very first full-fledged ab-initio Car-Parrinello molecular dynamics (CP) simulations [6] for a FE material, choosing the Purdue-Burke-Emzerhof [7] exchange-correlation functional. The clear advantage of such simulations over the effective Hamiltonian method is that the fully anharmonic potential is used for the dynamics. In the latter method results depend crucially on the correct identification of representative local modes while in the CP simulations no such identification is necessary. As a case study, we have chosen PbTiO$_3$, a relatively simple material, displaying only one phase transition, at T$_c$ = 763 K, to a tetragonal FE phase. This is characterized by a small uniaxial strain (6%), accompanied by microscopic displacements of the ions out of their high-symmetry sites: the structure is shown in Fig. 1. The phase transition is soft-mode driven, and is predominantly displacive.

Size and length of the simulation are crucial issues in our work, which is at the edge of present computational feasibility. In this preliminary report we present results for the thermal expansion, as obtained from a 135-atom simulation, and for the FE transition, as obtained from a 40-atom simulation. In the former case we get results in good agreement with the experiment, and we present reasons for trusting convergence with respect to both size and length of the simulation. In the latter case the results, although encouraging, are only semiquantitative due to a clear lack of size convergence. Further simulations for the 135-atom system are in progress.

II CALCULATIONS

We perform variable-cell CP simulations at constant pressure and temperature, using the Parrinello-Rahman scheme [8] and a Nosé-Hoover thermostat [9]. We adopt a pseudopotential approach, where the pseudopotentials are chosen as ultrasoft [10] for Ti and O, and norm-conserving for Pb. The semicore electrons of Pb ($5d^{10}$) and Ti ($3s^2 3p^6 3d^2$) are explicitly dealt with in the electronic structure calculations (44 electrons per 5-atom cell). A kinetic-energy cut-off of 40 Ry is used for the plane-wave basis set and of 200 to 240 Ry for the charge density. The Γ-point-only Brillouin-zone sampling is used throughout.

Since the CP simulations are rather computer-demanding, we have preliminarily performed some convergence tests by means of T = 0 conventional "total energy" calculations, with a 5-atom cell and Brillouin-zone sampling of increasing accuracy. As a check of internal consistency, we have also performed T = 0 Γ-point-only calculations for the 2×2×2 and 3×3×3 supercells, explicitly verifying that the results agree with those of the 5-atom cell within the appropriate sampling (k-point folding). These results are summarized in Table 1.

	tetragonal			cubic	
	a (Å)	c/a	d (Å)	a (Å)	ΔU_{c-t} (eV)
8 k points	3.89	1.10	0.54	3.95	0.87
2×2×2 (Γ-point)	3.88	1.09	0.49	3.94	0.83
27 k points	3.89	1.08	0.45	3.93	0.32
3×3×3 (Γ-point)	3.89	1.07	0.42	3.93	0.32
64 k points	3.90	1.07	0.44	3.93	0.25
Experiment	3.90	1.06	0.47	3.97 (763 K)	

TABLE 1. Computed structural parameters at T = 0 as a function of the Brillouin-zone sampling. The parameter d is displacement of the O octahedron along the tetragonal axis. The last column gives the energy difference between the cubic and tetragonal phase, in eV per 5-atom cell. Experimental values from Ref. [11].

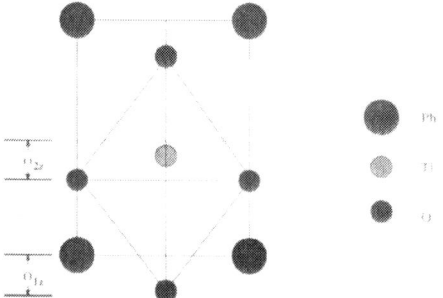

FIGURE 2. Coordinates used in defining our order parameter, Eq. (1). At variance with Fig. 1, for the sake of clarity here the displacements and macroscopic strain are exaggerated and *not* drawn to scale.

In our finite-temperature CP simulations, size effects enter in two distinct ways: the supercell size introduces spurious spatial correlations in the thermal fluctuations of the nuclear coordinates, while the Γ-point sampling introduces errors in the computed ground electronic state. The latter errors are fairly small for the $3\times3\times3$ supercell, as demonstrated by the $T=0$ results.

We perform CP simulations over $2\times2\times2$ supercells (40 atoms, 352 electrons) and $3\times3\times3$ (135 atoms, 1188 electrons). The latter ones are particularly demanding, and run at the Pittsburgh Supercomputing Center using 64 processors and taking about 90sec/CP step. We find that in order obtain good averages we need at least 10 psec long simulations. This means a trajectory of 40,000 CP time steps ($\Delta t = 10$ a.u.). The $3\times3\times3$ results reported here are only preliminary; other simulation runs are in progress at the time of writing.

As already stated, this material is predominantly soft-mode: therefore the phase transition can be studied by monitoring the temperature dependence of the soft mode. To this aim, we follow the approach of Fabris et al. [12], who addressed zirconia within an empirical model-potential scheme and a 96-atom supercell. What makes the method appealing is the quality of the results obtained with a supercell of moderate size: we have therefore implemented the method within our CP simulation. One needs to define an order parameter, which we have chosen as the average displacement of the oxygen sublattices with respect to the lead sublattice along the tetragonal axis. There are two oxygen planes in a unit cell and the instantaneous order parameter may then be expressed as:

$$\delta(t) = \frac{1}{2}(O_{1z} + O_{2z}), \qquad (1)$$

where O_{1z} and O_{2z} are the displacements of the two oxygen planes from their respective positions in the perfect perovskite structure (see Fig. 2) averaged over the supercell.

The order parameter is calculated from the instantaneous positions during the simulations, and its time-derivative is used to compute the velocity-velocity correlation function:

$$S(t) = \langle \dot{\delta}(t)\dot{\delta}(0) \rangle, \qquad (2)$$

where we average over the time evolution. The Fourier transform $S(\nu)$ of Eq. (2) is then plotted and the soft-mode peak is identified by its temperature dependence.

III RESULTS AND DISCUSSION

First we address our thermal-expansion results, as obtained from both 2×2×2 and 3×3×3 supercells. A plot of the temperature dependence of the lattice parameters in given in Fig 3. While the 3×3×3 results show good agreement with the experimental trend [13], the smaller supercell shows lack of convergence by displaying a qualitatively incorrect trend: the tetragonality (c/a) of the cell increases. Our finding of the correct thermal expansion gives confidence that the 3×3×3 size is large enough to accommodate the relevant correlation lengths, while the previously discussed T = 0 results also show that the Brillouin-zone-sampling error is negligible.

Next, we address the microscopic quantities: nuclear displacements and order parameter. We measure them in reduced units (fractions of the cell parameters).

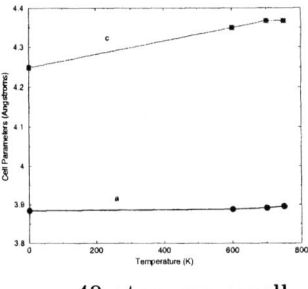

40-atom supercell

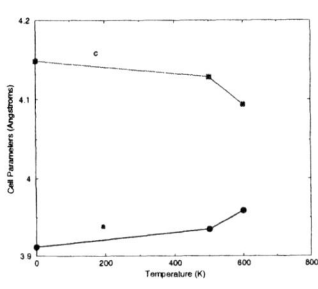

135-atom supercell

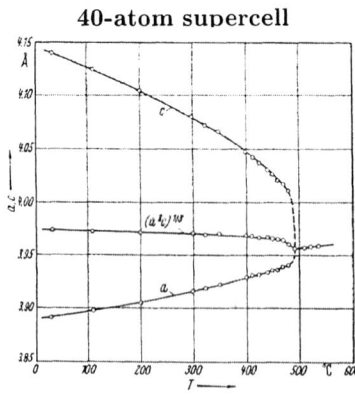

Experiment

FIGURE 3. Temperature dependence of the cell parameters from variable-cell CP simulations on 40-atom and 135-atom supercells and that from experiment. Note that the experimental plot is in degrees centigrade and starts from 0° C.

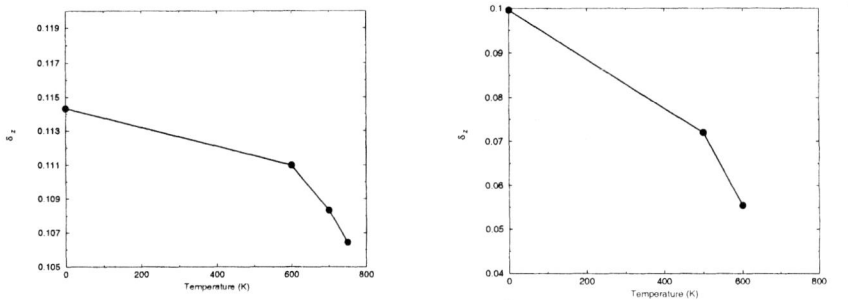

FIGURE 4. Temperature dependence of the order parameter. Left panel: 40-atom simulations. Right panel: 135-atom simulations.

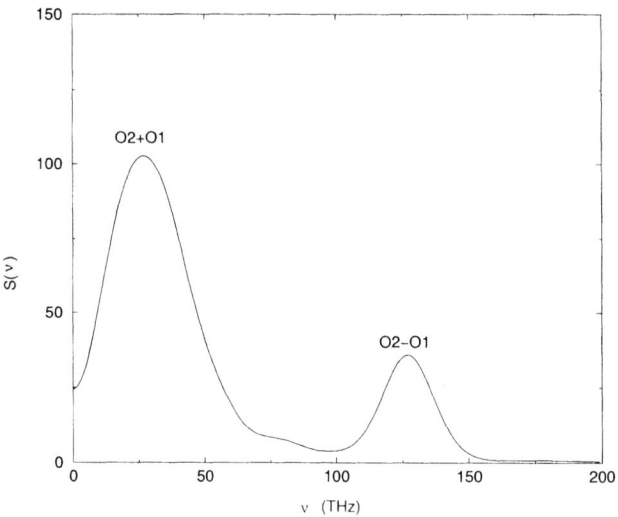

FIGURE 5. Velocity-velocity correlation function $S(\nu)$ from the 40-atom simulation.

In Fig. 4 we display the thermal behaviour of the order parameter obtained for both supercell sizes. Notice that the order parameter approaches zero (i.e. its value in the cubic phase) at a *lower* temperature, of the order of 100 K or possibly more, in the 3×3×3 supercell than in the 2×2×2 one.

The velocity-velocity correlation function for the 2×2×2 simulation at 700 K is shown in Fig. 5. The very perspicuous low-energy peak is present at all our simulation temperatures, while its position shifts dramatically with temperature, going soft. We therefore assign the lowest-frequency peak to the displacement of the O sublattices as a whole against the Pb atoms (soft mode), while we assign the other structures present in $S(\nu)$ to stretching between different O planes. In a second-order purely displacive transition the square of the soft-mode frequency

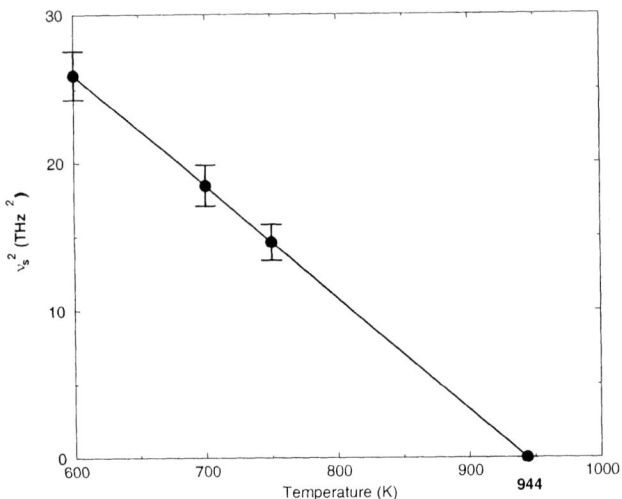

FIGURE 6. Square of the soft-mode frequency from three simulations, and their extrapolation to the transition temperature, for a 40-atom sample. Notice that, despite this size is insufficiently converged, the behavior is very accurately linear.

goes linearly with temperature near the transition temperature [14]: extrapolation to zero provides $T_c = 944$ K as shown in Fig. 6. Since the actual transition is weakly first-order—and not purely second-order—the calculated value, as obtained by linear extrapolation, is expected to overestimate (by a fairly small amount) the actual T_c value. Much more important, however, is the size effect. We have shown above that a 2×2×2 supercell is too small to correctly account for thermal fluctuations in this system, while we have also shown that 3×3×3 is expected to be a reasonable size. The calculations on the larger system are presently in progress; however, looking at Fig 4, it is easy to predict that the calculated value of T_c for the 3×3×3 supercell will be significantly lower than the present value of 944 K, thus hopefully improving the agreement with the experimental value of 763 K. So far, the only theoretical value is the one for Ref. [2], who report a value of 660 K from a model-Hamiltonian approach based on local-density functional first-principle calculations.

In conclusion, we have presented the preliminary results of the very first ab-initio molecular dynamics study of a ferroelectric material. By addressing the thermal expansion and the transition temperature of $PbTiO_3$, we have given evidence that the workload, although heavy, is within the capability of todays' supercomputers. The results discussed here are very encouraging; further simulations are running at the time of writing.

Work partially supported by the Office of Naval Research through grant N00014-01-1-0407.

REFERENCES

1. W. Zhong, D. Vanderbilt, and K. Rabe, Phys. Rev. Lett. **73**, 1861 (1994); Phys. Rev. B **52**, 6301 (1995).
2. U. V. Waghmare and K. M. Rabe, Phys. Rev. B **55**, 6161 (1997).
3. H. Krakauer, R. C. Yu, C. Z. Wang, and C. LaSota, Ferroelectrics **206**, 133 (1998).
4. H. Krakauer, R. C. Yu, C. Z. Wang, K. M. Rabe, and U. V. Waghmare, J. Phys.: Condens. Matter **11**, 3779 (1999).
5. S. Tinte, M. G. Stachiotti, M. Sepliarsky, R. L. Migoni, and C. O. Rodriguez, J. Phys.: Condens. Matter **11**, 9679 (1999).
6. R. Car and M. Parrinello, Phys. Rev. Lett. **55**, 2471 (1985).
7. J. P. Perdew, K. Burke and M. Emzerhof, Phys. Rev. Lett. **77**, 3865 (1996).
8. M. Parrinello and A. Rahman, J. Appl. Phys **52**, 7182 (1981).
9. S. Nosé, J. Chem. Phys. **81**, 511 (1984).
10. D. Vanderbilt, Phys. Rev. B **41**, 7892 (1990).
11. G. Shirane, R. Pepinsky, and B. C. Frazer, Acta Crystallogr. **9**, 131 (1956).
12. S. Fabris, A. T. Paxton and M. W. Finnis, Phys. Rev. B **63** 094101 (2001).
13. G. Shirane, and S. Hoshino, T. Phys. Soc. Japan **6**, 265 (1951).
14. A. D. Bruce, Adv. Phys. **29**, 117 (1980).

Ferroelectric instabilities and self-consistent mechanism for the isotopic substitution in KDP

S. Koval*, J. Kohanoff[†], R. L. Migoni* and E. Tosatti**[‡]

*Instituto de Física Rosario, Universidad Nacional de Rosario, 27 de Febrero 210 Bis, 2000 Rosario, Argentina
[†]Atomistic Simulation Group, The Queen's University, Belfast BT7 1NN, United Kingdom
**International Centre for Theoretical Physics, Strada Costiera 11, I-34014 Trieste, Italy
[‡]International School for Advanced Studies (SISSA), and Istituto Nazionale Fisica della Materia (INFM/DEMOCRITOS), Via Beirut 4, I-34014 Trieste, Italy

Abstract.
We performed *ab initio* calculations to study ferroelectric (FE) instabilities and isotope effects in the H-bonded ferroelectric KH_2PO_4 (KDP). We demonstrate that the source of the FE instability is the hydrogen off-centering. This ordering, produces an electronic charge redistribution within the PO_4 tetrahedral units, which polarize along *c*. Cluster distortions following the H off-centered relaxation pattern in a mean-field paraelectric (PE) phase, lead to instabilities which are significant only when the heavy ions P and K are also allowed to relax. Subsequent quantization in small clusters, leads to tunneling only for distortions including heavy ions relaxations. This explains the H double-site occupancy observed experimentally in the PE phase, and is also in agreement with the P-atom multi-site distribution detected experimentally in deuterated KDP (DKDP). Mass changes due to deuteration at fixed structural parameters cannot account for the huge isotope effect. However, the main effect of deuteration is a depletion of the proton probability density at the O-H-O center, which in turn weakens the proton-mediated covalency in the bridge. A lattice expansion follows then, which is coupled self-consistently with the proton off-centering. This self-consistent mechanism is illustrated with a non-linear model deduced from the *ab initio* calculations, and allows us to explain the huge isotope effect observed and the importance of geometrical effects proved by high-pressure experiments.

INTRODUCTION

Potassium dihydrogen phosphate (KH_2PO_4, or KDP) crystals are a key component widely used in optoelectronic devices. Besides the obvious technological interest, KDP is also interesting from a fundamental point of view. KDP is a prototype ferroelectric (FE) crystal belonging to the family of hydrogen-bonded ferroelectrics extensively studied in the past.[1] Their molecular units are linked by hydrogen bonds, and ferroelectricity appears to be connected to the behavior of the protons in these H-bonds. A characteristic feature of this family is the large increase in the Curie temperature T_c upon deuteration (from $\simeq$ 122 K in KDP to $\simeq$ 229 K in DKDP). The origin of this huge isotope effect is still controversial, and has been mostly analysed in terms of the quantum tunneling model proposed in the early sixties by Blinc,[2] modified by the inclusion of the coupling between protons and the K-PO_4 dynamics.[3] These models have been validated *a posteriori* on the basis of their predictions, but direct experimental indications of tunneling are only very recent.[4] In addition, the connection between tunneling and iso-

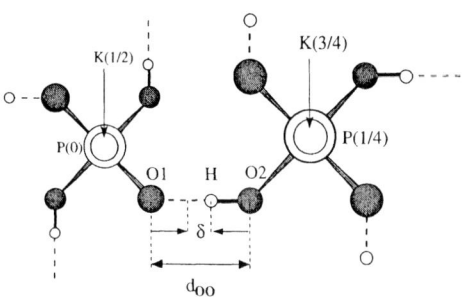

FIGURE 1. Schematic view of the internal structure of KDP along the tetragonal axis. P and K atoms are aligned (coordinates relative to c are indicated in brackets). Covalent and H-bonded hydrogens are attached to corresponding oxygens by full and broken lines, respectively.

tope effects remains unclear. On the contrary, there is experimental evidence,[5, 6, 7, 8] that the geometrical modification of the hydrogen bonds and the lattice parameters upon deuteration (Ubbelohde effect [9]) is intimately connected to the mechanism of the phase transition. The distance δ between the two collective equilibrium positions of the protons (see Fig. 1) was shown to be remarkably correlated with T_c.[8] These findings stimulated new theoretical work where virtually the same phenomenology could be explained without invoking tunneling.[10].

It is the purpose of this work to investigate, using electronic structure calculations within the Density Functional Theory (DFT), the aspects of proton ordering, tunnneling and isotope effects in KDP. To this aim we study the dependence of the system energetics and charge distributions upon the proton position in the H-bonds under various conditions: allowing and forbidding K and P ions displacements, and considering global and small cluster distortions. The next quantization of local motions will allow us to study nuclei quantum effects.

The paraelectric (PE) structure of KDP is body-centered tetragonal with two KH_2PO_4 units per lattice site. We use the conventional *bct* cell doubled along the tetragonal c axis (64 atoms) to describe homogeneous distortions, and a conventional *fct* cell doubled along the c-axis (128 atoms) for local distortions. The internal structure is depicted in Fig. 1. Above T_c the protons occupy with equal probability two equivalent off-center positions in the H-bond,[11] while below T_c they order in such a way that each PO_4 group has two covalently bonded and two H-bonded hydrogen atoms.

AB INITIO METHODS

We use two different types of pseudopotential DFT approaches: a method employing a basis set of confined pseudoatomic orbitals (SIESTA),[12] and a plane wave (PPW) method.[13] For the first approach we choose a double-zeta basis set with polarization functions, and an orbital confinement energy $E_c = 50$ meV. In the PPW method, we set the energy cutoff to 150 Ry. In both cases exchange-correlation terms are computed using a gradient-corrected (GGA) functional,[14] and norm-conserving pseudopotentials [15] are employed to represent the interaction between ionic cores and valence electrons. We also include nonlinear core corrections for a proper description of the K ion. The electronic Brillouin zone sampling reduces to the Γ-point, which proved sufficient for the large supercells used.

PROTON ORDERING AND POLARIZATION

We first analyse the relation between proton ordering and polarization. We start from the average experimental structure of the KDP PE phase at T_c+5 K,[11] which has the hydrogens centered in the O-H-O bonds (TS phase). We make this choice, instead of the optimized structure,[16] because minimal differences which appear in the calculations are capable to spoil actual instabilities related to the H positions in the bonds, which are extremely sensitive to the d_{OO} distance (see Fig. 1). By relaxing the atomic positions with fixed tetragonal lattice, the H atoms move collectively off-center towards the O2 oxygens, and away from O1, as indicated in Fig. 1. This is the FE phase with tetragonal structure (FP_t). We plot in Fig. 2 the charge density difference $\Delta\rho(\mathbf{r}) = \rho_{FP_t}(\mathbf{r}) - \rho_{TS}(\mathbf{r})$ between both phases, in the planes determined by the atoms P-O1$\cdots$H (Fig. 2a) and P-O2-H (Fig. 2b). We observe that the collective H motion induces a charge delocalization from O2 towards the O2-H bond, which increases its covalency, while the O1-H bond weakens becoming a hydrogen-bond. In addition, there is charge flow from the O2-P to the O1-P bond. This picture also emerges from the analysis of Mulliken populations.[16, 17] This charge redistribution is reflected in an increase of the P-O2 and a decrease of the P-O1 distances. The P-atoms are thus driven off-center in the PO_4 tetrahedra, which further polarizes, and unbalanced electrostatic forces induce a displacement of the K^+ ions along the c-axis, towards the charge-excess (O1) side of PO_4 units.[16]

In order to identify the driving mechanism of the ferroelectric instability, we investigate the *ab initio* potential energy surface (PES) as a function of the proton off-centering parameter $\delta = d_{OO} - 2d_{OH}$, and the K-P relative displacement along c, quantified as $\gamma = d_{PP} - 2d_{KP}$, which is shown to be a measure of polarization.[18] We fully relax the oxygen positions for each chosen (δ, γ) pair, thus obtaining a two-dimensional double-well collective PES with a saddle point at $\delta = \gamma = 0$, whose contours are represented in Fig. 3. According to this PES, the crystal is stable against polarization ($\gamma \neq 0$) unless the protons are ordered off-center. Cuts of this PES at different values of γ indicate that, even for vanishing γ, the energy minimum corresponds to a finite δ, i.e. protons are always collectively unstable at the H-bond centers. [19, 21] Therefore, we confirm that the source of the ferroelectric instability is, indeed, the H off-centering.

LOCAL INSTABILITIES AND QUANTIZATION IN SMALL CLUSTERS

We now address the microscopic origin of the observed proton double-occupancy in the PE phase,[11] which is an indication of the order-disorder character of the transition.

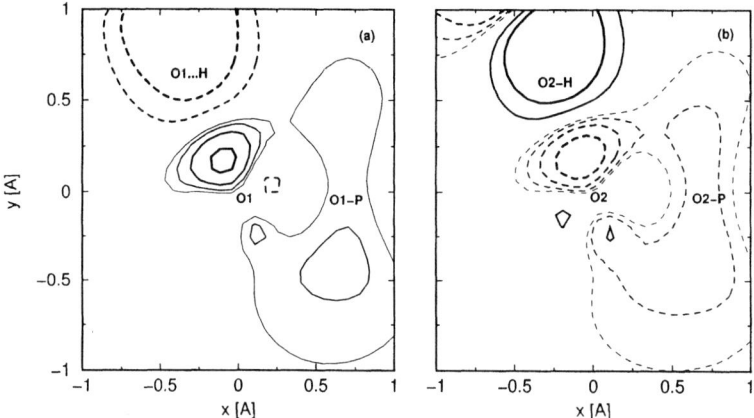

FIGURE 2. Differential charge density contours $\Delta\rho(\mathbf{r})$ in the planes containing the following atoms: (a) P-O1···H, (b) P-O2-H. Labels O1 and O2 denote the positions of the respective nuclei, positioned at (0,0). Labels O2-P and O1-P indicate the position of the center of the corresponding bonds. The same convention is used for the O2-H and O1···H bonds. Positive (negative) contours are in solid (dashed) lines. The thickest lines represent an absolute value of 2.96×10^{-3} eÅ^{-3}. The thinner lines are obtained by successively halving this value, down to 3.70×10^{-4} eÅ^{-3}.

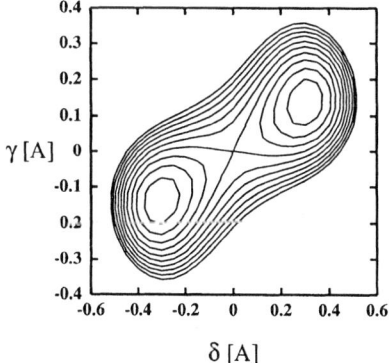

FIGURE 3. Equispaced energy contours of the two-dimensional PES as a function of δ and γ (step = 13.6 meV/ KH$_2$PO$_4$ unit). The minima at $(\delta, \gamma) \simeq \pm(0.3, 0.15)$Ålie $\simeq$ 50 meV below the saddle point at (0,0).

This phenomenom can be ascribed either to static or thermal activated dynamic disorder, or to tunneling between the two sites. Any of these possibilities requires the search for instabilities with respect to correlated but localized H motions in the PE phase, including also the possibility of heavy-ions relaxation. To this purpose we consider distortions in increasingly larger clusters embedded in a host PE matrix. For the reasons explained above, the host is modeled by protons centered between oxygens, and experimental structural parameters (including the O-O distances) for KDP at T_c^{KDP}+5 K (127 K). [11] In order to asses the effect of the volume increase seen upon deuteration, we also analyse the analogous case of D in DKDP by now expanding the host structural parameters to the corresponding experimental values at T_c^{DKDP}+5 K (234 K). [11]

We analyse results for different clusters comprising N hydrogens (deuteriums): (a) N=1 H (D) atom, (b) N=4 H (D) atoms which connect a PO_4 group to the host, (c) N=7 H (D) atoms localized around two PO_4 groups, and (d) N=10 H (D) atoms localized around three PO_4 groups. For all clusters, we consider correlated motions with the pattern shown in Fig. 1, which are the most favorable for exhibiting FE instabilities, as it was illustrated in the previous section. This correlated pattern, is retained as a single collective coordinate δ_c and two cases are considered: i) in the first case, we allow for the motion of H atoms alone, all other atoms being kept fixed, ii) secondly, we also allow for the relaxation of the heavy ions K and P, which will follow the ferroelectric mode pattern [20], as expected. Subsequent quantization of the cluster motion in the corresponding effective potential, would allow us to determine the importance of tunneling in the disordered phase. Although computing limitations don't allow us to consider larger clusters than those mentioned, we are assuming here that quantum fluctuations ocurring at short range in the PE phase will be sufficiently revealing, specially far away from critical points.

In Fig. 4, we show for the clusters considered, the total energy variation as a function of $\delta_c/2$. For the case of H motions alone, we do not observe any instability for N=1 and N=4, both in KDP and DKDP. A small barrier of $\sim$ 6 meV appears in DKDP for the N=7 move, as shown in Fig. 4(b) (open squares). This barrier grows up to $\approx$ 25 meV for the N=10 cluster in DKDP (open circles). However, quantum mechanical calculations of the cluster levels, which will be described bellow, yield ground states (GS) above the barriers and consequently, the absence of tunneling in these cases (see Fig. 4). In KDP, even the larger cluster considered is very stable, as indicated by the open circles in Fig. 4(a). The result for KDP suggests to rule out this type of motions in the PE phase, because they are incompatible with double-site occupancy.

The next step, was to considered also the motion of the heavy atoms in the above correlated local motions. Now the situation changes drastically, as shown by the solid lines and full symbols in Fig. 4 (a) and (b). In fact, clusters involving two or more PO_4 units - cases (c) and (d) above - exhibit instabilities in both KDP and DKDP, with a significant barrier in DKDP for case (d), of the order of $\approx$ 150 meV. We note here, that the instability appears in clusters which are sufficiently large, thus providing a measure of the FE correlation length. Moreover, the instabilities are much stronger (and the correlation lenght accordingly shorter) in the expanded DKDP lattice, than in KDP.

We treat these clusters quantum-mechanically, by solving the Schrödinger equation for the collective coordinate $\delta_c/2$, which is done for each cluster in the corresponding effective potentials of Fig. 4. As a local collective motion, the cluster has an effective

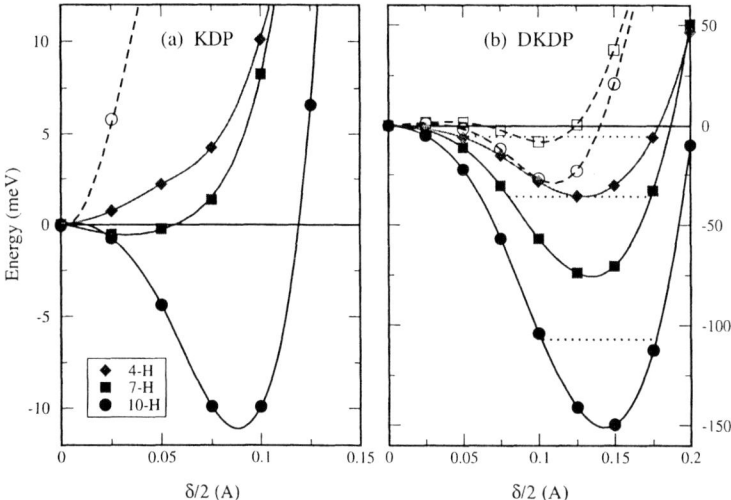

FIGURE 4. *Ab initio* energy profiles for correlated local distortions in (a) KDP and (b) DKDP. Reported are clusters of: N=4 H(D) (diamonds), N=7 H(D) (squares), and N=10 H(D) (circles). Empty symbols and dashed lines indicate that only the H(D) atoms move. Motions that involve also heavy atoms (P and K) are represented by filled symbols and solid lines. Dotted lines: GS energies (only negative ones, signaling tunneling)

mass which is calculated as $\mu = \sum_i m_i a_i^2$, where i runs over the displacing atoms and m_i are there corresponding atomic masses. a_i is the i-atom displacement at the minimum from their positions in the PE phase, relative to the H(D) displacement. The hydrogen (deuterium) effective masses calculated for these correlated motions in different clusters are about $\mu_H = 2.3$ ($\mu_D = 3.0$) proton masses (m_p) in KDP (DKDP), respectively. The calculation of the GS energy in the *heavy clusters*, leads now to quantized levels bellow the barriers, as shown by dotted lines in Fig. 4 (b), for all clusters in DKDP. This is a clear sign for tunneling arising from correlated D motions including heavy ion relaxations. However, in KDP, even the largest cluster considered (N=10), has the GS level above the barrier. The onset of tunneling at a critical cluster size, provides a rough indication of the correlation volume: it comprises more than 10 hydrogens in KDP, but no more than 4 deuteriums in DKDP. We clearly observe, then, that the dynamics of the order-disorder transition would involve fairly large H(D)-clusters together with heavy-atom (P and K) displacements. Thus, the observed proton double-occupancy is explained in our calculation by the tunneling of large and *heavy* clusters. [21] The last conclusion is confirmed by the multi-site distribution observed for the P atoms. [22]

QUANTUM DELOCALIZATION AND THE GEOMETRICAL EFFECT

Geometrical Effect and Tunneling

Let us now, address the origin of the huge isotope effect on T_c, observed in KDP. For forty years, after the pioneering work of Blinc,[2] the central issue in KDP has been whether tunneling is or not at the root of the large isotope effect, a fact that was never rigorously confirmed. Moreover, a crucial set of experiments against the tunneling picture was done recently by Nelmes and coworkers: by applying pressure, and tunning conveniently the D-shift parameter δ, they brought T_c^{DKDP} almost in coincidence with T_c^{KDP}, in spite of the mass difference between D and H in both systems. [7, 8, 23] This seems to indicate the preponderance of the modification of the H-bond geometry by deuteration – the *geometrical effect* – as the mechanism that accounts for the isotope effects in the transition.

Our calculations show that as the cluster size grows ($N \to \infty$), the tunnel splitting Ω tends to vanish. On the other hand, only large clusters are expected to be relevant for the nearly second-order FE transition in these systems.[24] Thus, for large tunneling clusters the potential barrier is sufficiently large and the GS levels are deep enough (see Fig. 4), that the relation $\hbar\Omega_{H(D)} \ll K_B T_c$ is fulfilled so much for D as for H. So, according to the tunneling model, the above relation leads to the fact that the simple change of mass upon deuteration at fixed potential could not explain the near duplication of T_c. In fact, let us consider the largest cluster (N=10) in Fig. 4 for DKDP, which is larger than the crossover lenght in this system. The GS level for the deuterated case (calculated with an effective mass of $\mu_D = 35.4\ m_p$) is around $E_{GS} = -107$ meV, well bellow the central barrier, and the tunnell splitting ammounts to $\hbar\Omega_D = 0.34$ K. Changing the mass at fixed potential ($\mu_H = 25.3\ m_p$), leads to a tunnell splitting only slightly larger $\hbar\Omega_D = 1.74$ K. As $T_c^{DKDP} \approx 229K$, the relation $\hbar\Omega_{H(D)} \ll K_B T_c$ clearly holds, and the change in Ω at fixed potential should account only for a small change in T_c. This is in agreement with the high-pressure experiments mentioned above,[7, 8, 23] where at fixed structural conditions, the isotope effect in T_c appears to be very small.

On the other hand, the geometric effect in the H-bond at fixed potential is very small too. Actually, the plot of the proton and deuteron wave functions (WF) in the DKDP potential for the N=7 cluster (see Fig. 5 (a)), shows very slight differences between both WF. As a matter of fact, the distance between peaks as a function of the effective mass at fixed potential almost remain unchanged, as can be seen by the square symbols in Fig. 5 (c).

By contrast, the proton WF for the N=7 cluster in the KDP potential, exhibits a broad single peak, as it's shown in Fig. 5 (a). But, how can we explain such a big geometric change in going from DKDP to KDP? The first observation after this question, comes from what is apparent in Figure 4: energy barriers in DKDP are much larger than those in KDP, implying that quantum effects are significantly reduced in the expanded DKDP lattice. In fact, the proton WF in KDP, which is a function of the collective coordinate δ_c, have more weight around the middle of the H-bond ($\delta_c \approx 0$) than in DKDP (Fig. 5 (a)). That is, zero-point motion pushes proton to be at the H-centered position between oxygens, more effectively than deuteron. This affects the covalency of the bond, which becomes stronger as the proton moves to the H-bond center. The geometric change of the O-H-O bridge, produced by the mixed effect of quantum delocalization and the gain in covalency, affects in turn the crystal cohesion. Thus, the increased probability of the proton to be midway between oxygens, strengthens the O-H-O covalent grip and pulls the oxygens together, causing a small shrinking of the lattice. This shrink has the effect of decreasing the potential depth, making the proton even more delocalized, an so on in a self-consistent way. This self-consistent procedure is finally identified as the phenomenon that shrinks the lattice from the larger classical value to the smaller value found for KDP. This phenomenon, triggered by tunneling, leads to an enhance of the geometrical effect. The overall self-consistent effect is eventually much larger than the deuteration effect obtained at fixed potential.

To estimate an upper limit to that effect, we considered the comparison of the lattice parameters and the bridge lenghts by making classical electronic calculations. This was done in two different cases: one with the hydrogens forced to stay in the middle of the H-bond, and the other, with the hydrogens fully off-centered in the FE state of KDP. In the last case, the distance between oxygens is $d_{OO} \approx 2.51$Å, dropping to $d_{OO} \approx 2.42$Å when H is centered. In adition, the lattice volume shrinks about 2.3 %. Thus, the proton centering acts as a very strong attraction center, pulling the two oxygens together. We estimate that, at the equilibrium volume, the proton centering creates an equivalent pressure of ≈ 20 Kbar. In the true high-temperature PE phase, the protons are not centered in the middle of the H-bonds, but they are equally distributed on both sides of the bond, thus reducing the magnitude of the effect.

The Huge Isotope Effect: a Nonlinear Self-consistent Phenomenon

In the previous subsection, we discussed how a self-consistent mechanism combining quantum delocalization, the modification of the covalency in the bond, and the effect over the lattice parameters, can account for the large geometric effect observed by deuteration. Actually, this mechanism is now capable of explaining, at least qualitatively, the increase in the order parameter and T_c with deuteration. This self-consistent mechanism has obviously its origin in the difference in tunneling induced by different masses, but is largely amplified through the geometric modification of the lenghts and energy scales.

We constructed a simple model, to demonstrate the non-linear self-consistent behaviour descripted above in the mechanism of the isotope substitution in KDP. To this purpose, we considered the Schrödinger equation for the clusters, and added to the effective hydrogen potential, a quadratic WF-dependent term of the form $V_{\text{eff}}(x) = V_0(x) - k|\Psi(x)|^2$, where $x = \delta_c/2$ and $V_0(x)$ is a quartic double-well similar to those of Fig. 4. The quadratic term $|\Psi(x)|^2$ serves as a non-linear feedback in the model: when proton delocalizes, it has more weight to be in the middle of the H-bond ($|\Psi(x)|^2$ increases at the center), producing a decrease in the barrier for the effective potential, which further delocalizes the proton, and so on.

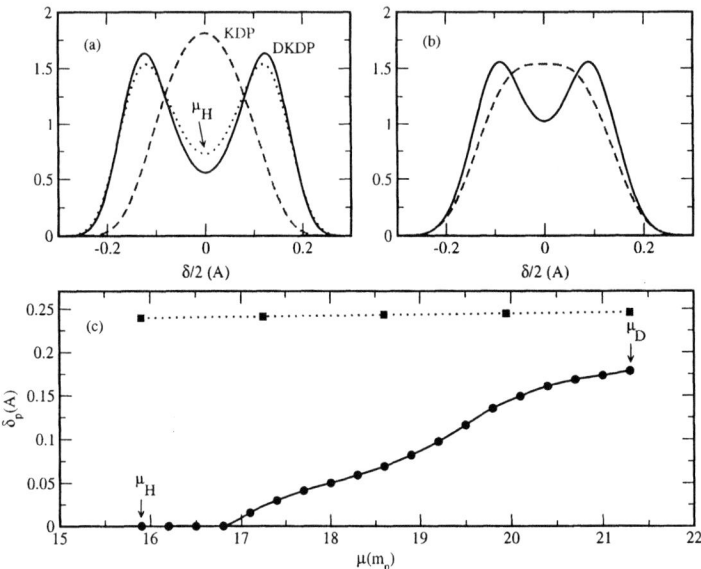

FIGURE 5. WF in the 7-H(D) cluster PES for (a) ab initio and (b) self-consistent model calculations. Solid (dashed) lines are for D (H). Dotted line is for H in the DKDP PES. (c) WF peak separation δ_p as a function of the cluster effective mass μ (given in units of the proton mass) for the self-consistent model (circles) and for fixed DKDP potential (squares). Lines are guides to the eye.

The parameter k and those of the bare potential $V_0(x)$, are choosen so as to qualitatively reproduce the WF profiles in the cases of KDP (broad single peak) and DKDP (double peak). Once these parameters are fixed, the WF self-consistent solutions depend only on the effective mass μ. Figure 5 (b) shows the WF corresponding to μ_D (solid line) and μ_H (dashed line), which are similar to those calculated from the ab initio potentials for the N=7 cluster (Fig. 5 (a)).

In Fig. 5 (c), we show the distance between peaks δ_p in the WF as a function of μ. For the self-consistent solution, starting from the finite value for μ_D (DKDP), δ_p decreases remarkably towards lower μ values, until it vanishes near μ_H (KDP) (see cicles in Fig. 5 (c)). Now, this strong dependence of δ_p on the mass, is in striking contrast with the very weak dependence obtained at fixed DKDP potential and geometry (square symbols). Such a large mass dependence, can now explain the large isotope effect found in KDP, via an amplified and self-consistent geometrical modification of the H-bond.[21]

CONCLUSIONS

We confirm the H off-centering as the source of the FE instability. The double occupancy in the PE phase is explained by cluster tunneling involving also heavy ions, in agreement with the multi-site distribution observed for the P ion.[22] At fixed lattice geometry, the mere mass change cannot explain the huge isotope effect. The main effect of replacing deuterons by protons is an enhancement of the quantum delocalization at the bond center. This is combined with a gain of covalency in the H-bond, which shrinks the lattice, which further delocalizes the proton, and so on in a selfconsistent way. In the end, the isotope effect is dominated by the geometrical effect, which is triggered by tunneling and is in agreement with experiments.

ACKNOWLEDGMENTS

We aknowlegde helpfull discussions with R.J. Nelmes, M.I. McMahon, R. Resta, A. Bussmann-Holder, G. Colizzi, G. Reiter, M.G. Stachiotti and D. Marx. R.M. and S.K. thank support from CONICET, Argentina, and from ICTP, Trieste. S. K. also thanks support from Fundación Antorchas, Argentina. E.T.'s work was also supported by MIUR COFIN01, and by INFM/G.

REFERENCES

1. M. E. Lines and A. M. Glass, *Principles and Applications of Ferroelectric and Related Materials* (Clarendon Press, Oxford, 1977).
2. R. Blinc, J. Phys. Chem. Solids **13**, 204 (1960).
3. K. Kobayashi, J. Phys. Soc. Jpn. **24**, 497 (1968); E. Matsushita and T. Matsubara, Prog. Theor. Phys. **67**, 1 (1982); T. Matsubara and E. Matsushita, Prog. Theor. Phys. **71**, 209 (1984); M. Kojyo and Y. Onodera, J. Phys. Soc. Jpn. **57**, 4391 (1988); A. Bussmann-Holder and K.H. Michel, Phys. Rev. Lett. **80**, 2173 (1998).
4. G.F. Reiter *et al.*, Phys. Rev. Lett. **89**, 135505 (2002)).
5. M. Ichikawa, K. Motida and N. Yamada, Phys. Rev. B **36**, 874 (1987).
6. Z. Tun *et al.*, J. Phys. C: Solid State Phys. **21**, 245 (1988).
7. R.J. Nelmes, J. Phys. C: Solid State Phys. **21**, L881 (1988).
8. M.I. McMahon *et al.*, Nature (London) **348**, 317 (1990).
9. J.M. Robertson and A.R. Ubbelohde, Proc. R. Soc. London A **170**, 222 (1939).
10. H. Sugimoto and S. Ikeda, J. Phys.: Condens. Matter. **8**, 603 (1996)
11. R.J. Nelmes, Z. Tun and W.F. Kuhs, Ferroelectrics **71**, 125 (1987).
12. P. Ordejón, E. Artacho and J.M. Soler, Phys. Rev. B **53**, R10441 (1996); D. Sánchez-Portal *et al.*, Int. J. Quantum Chem. **65**, 453 (1997).
13. C. Cavazzoni and G.L. Chiarotti, Computer Phys. Commun. **123**, 56 (1999).
14. J.P. Perdew, K. Burke and M. Ernzerhof, Phys. Rev. Lett. **77**, 3865 (1996).
15. N. Troullier and J.L. Martins, Phys. Rev. B **43**, 1993 (1991).
16. S. Koval *et al.*, Comput. Mater. Science **22**, 87 (2001).
17. Q. Zhang *et al.*, Phys. Rev. B **65**, 024108 (2002).
18. J. Kohanoff, S. Koval and R. Migoni (unpublished).
19. S. Koval, J. Kohanoff and R. L. Migoni, Ferroelectrics **268**, 239 (2002).
20. R. J. Nelmes, Ferroelectrics **71**, 87 (1987).
21. S. Koval, J. Kohanoff, R. L. Migoni and E. Tossati, Phys. Rev. Lett. **89**, 187602 (2002).
22. M.I. McMahon *et al.*, Europhys. Lett. **13**, 143 (1990).
23. R.J. Nelmes *et al.*, Ferroelectrics **124**, 355 (1991).
24. G.A. Samara, Ferroelectrics **5**, 25 (1973)

First-principles calculations of K_2SeO_4 dielectrics

Razvan Caracas and Xavier Gonze

Université Catholique de Louvain, Unité de Physico-Chimie et de Physique de Matériaux, pl. Croix du Sud 1, B-1348 Louvain-la-Neuve, Belgium

Abstract. We study from first principles the groundstate properties of the high-symmetry phases of K_2SeO_4: two ideal ordered high-temperature hexagonal ones and an orthorhombic one. We analyze in detail the dielectric properties of these phases and discuss certain aspects of the zone-center dynamical properties.

INTRODUCTION

K_2SeO_4 is the prototype material of the A_2BX_4 family of dielectrics. Such compounds exhibit a rich sequence of phase transitions with normal, commensurate and incommensurate structures [1, 2, 3, 4]. All these structures have in common a high-temperature (disordered) hexagonal phase, and are obtained by different slight distortions of it. The hexagonal phase has been experimentally identified only in K_2SeO_4 and K_2SO_4: the hypothetical phase transition temperature is probably higher than the melting point for other materials, like Rb_2ZnBr_4 [5] or Rb_2ZnCl_4 [6]. Under cooling the hexagonal structure transforms first to an orthorhombic Pbnm structure and further to an incommensurately modulated (IC) structure, whose average structure is the previously mentioned orthorhombic one. Then, the modulation wavevector locks-in and the structure transforms to a commensurately modulated superstructure with orthorhombic symmetry. The different compounds belonging to the A_2BX_4 family of dielectrics exhibit different other orthorhombic, monoclinic and triclinic structures, some of them with ferroelectric properties.

K_2SeO_4 follows the phase transition sequence : hexagonal, orthorhombic, incommensurately modulated, and orthorhombic, with 545K, 130K and 93 K temperature transitions.

In the experimental studies dealing with the hexagonal structure, the determination of the space group and the ordering state has received much attention [7, 8, 9]. These data show the $P6_3/mmc$ space group, with a structure similar to the α-K_2SO_4. It is assumed that the SeO_4 tetrahedra are orientationally disordered, with their apex pointing independently up and down with respect to the hexagonal c axis.

The normal-incommensurate-commensurate transitions played a central role in different theoretical studies. These used empirical and/or semi-empirical potentials (with one exception, Ref. [10] based on a Gordon-Kim model) in order to describe the dynamics of the phase transitions: neither the ground-state features, nor the hexagonal structures are discussed.

In this study we report results from first principles calculations, using the local density approximation (LDA) within the density functional theory (DFT). We briefly describe the crystal structure of the hexagonal and orthorhombic phases, and present the corresponding electronic, dielectric and zone-center dynamical properties. The paper ends with the conclusions.

COMPUTATIONAL DETAILS

All the calculations are based on the local density approximation (LDA) of the density functional theory (DFT) [11, 12]. We use the ABINIT package (ABINITv2.x, 1999-2000 [13] and ABINITv3.x 2001-2002 [14]). The ABINIT software is based on planewaves and pseudopotentials. The three elements are represented by Troullier-Martins pseudopotentials [15]. The considered valence electrons for K, Se and O are $3p^64s^1$, $4s^24p^4$ and $2s^22p^4$ respectively.

The structural relaxation was conducted using the Broyden-Fletcher-Goldfarb-Shanno minimization [16], modified to take into account the total energy as well as the gradients.

The dynamical matrices, Born effective charges and dielectric permittivity tensors, were computed within density-functional perturbation theory, using the responses to atomic displacements and homogeneous electric fields [17, 18, 19, 20].

CRYSTAL STRUCTURE

Hexagonal phase

According to the experimental results [21, 22], the hexagonal phase is disordered, with average P6$_3$/mmc space group and two molecular units in the unit cell. The SeO$_4$ tetrahedra may point up or down the c hexagonal axis (Fig. 1). The two possible orientations combined together with the possible symmetries, we obtain two ideal ordered structure, that describe the two extreme cases of the disordered state. The up_down structure, where the two tetrahedra of the primitive unit cell have opposite orientations and the up_up structure, where the tetrahedra have similar orientations. The space group of the up_down structure is P$\bar{3}$m1 and of the up_up structure P6$_3$mc, both groups being subgroups of the P6$_3$/mmc space group.

To start the determination of the up_down structure, we use the experimental lattice constants of the K$_2$SeO$_4$ [21] and the atomic positions of the up_down K$_2$SO$_4$ hexagonal structure [23], then we fully optimize both cell parameters and internal atomic positions. The up_up structure is built by switching the orientation of one of the two tetrahedra from the up_down structure.

The a and c theoretical lattice constants are 6.1186 Å, 7.9443 Å, for the up_down and 5.9125, 9.0025 Å, for the up_up structure, respectively. The unit cell volumes vary by about 6% between the two structures: 257.50 Å^3 (up_down) vs. 272.54 Å^3 (up_up). The electronic LDA energy of the up_down structure is 9 mHa (0.245 eV) per unit cell lower than the energy of the the up_up structure.

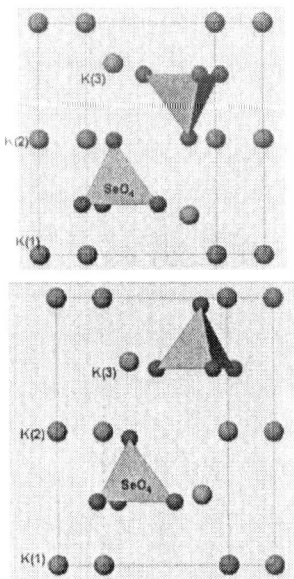

FIGURE 1. The two theoretical structures of the high-temperature phase of K_2SeO_4 having the SeO_4 pointing up or down the c hexagonal axis

The experimental lattice constants of the disordered phase are a=6.14 Å, c=8.90 Å, with volume of 290.6 Å^3. We remark that they are similar to the a of the up_down and the c of the up_up theoretically determined structures, (with deviations of about 1% from the experimental parameters). Beyond the expected error due to LDA (usually the theoretical volume is smaller than the experimental one by a few percent), this similarity may arise from a dynamical disorder state of the SeO_4 tetrahedra at high temperatures. The lattice constants of the final disordered structure should allow all possible orientations of the selenate groups, that leads to the choice of the maximum values of these parameters.

Orthorhombic structure

The orthorhombic structure represents a distorted c x $a\sqrt{3}$ x b hexagonal up_down superstructure. Its space group is Pnam. The theoretical a=7.709 Å and b=10.383 Å lattice constants are underestimated with respect to the experimental ones, a=7.661 Å and b=10.466 Å, by less than 1 %. The theoretical c parameter, 6.425 Å, is overestimated by about 7 % (this large overestimation may be an artifact of the LDA).

The distortion with respect to the hexagonal structure consists of complex displacements of the atoms, mainly K, in the hexagonal (001) plane, combined with about $c/2$ shifts of half of the K-SeO$_4$ chains along the c hexagonal axis. The orthorhombic theoretical unit cell volume is 1.997 of the up_down hexagonal unit cell volume. During the phase transition the structure is elongated along the hexagonal a axis and compressed along the hexagonal [2 1 0] and [0 0 1] directions. The orthorhombic phase has a 23.84 mHa (0.649 eV) lower LDA energy than the hexagonal up_down structure.

ELECTRONIC PROPERTIES

The structure has a mixed ionic-covalent character, formed by isolated K cations and SeO_4 tetrahedral anionic groups.

K_2SeO_4 is an insulator, with the electronic band structure characterized by weakly dispersive bands. The differences between the electronic structure of the different analyzed phases, are quite small: they involve the dispersion ranges of the different groups of bands and the gaps between them. The changes are on the order of less than 1 eV. The 0K-LDA gap, direct in Γ, is 3.5, 3.7 and 3.7 eV for the experimental orthorhombic, theoretical up_down and up_up hexagonal structures respectively. The experimental gap obtained from absorbtion edge measurements [24], decreases linearly with increasing temperature. At the room-T (orthorhombic structure), is about 6.0-6.1 eV. At 800 K (the hexagonal phase) the interpolated value is 5.4 to 5.6 eV. As often observed, the LDA gap is smaller than the experimental gap.

There is a one-to-one correspondence between each group of bands and the K orbitals or the SeO_4 molecular orbitals. We may use the irreducible representations of the $\bar{4}3m$ point group to analyze the hybridization between the different SeO_4 orbitals. The five s orbitals (1 Se4s + 4 O2s) and fifteen p orbitals (3 Se4p + 12 O2p) in each SeO_4 group are hybridized and decomposed as $2A_1+T_2$ and $A_1+E+3T_2+T_1$ respectively. The s bands belonging to the SeO_4 group lie in the low-energy zone of the valence bands. The K3p orbitals lie relatively high in energy between the s SeO_4 bands. The p SeO_4 bands lie in the high-energy zone of the valence bands and the first conduction bands.

DIELECTRIC PROPERTIES

The computed electronic dielectric tensors, ε^∞, do not vary significantly between the three structures. The diagonal values of the ε^∞ are [2.37 2.37 2.43], [2.37 2.37 2.28] and [2.40 2.38 2.36] for the hexagonal up_down, hexagonal up_up and orthorhombic structures, respectively. The values for the orthorhombic structure are closer to the up_down than the up_up structure. It is likely that these values are overestimated, as a consequence of the well-known problem in reproducing ε^∞ in LDA [25] and the previously mentioned gap underestimation.

Next, we compute the Born effective charges. Table 1 shows the values obtained for the atoms in the hexagonal up_down and up_up structures. The Born effective charges of the atoms in the orthorhombic structure are similar to the values from the hexagonal ones. All the atoms present differences with respect to the nominal charges. The Born effective charges of the K are slightly higher than the nominal charges, while those of Se and O are lower than the nominal charges (+6 and -2). The differences with respect

TABLE 1. Calculated Born effective charge tensors for the atoms in the hexagonal up_down and up_up structures. The columns(lines) correspond to atomic displacements(polarization) directions. For each atom, the fourth line gives the eigenvalues of the tensor. The first three column of numbers correpond to the up_down structure, while the last three to the up_up structure. The labels of the atoms correspond to the symmetry independent sites, with the exception of the O atoms, where O(1) designate the apex atom of the SeO_4 group and the O(2), O(3) and O(4) its basis.

Atom	P	x	y	z	P	x	y	z
K1	x	1.215	0.014	0.001	x	1.186	0.000	0.000
	y	0.014	1.231	0.000	y	0.000	1.186	0.000
	z	0.001	0.000	1.334	z	0.000	0.000	1.271
		1.207	1.239	1.334		1.186	1.186	1.271
K2	x	1.234	0.012	0.000	x	1.186	0.000	0.000
	y	0.012	1.247	0.000	y	0.000	1.186	0.000
	z	0.000	0.000	1.047	z	0.000	0.000	1.271
		1.254	1.227	1.047		1.186	1.186	1.271
K3	x	1.126	0.008	0.000	x	1.171	0.000	0.000
	y	0.008	1.135	0.000	y	0.000	1.171	0.000
	z	0.000	0.000	1.281	z	0.000	0.000	1.090
		1.121	1.140	1.281		1.171	1.171	1.090
Se	x	3.046	0.001	0.000	x	2.983	0.000	0.000
	y	0.001	3.048	0.000	y	0.000	2.983	0.000
	z	0.000	0.000	3.012	z	0.000	0.000	2.863
		3.046	3.048	3.012		2.983	2.983	2.863
O1	x	-0.941	0.010	0.000	x	-0.958	-0.001	0.000
	y	0.010	-0.929	0.000	y	-0.001	-0.955	0.000
	z	0.000	0.000	-1.945	z	0.000	0.000	-2.031
		-0.947	-0.923	-1.945		-0.955	-0.958	-2.031
O2	x	-1.692	0.372	-0.317	x	-1.664	0.350	-0.272
	y	0.372	-1.263	0.183	y	0.350	-1.260	0.157
	z	-0.317	0.183	-1.181	z	-0.272	0.157	-1.064
		-1.028	-1.048	-2.059		-0.956	-1.058	-1.974
O3	x	-1.706	-0.389	0.319	x	-1.663	-0.350	0.273
	y	-0.389	-1.290	0.187	y	-0.350	-1.260	0.157
	z	0.319	0.187	-1.179	z	0.273	0.157	-1.065
		-1.028	-1.057	-2.089		-0.956	-1.058	-1.974
O4	x	-1.057	-0.015	-0.002	x	-1.057	0.000	0.000
	y	-0.015	-1.939	-0.370	y	0.000	-1.865	-0.315
	z	-0.002	-0.370	-1.179	z	0.000	-0.315	-1.063
		-1.028	-1.057	-2.090		-0.956	-1.058	-1.974

to the nominal charges arise from the covalent bonding in the SeO_4 tetrahedra (for Se and O) and from some weak covalent component of the K-O bonding (for K) [26]. The cations show relatively isotropic effective charges, that are slightly larger than the isotropic effective charges obtained by interpolation of the infrared spectra [27]. The O present anisotropic effective charges. Their anisotropy is less pronounced than that for the O atoms in the SiO_4 tetrahedra from silica [17]. The differences between the two hexagonal structures are more visible for the Se atoms, that have higher charges in the up_down than the up_up structures. This difference is compensated by higher charges on the O atoms.

ZONE-CENTER LATTICE DYNAMICS

For the hexagonal structures, on the basis of group theory, we obtain decompositions of the vibration modes as $5A_{1g} + 1A_{2g} + 6E_g + 1A_{1u} + 7A_{2u} + 8E_u$ and $6A_1 + 1A_2 + 6B_1 + 6B_2 + 7E_1 + 7E_2$ for the up_down and up_up structures respectively.

The up_down hexagonal structure exhibits in the zone center one unstable mode (LO=49.54i cm^{-1} and TO=51.28i cm^{-1}), with E_u character. The real space eigendisplacements of this vibration mode are dominated by the O and K(2) movements. They tend to rotate the SeO$_4$ tetrahedra and to displace the K(2) atoms parallel to the cartesian axes, in the (001) plane. This mode may be the zone-center remnant of a zone-boundary soft mode responsible for the hexagonal-orthorhombic phase transition. A full Brillouin-zone analysis would be needed to establish its precise role during the phase transitions.

The up_up hexagonal structure exhibits in the zone center two unstable modes. The first mode has E_2 character (59.97i cm^{-1}) and the second mode has E_1 character (LO=40.02i cm^{-1} and TO=32.62i cm^{-1}). They are both dominated by the oxygen displacements that tend to rotate the SeO$_4$ tetrahedra around directions close to [1$\bar{1}$0] or [110], depending on the tetrahedra position.

For the orthorhombic structure we obtain a decomposition of the vibration modes as $13A_g + 8B_{1g} + 13B_{2g} + 8B_{3g} + 8A_u + 13B_{1u} + 8B_{2u} + 13B_{3u}$. The A_u modes are silent, all other modes being Raman (g-modes) or infrared (u-modes) active.

The orthorhombic structure presents two instabilities in Γ. The first unstable mode has A_u character and the second has a B_{1g} character. Both unstable modes represent, in the real space, mixings between alternating displacements of K along the c axis and tetrahedral E-type modes of the SeO$_4$ groups. We will emphasize here some dielectric properties related to the u-modes in the three structures.

Tables 2 and 3 shows the IR active modes in the zone center for, respectively, the hexagonal up_down and hexagonal up_up structures, as well as the corresponding oscillator strengths.

The oscillator strengths $S_{m,\alpha\beta}$ corresponding to the mth mode, are related to the eigendisplacements $U_m(\kappa\alpha)$ and the Born effective charges $Z^*_{\kappa,\alpha\alpha'}$ of atom κ according to [19]:

$$S_{m,\alpha\beta} = (\sum_{\kappa,\alpha'} (Z^*_{\kappa,\alpha\alpha'} U^*_m(\kappa\alpha')))(\sum_{\kappa',\beta'} (Z^*_{\kappa,\beta\beta'} U_m(\kappa'\beta'))) \qquad (1)$$

The displacements are normalized using the masses. The oscillator strengths determine the contribution of the phonons to the low-frequency dielectric permittivity tensor via:

$$\varepsilon^\infty_{\alpha\beta} + \frac{4\pi}{\Omega_0} \sum_m (\frac{S_{m,\alpha\beta}}{\omega^2_m}) \qquad (2)$$

For the hexagonal up_down structure the highest frequency A and E modes exhibit the largest effective charges and oscillator strengths. However the largest contribution to the static dielectric tensor is provided by the low frequency modes. In the up_up structure,

TABLE 2. Symmetry characteristics, LO-TO splitting and oscillator strengths (OS) for the IR-active zone-center vibrational modes in the hexagonal up_down structure. The oscillator strengths are expressed in 10^{-4} atomic units (1 a.u. = 0.342036 m^3/s^2). The values given for the oscillator strengths for the A_{2u} modes are the zz components and for the E_u modes the xx and yy components along cartesian directions.

Mode	TO	LO	OS	
A_{2u}	130	131	0.0791	
	150	250	0.7208	
	220	240	0.6160	
	412	437	1.4082	
	869	873	0.5899	
	956	991	4.4024	
E_u	51i	40i	0.0069	0.0209
	95	101	0.1832	0.5472
	104	138	0.2128	0.0486
	186	204	0.0546	0.3297
	327	327	0.0033	0.0098
	408	412	0.3061	0.9189
	883	927	1.2601	3.6249

there is no general trend concerning the effective charges and the oscillator strengths. The largest contributions to the static dielectric tensor is provided by the low frequency A_1 and largest frequency E_u modes.

Further, using the information contained in the dynamical matrix in Γ we are able to determine the self-force constants (SFCs) and some of the short-range interatomic force constants (IFC). We define the SFC as the force acting on an isolated atom during its displacement along the cartesian direction, all the other atoms in the lattice being fixed. The eigenvalues of the SFCs tensors for the hexagonal up_down, hexagonal up_up and orthorhombic structures are listed respectively in Tables 4, 5, 6.

The SFCs values are similar between the three investigated structures. The largest SFCs are on the Se atoms, while the lowest ones are on the K atoms. The values of the SFCs on the K(1) atoms in the up_down structure are much larger than the values of all the other K atoms in all three structures. This might be due to the slightly changed local environment in the up_down structure (in terms of surrounding O atoms positions).

Certain atoms may present different off-diagonal negative values, that arise mainly from the departure of the atomic position from the high-symmetry positions. However, the possible tendency of displacement of one isolated atom in, e.g the xz plane, is cancelled by the symmetry. We observe from the above mentioned tables that in all the three structures the eigenvalues of the self-force constants for all the atoms are positive. This shows that the only possible way to reduce the energy is the collective displacement of atoms, corresponding to phonon modes.

TABLE 3. Symmetry characteristics, LO-TO splitting and oscillator strengths (OS) for the IR-active zone-center vibrational modes in the hexagonal up_up structure. The oscillator strengths are expressed in 10^{-4} atomic units (1 a.u. = 0.342036 m^3/s^2). The values given for the oscillator strengths for the A_{2u} modes are the zz components and for the E_u modes the xx and yy components along cartesian directions.

Mode	TO	LO	OS	
A1	135	140	0.2943	
	164	197	0.8404	
	413	435	1.2804	
	857	860	0.5061	
	918	955	4.3805	
E1	40i	33i	0.0640	0.0222
	110	125	0.1387	0.4951
	149	151	0.4613	0.1646
	335	335	0.0104	0.0036
	411	413	0.0082	1.3456
	889	928	3.5492	1.2022

TABLE 4. Self-force constants on the atoms in the up_down hexagonal structure. Values represent the eigenvalues of the force constants, expressed in Ha/bohr.

K(1)	0.051	0.051	0.064
K(2)	0.012	0.012	0.025
K(3)	0.016	0.016	0.032
Se	0.623	0.623	0.718
O(1)	0.038	0.038	0.467
O(2)	0.409	0.046	0.046

TABLE 5. Self-force constants on the atoms in the the up_up hexagonal structure. Values represent the eigenvalues of the force constants, expressed in Ha/bohr.

K(1)	0.014	0.014	0.030
K(2)	0.014	0.014	0.030
K(3)	0.027	0.027	0.035
Se	0.643	0.643	0.650
O(1)	0.037	0.037	0.449
O(2)	0.398	0.047	0.048

TABLE 6. Self-force constants on the atoms in the orthorhombic structure. Values represent the eigenvalues of the force constants, expressed in Ha/bohr.

K(1)	0.033	0.016	0.008
K(2)	0.038	0.029	0.017
Se	0.676	0.630	0.619
O(1)	0.393	0.049	0.045
O(2)	0.397	0.051	0.043
O(3)	0.437	0.048	0.039

The accurate determination of the IFCs is possible only for the short-range interactions, between atoms belonging to same unit cell. We may compute the IFCs between the Se and the O atoms that build the SeO_4 tetrahedra.

The eigenvalues of the Se-O IFCs tensors for the theoretical hexagonal up_down and up_up structures are respectively [-0.413 -0.057 -0.057] and [-0.377 -0.056 -0.056] for the Se-O bond that is parallel to the threefold axis and [-0.347 -0.060 -0.059] and [-0.350 -0.060 -0.054] for the three other bonds. Within each set of eigenvalues, the largest value may be assigned to the "longitudinal" IFC and the two lowest values to the "transverse" IFCs. The differences between IFCs of the two hexagonal structures arise mainly from the differences in the Se-O bond lengths. For the orthorhombic structure, the eigenvalues of the four, symmetry independent, IFCs are: [-0.382 -0.058 -0.054], [-0.345 -0.060 -0.057], [-0.345 -0.060 -0.057] and [-0.346 -0.058 -0.057]. The four IFCs are very similar to each other and also similar to the IFCs of the hexagonal phases. This similarity reflects the constancy of the properties of the Se-O bond during the phase transitions.

CONCLUSION

We determine within the local density approximation of the density functional theory structural, electronic, dielectric and zone-center dynamical properties of the hexagonal and orthorhombic phases of K_2SeO_4

K_2SeO_4 consists of SeO_4 anionic tetrahedral groups and isolated K cations. Within the anionic groups the Se and O are covalently bonded.

We compute two ideal ordered structures of the high-temperature hexagonal phase: the first one ($P6_3mc$) with the SeO_4 tetrahedra pointing in the same direction, and the second one ($P\bar{3}m1$) with the SeO_4 tetrahedra pointing in opposite directions. The lattice parameters of the disordered experimental high-T hexagonal structure may be strongly influenced by anharmonic effects present at high temperatures.

The Born effective charges are similar for the three analyzed structures. They do not present large deviations from the nominal charges, and confirm the ionic character of the K atoms and the covalent bonding within the SeO_4 tetrahedral groups.

The hexagonal up_up structure presents two unstable modes in Γ and the hexagonal up_down structure presents only one unstable mode in Γ. The orthorhombic structure presents one g and one u unstable modes in the zone-center. We also report the oscillator strengths the IR active modes in the two hexagonal structures.

The self-force constants on all the atoms and the interatomic force constants for the atoms belonging to the same tetrahedra are very similar between the three analyzed structures.

ACKNOWLEDGMENTS

We aknowledge help from computer scientists B. van Renterghem and J.-M. Beuken. XG acknowledges financial support from the National Fund for Scientific Research (Belgium). Support also came from the FRFC project No. 2.4556.99 "Simulation numérique et traitement des données".

REFERENCES

1. J.D. Axe, M. Iizumi and G. Shirane, in *Incommensurate Phases in Dielectrics*, R. Blinc and A.P. Levanyuk (eds.), pp.1-48, North Holland, Amsterdam (1986)
2. H.Z. Cummins, Phys. Rep. **185**, 211 (1990).
3. Y. Ishibashi, in *Incommensurate phases in dielectrics 2*, eds. R. Blinc and A.P. Levanyuk, Elsevier Science Publishers B.V., pp. 49-69 (1986).
4. R. Caracas, J. Appl. Cryst. **35**, 120 (2002), http://www.mapr.ucl.ac.be/~crystal/.
5. M.S. Novikova, R.A. Tamazyan, and I.P. Aleksandrova, Crystallogr. Rep. **40**, 31 (1995).
6. N.G. Zamkova and V.I. Zinenko, J. Experim. Theor. Phys. **80**, 713 (1995).
7. K. Inoue, K. Suzuki, A. Sawada, Y. Ishibashi and Y. Takagi, J. Phys. Soc. Jpn. **46**, 609 (1979).
8. K.S. Aleksandrov, Crystallogr. Rep. **38**, 67 (1993).
9. T.M. Chen and R.H. Chen, J. Solid State Chem. **111**, 338 (1994).
10. H.M. Lu and J.R. Hardy, Phys. Rev. Lett. **64**, 661 (1990).
11. P. Hohenberg and W. Kohn, Phys. Rev. **136**, B864 (1964).
12. W. Kohn and L.J. Sham, Phys. Rev. **140**, A1133, (1965).
13. X. Gonze, R. Caracas, P. Sonnet, F. Detraux, P. Ghosez, I. Noiret, J. Schamps, American Institute of Physics Conference Proceedings **535**, edited by H. Krakauer, pp. 13-20, Melville, New York (2000).
14. X. Gonze, J.-M. Beuken, R. Caracas, F. Detraux, M. Fuchs, G.-M. Rignanese, L. Sindic, M. Verstraete, G. Zerah, F. Jollet, M. Torrent, A. Roy, M. Mikami, Ph. Ghosez, J.-Y. Raty and D.C. Allan, Comput. Mater. Sci. **25**, 478 (2002).
15. N. Troullier and J.L. Martins, Phys. Rev. B **43**, 1993 (1991). URL: http://www.abinit.org/ABINIT/Psps/LDA_TM/lda.html
16. W.H. Press, B.P. Flannery, S.A. Teukolsky and W.T. Vetterling *Numerical Recipes. The Art of Scientific Computing (FORTRAN Version)*, Cambridge University Press, Cambridge (1989).
17. X. Gonze, D.C. Allan, and M.P. Teter, Phys. Rev. Lett. **68**, 3603-3606 (1992).
18. X. Gonze, Phys. Rev. B **55**, 10337 (1997).
19. X. Gonze and C. Lee, Phys. Rev. B **55**, 10355 (1997)
20. S. Baroni, S. de Gironcoli, A. Dal Corso, and P. Giannozzi, Rev. Mod. Phys. **73**, 515-562 (2001).
21. V.I. Zinenko and N.G. Zamkova, Phys. Rev. B **57**, 211 (1998).
22. N.E. Massa, F.G. Ullman and J.R. Hardy, Phys. Rev. B **27**, 1523 (1983).
23. A.J. van den Berg and F. Tuinstra, Acta Cryst. B**34**, 3177 (1978).
24. S. Pacesova, B. Brezina, and L. Jastrabik, phys. stat. sol. (b) **116**, 645 (1983).
25. X. Gonze, Ph. Ghosez, and R.W. Godby, Phys. Rev. Lett. **74**, 4035-4038 (1995).
26. Ph. Ghosez, J.-P. Michenaud, and X. Gonze, Phys. Rev. B **58**, 6224-6240 (1998).
27. P. Echegut, F. Gervais and N.E. Massa, Phys. Rev. B **34**, 278 (1986).

Point Defects and Physical Properties of Ferroelectrics: Lithium Niobate

G. Malovichko[1] V. Grachev[1] and O. Schirmer[2]

[1]*Physics Department, Montana State University, Bozeman 59717 Montana, USA*
[2]*Physics Department, University of Osnabrück, 49069 Osnabrück, Germany*

Abstract. The strong dependence of the properties of ferroelectric lithium niobate crystals on the concentration of intrinsic (non-stoichiometric) and extrinsic (impurity) defects is analyzed. Spectra of optical absorption and magnetic resonance spectroscopy are compared for crystals with different compositions. The results show clearly that the possibility to vary the concentration of intrinsic defects offers extraordinary informative opportunities for the investigation of the fundamental physics of the material.

INTRODUCTION

The fast development of opto- and acoustoelectronics as well as photonics requires new ferroelectric materials with improved characteristics. Sometimes the requested quality or parameters can be successfully achieved with a well-known, commonly used material, if its defect system is changed in a prescribed way.

In 1960-1980, the conventional, congruent lithium niobate ($LiNbO_3$, LN) was considered to be a well-known and well-characterized crystal. Conventional LN crystals, grown from a congruent melt with lithium deficiency ($X_{melt} = X_{Crystal} \approx 48.4\%$, where $X = [Li]/([Li]+[Nb])$, contain a rather large percentage of intrinsic (non-stoichiometric) defects.

About 10 years ago, our group found a method to obtain LN crystals with a strongly reduced concentration of intrinsic defects [1, 2]. These samples are often called Li-rich, nearly stoichiometric or stoichiometric crystals. The most perfect among them represent regularly ordered crystals (ROC) [3] or are very close to them. We predicted that such crystals should have properties, which are quite different from those of congruent crystals [4, 5]. This suggestion had considerable impact leading to some kind of revolution in understanding the new and very rich possibilities of stoichiometric materials. Strong changes of crystal properties with the decrease of the concentration of intrinsic defects were found: a tremendous narrowing of spectral lines and corresponding increase of spectral resolution, 5-100 times lower electric field for domain switching, two order slower rate of carrier recombination, two order higher sensitivity to holographic recording, non-linear dependence of the ultraviolet absorption edge on crystal composition, different dynamical mechanical properties etc. [6, 7, 8, 9, 10, 11, 12, 13, 14, 15, 16].

Several ways to obtain crystals with low concentrations of intrinsic defects have been developed up to now: growth from melts with Li excess (up to X_{melt}=60%) [17, 18], post-growth vapor transport equilibration (VTE) treatment [19, 20], growth from melts, to which potassium has been added [1, 21] (later on labeled $LN_{(K)}$), pulse laser deposition, micropulling down, laser heated pedestal growth etc.

Here we report results of our study of LN crystals by Electron Paramagnetic Resonance (EPR), Electron Nuclear Double Resonance (ENDOR) and optical absorption; these are the most sensitive and informative tools for the study of defect structures. Our goals are to demonstrate the strong dependence of the properties of LN on intrinsic and extrinsic defects and to present spectroscopic data, which could be used as cornerstones for theoretical calculations.

COMPARISON OF SPECTRA OF CONGRUENT AND STOICHIOMETRIC CRYSTALS

LN crystals with a very wide range of compositions (from $x_C \approx 46\%$ to $x_C \approx 50\%$), doped with different divalent and trivalent impurities, were studied. Below we shall present typical data mainly for congruent $x_C = 48.4\%$ and stoichiometric $LN_{(K)}$ with $x_C \approx 50\%$ doped with Cr^{3+} or Fe^{3+} as paramagnetic probes.

Optical absorption spectra of congruent and $LN_{(K)}$ samples doped with 1 wt.% of Cr are presented in Fig. 1. We observed also a strong blue shift of the fundamental absorption edge for $LN_{(K)}$ doped with 0.01 and 0.1 wt.% of Cr. This reflects a decrease of the intrinsic defect content in $LN_{(K)}$ crystals at low levels of chromium doping.

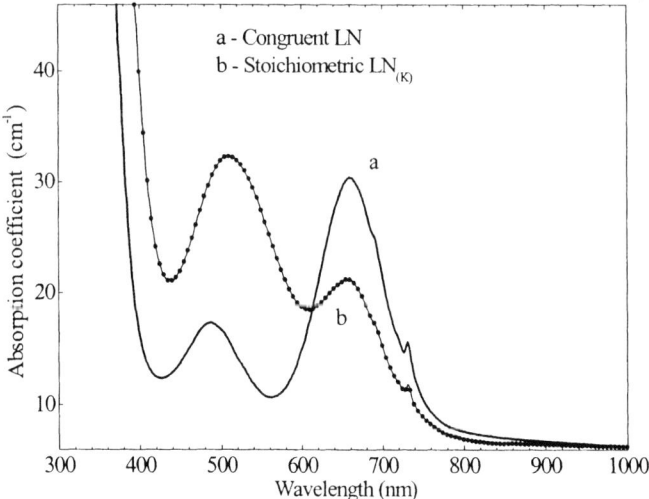

FIGURE 1. Optical absorption spectra of Cr doped crystals at room temperature. Light propagation along **y** axis, light polarization **E**∥**z**.

Three broad bands centered approximately at 340, 490 and 660 nm definitely increase with the rise of the Cr concentration. We supposed that the band at 660 nm could serve as the reference band for the characterization of the concentration of the well-known Cr_1^{3+} centers, which are always observed in congruent crystals. The appearance of a new band near 500-550 nm in the $LN_{(K)}$ sample has to be attributed to other Cr^{3+} centers. The concentration of the centers responsible for the absorption at 530 nm in stoichiometric $LN_{(K)}$ sample is several times higher than in congruent sample. To determine definitely the origin of the new optical band and the structure of centers responsible for it, we carried out EPR (Fig. 2) and ENDOR studies of these samples.

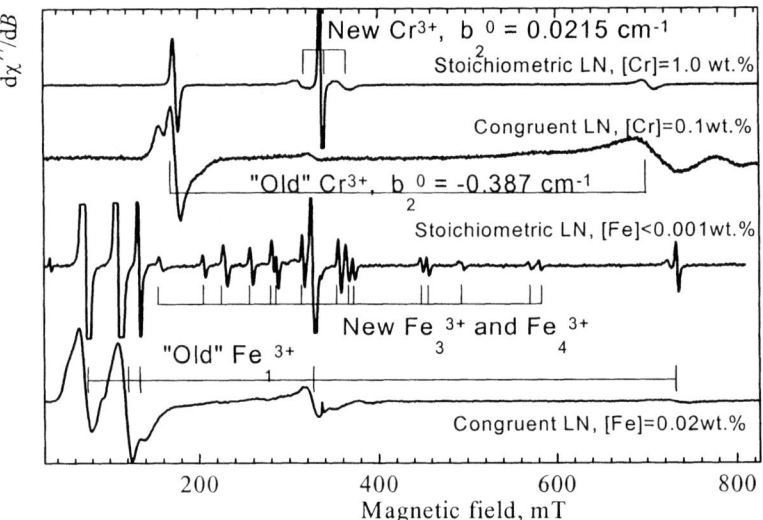

FIGURE 2. The EPR spectra of Cr^{3+} and Fe^{3+} in stoichiometric and congruent LN. T=5 K, ν=9.4 GHz.

In congruent LN crystals doped with chromium only the centers with zero-field splitting $\Delta = |2b_2^0| \approx 0.8$ cm^{-1} were observed. In stoichiometric crystals a new group of three EPR lines with a twenty times smaller splitting was registered. The new lines belong to another Cr^{3+} center, we labeled it Cr_{10}^{3+}. Since a similar appearance of two new Fe^{3+} centers was also registered for iron doped stoichiometric crystals (Fig. 2), we can conclude that the structure of impurity centers depends on the concentration of intrinsic defects.

The temperature dependence of the axial crystal field parameters b_2^0 for Cr_1^{3+} and Cr_{10}^{3+} centers has an opposite behavior (Fig. 3). In ferroelectric crystals the temperature dependence of the spin-Hamiltonian parameters can often be caused by the lattice transformation due to the phase transition. LN has such a transition from the ferroelectric R3c to the paraelectric R$\bar{3}$c one at $T_C \approx 1200$ C. The observed smooth dependence $b_2^0(T)$ for Cr_1^{3+} can originate from this transition [22]. However, in the temperature range between 4.2 and 400 K there are practically no changes of the positions of lattice ions, therefore, the disappearance of the axial crystal field for Cr_{10}^{3+} is related to the ion movements in the nearest surrounding of the Cr impurity.

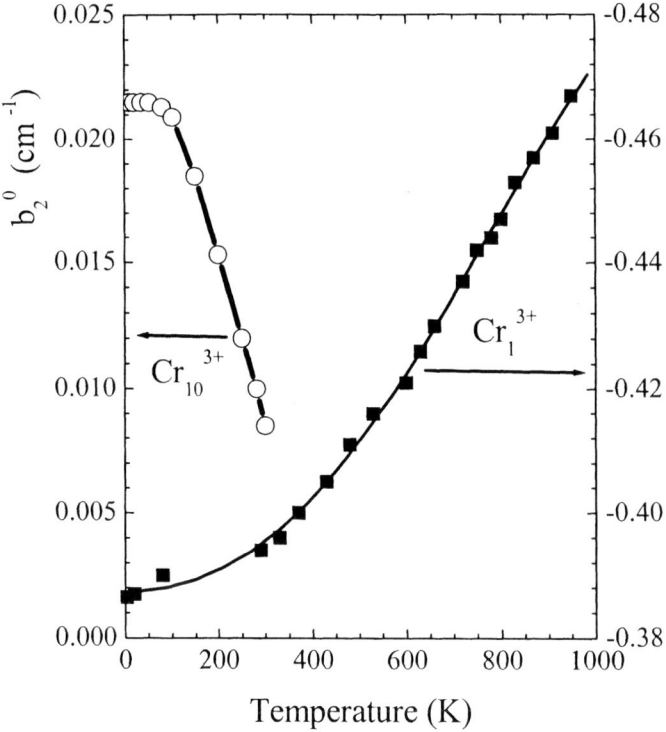

FIGURE 3. Temperature dependence of the axial crystal field parameter b_2^0 for Cr_1^{3+} and Cr_{10}^{3+}.

The ENDOR data gave us information about the hyperfine interactions of the chromium electrons with the surrounding nuclei and helped to determine the structures of these two centers [23, 24] in a final way. The models of the Cr_1^{3+} and Cr_{10}^{3+} centers are completely different (Fig. 4). In the first case Cr^{3+} substitutes Li^+ (Cr_{Li}^{3+}) and has an excess of positive charge, which can be compensated by intrinsic defects, like Nb and Li vacancies. These vacancies can be located very far from Cr^{3+} (distant compensation, the case of axial center Cr_1^{3+}) or in the nearest neighborhood (local compensation, low-symmetry centers [25]). In the second case the Cr^{3+} ion replaces Nb^{5+} and is negatively charged relative to the regular lattice. The used procedure for the growth of $LN_{(K)}$ crystals reduces the concentration of non-stoichiometric defects up to a very low level, and the Cr_1^{3+} centers bind the rest of conventional intrinsic defects. This creates the conditions, under which a chromium ion, looking for the charge compensator, enters the crystals together with some of the other available extrinsic defects. In our case Cr_{Nb}^{3+} associates with a H^+ ion, since three groups of lines around the Larmor frequency of hydrogen were found in the ENDOR spectra for Cr_{10}^{3+} [24].

Different groups of the ENDOR lines correspond to different locations of H$^+$ in the surrounding of Cr$_{Nb}^{3+}$. It means that both Cr$_{Li}^{3+}$ and Cr$_{Nb}^{3+}$ create families of similar centers; members of the families have different locations of their charge compensators.

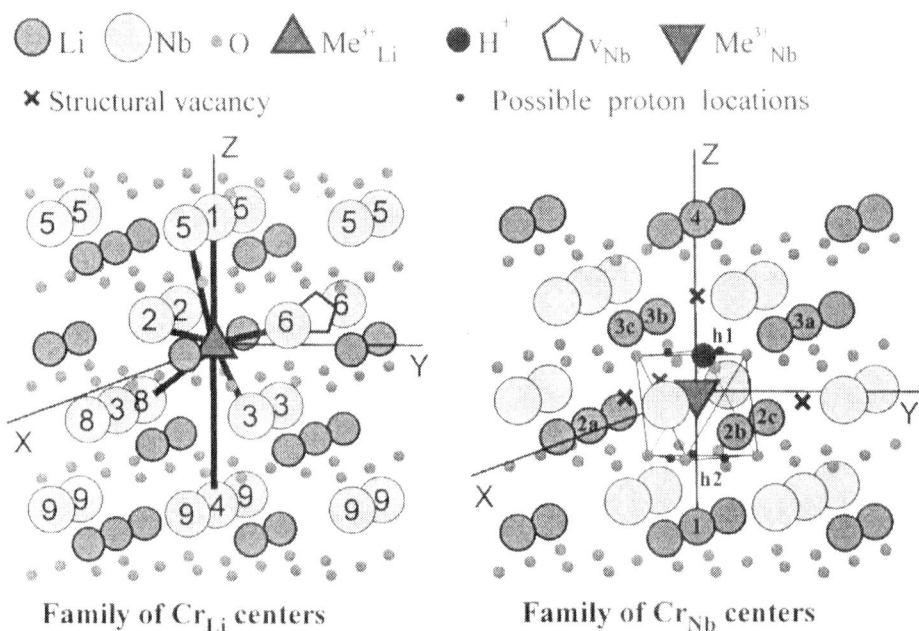

FIGURE 4. Models of chromium centres in lithium niobate. Thick lines on the left part connect Cr^{3+} with possible positions of the charge compensator. Black small circles and black crosses on the right part are possible locations of H$^+$.

According to the simplest theory the parameter of the isotropic (contact) hyperfine interaction of paramagnetic electrons with the i-th nucleus is given by

$$a^i = \frac{8\pi}{3} g_e g_n \beta \beta_n \Phi^i,$$
$$\Phi^i = |\psi(\vec{R}^i)|^2_\downarrow - |\psi(\vec{R}^i)|^2_\uparrow . \quad (1)$$

Here g_e, g_n, β, β_n are the electron and nuclear g-factors and magnetons, respectively, and $|\psi(\vec{R}^i)|^2_\downarrow$ and $|\psi(\vec{R}^i)|^2_\uparrow$ the total densities of electrons with spin down and spin up at the point $\vec{R}^i$ of a nuclear location.

Based on the measured parameters of the hyperfine interactions a^i, we found that for the Cr$_l^{3+}$ center the values of the electron density on the Nb nuclei are considerably higher than on the Li nuclei. This can be easily explained, if the chromium ion substitutes Li or occupies an octahedral vacancy site and has Nb nuclei in the nearest cation neighborhood. It was found that the spin densities on the surrounding nuclei are unexpectedly high. They are positive and decrease with rising distance from the chromium ion (Fig. 5). The densities at the Nb nuclei are several times larger than on the Li ones; this is obviously related to the stronger binding of the chromium electrons to the larger positive charge of Nb^{5+}.

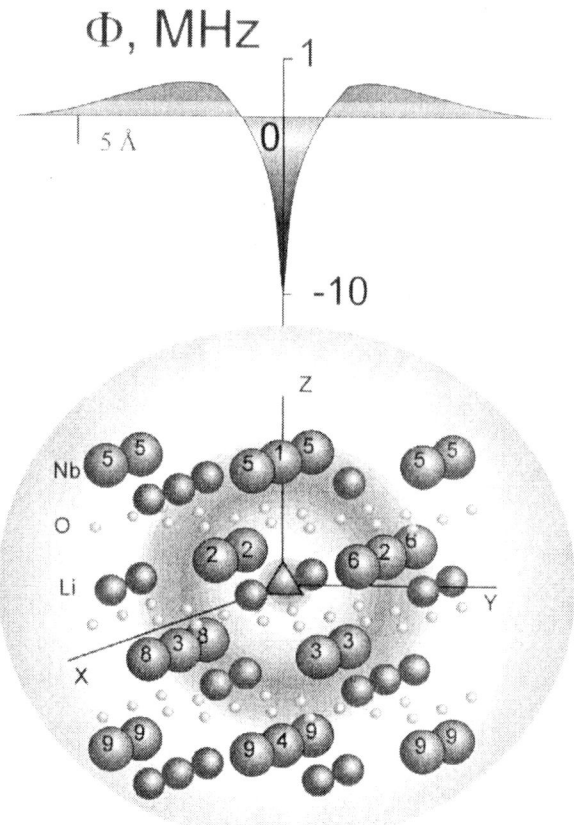

FIGURE 5. Distribution of electron spin density for the Cr_1^{3+} center derived from the ENDOR data. Numbers on circles denote shells of niobium nuclei.

In the case of Nb substitution (Cr_{10}^{3+}) the nearest cation surroundings are Li nuclei, and measured hyperfine interactions with these nuclei are the strongest ones. Rather high values of isotropic hyperfine interactions give us the evidence for a transfer of the electron density from chromium to neighboring nuclei.

DISCUSSION AND CONCLUSIONS

All microscopic and macroscopic properties of ferroelectric crystals depend to some extent on the concentration of intrinsic defects. For instance, the melting point of congruent LN is about 1240 C, however it decreases to about 100 C, if the crystal composition approaches the stoichiometric one. The temperature of the ferroelectric phase transition varies in the range from 1020 C (for Li-pure crystals) to 1190 C (Li-rich crystals). As a result of these opposite tendencies for stoichiometric composition the melting temperature is lower than the Curie temperature, and a stoichiometric crystal grows initially in the ferroelectric phase. The disappearance of intrinsic defects in stoichiometric crystals also leads to a measured decrease of lattice constants [5], a blue shift of the absorption band edge and a change of the electro-optical coefficients [26].

The refinement of the crystal lattice from non-stoichiometric defects leads also to a tremendous narrowing of the spectral lines of not only EPR (up to 10-40 times) and NMR, but also of Raman scattering (2 times) and luminescence (6-10 times) [27, 28], X-ray structural analyses and other techniques. This increases the resolution of spectroscopic methods considerably and, as a result, the quantity and quality of the obtained information.

The most successful way to study crystal properties proceeds from congruent crystals through nearly stoichiometric, stoichiometric and regularly ordered materials. In the regularly ordered materials the properties are not covered by the presence of intrinsic defects. The use of crystals with different x_C offers many advantages for the crystal investigation, especially, for the study of impurity centers. We found that positions occupied by impurity ions depend on the crystal nonstoichiometry. This interrelation of impurity locations and crystal composition is not very obvious and has to be studied further theoretically and experimentally. The found features give the opportunity to tailor the crystal properties by variation of x_C. In particular, it helped us to find crystals with optimal conditions for the observation of ENDOR signals.

We have shown that the EPR and ENDOR data give us very detailed and reliable information about the fundamental properties of LN crystals. The derived models of the impurity centers, the positions of the optical bands, the parameters of crystal fields and their temperature dependencies, the values of electron spin density, together with the changes of lattice constants and Curie temperature, can be efficiently used as corner stones for theoretical calculations.

The ideas developed in this work have a general character; therefore they should also be valid for other non-stoichiometric oxide ferroelectrics.

REFERENCES

[1] G.I.Malovichko, V.G.Grachev, L.P.Yurchenko, V.Ya.Proshko, E.P.Kokanyan, V.T.Gabrielyan. *Phys. Stat. Sol. (a)*, v. **133**, No 1, p. K29-K33 (1992).
[2] G.I. Malovichko, V.G. Grachev, E.P. Kokanyan, O.F. Schirmer, K. Betzler, B. Gather, F. Jermann, S. Klauer, U. Schlarb, M. Wöhlecke. *Appl. Phys.*, **A56**, 103 (1993).
[3] G.Malovichko, V.Grachev, O.Schirmer: *Appl. Phys.* **B**, **68**, 785-793 (1999).

[4] G.I.Malovichko, V.G.Grachev, O.F.Schirmer: *Solid State Commun.,* **89**, 195 (1994).
[5] G.Malovichko, O.Cerclier, J.Estienne, V.Grachev, E.Kokanyan, C.Boulesteix: *J. Phys. Chem. Solids,* **56**, 1285 (1995).
[6] G.Malovichko. Proc. of the 7-th Top.Meet. on *Photorefractive Materials, Effects and Devices,* Denmark, June 27-30 1999, *OSA TOPS,* **27**, 59-66.
[7] L. Hesselink, S. Orlov, A. Liu, A. Akella, D. Lande, R. Neurgaohkar: *Science,* **282**, 1089 (1998).
[8] H. Guenter, R. Macfarlane, Y. Furukawa, K. Kitamura, R. Neurgaonkar: *Appl. Optics,* **37**, 7611 (1998).
[9] V.Gopalan, T.E.Mitchell, Y.Furukawa, K.Kitamura: *Appl. Phys.Lett.,* **72**, 1981 (1998).
[10] A.Grisard, E.Lallier, K.Polgar, A.Peter. Abstracts of Fourth Annual Meeting of the COST Action P2 „*Application of Nonlinear Optical Phenomena*" and Workshop on LiNbO3, Budapest, 16-19 May 2001, p. 2.2 (2001).
[11] L. Kovacs, G. Ruschhaupt, K. Polgar, G. Corradi, M. Woehlecke: *Appl. Phys. Lett.* **70**, 2801 (1997).
[12] A. de Bernabe, C. Prieto, A. de Andres: *J. Appl. Phys.,* **79**, 143 (1995).
[13] F. Abdi, M. Aillerie, P. Bourson, M.D. Fontana, K. Polgar: *J. Appl. Phys.,* **84**, 2251 (1998).
[14] L. Kovacs, G. Ruschhaupt, K. Polgar, G. Corradi, M. Woehlecke: *Appl. Phys. Lett.* **70**, 2801 (1997).
[15] A. de Bernabe, C. Prieto, A. de Andres: *J. Appl. Phys.,* **79**, 143 (1995).
[16] F. Abdi, M. Aillerie, P. Bourson, M.D. Fontana, K. Polgar: *J. Appl. Phys.,* **84**, 2251 (1998).
[17] G.I.Malovichko, V.G.Grachev, V.T.Gabrielyan, E.P.Kokanyan: *Sov.: Phys. Solid State,* **28**, 1453 (1986).
[18] N.Iyi, K.Kitamura, F.Izumi, J.K.Yamamoto, T.Hayashi, H.Asano, S.Kimura: *J. Sol. State Chem.,* **101**, 340 (1992).
[19] P.F.Bordui, R.G.Norwood, C.D.Bird, G.D.Calvert: *J. Cryst. Growth,* **113**, 61 (1991).
[20] D.H.Jundt, M.M.Fejer, R.L.Byer: *IEEE J. QE,* **26**, 135 (1990).
[21] K.Polgar, A.Peter, L.Kovacs, G.Corradi, Zs.Szaller: *J. Cryst. Growth,* **177**, 211, (1997).
[22] V.G.Grachev, G.I.Malovichko: *Sov.: Phys. Solid State,* **27**, 424 (1985).
[23] V.Grachev, G.Malovichko. *Phys. Rev.,* **B62**, 7779 (2000).
[24] G.Malovichko, V.Grachev, A.Hofstaetter, E.Kokanyan, A.Scharmann, O.Schirmer. *Phys. Rev.,* **B65**, 224116 (2002).
[25] G.Malovichko, V.Grachev, E.Kokanyan, O.Schirmer. *Phys. Rev.,* **B59**, 9113 (1999).
[26] F. Abdi, M. Aillerie, P. Bourson, M.D. Fontana, K. Polgar. *J. Appl. Phys.,* **84**, 2251 (1998).
[27] F.Lhomme, P.Bourson, M.D.Fontana, G.Malovichko, M.Aillerie, E. Kokanyan.- *J. Phys.: Condensed Matter,* **10**, 1137-46 (1998).
[28] G.M. Salley, S.A.Basun, A.A.Kaplyanskii, R.S.Meltzer, K.Polgar, U.Happek. *J. Lumines.,* **87-9**, 1133 (2000).

Quantum chemical modeling of electron and hole polarons in ABO_3 perovskites

R. I. Eglitis*, E. A. Kotomin[†**], G. Borstel* and V.S. Vikhnin[‡]

*Fachbereich Physik, Universität Osnabrück, D-49069 Osnabrück, Germany
[†]MPI für Festkörperforschung, Heisenbergstrasse 1, D-70569 Stuttgart, Germany
[**]Institute for Solid State Physics, 8 Kengaraga str., Riga LV-1063, Latvia
[‡]A.F. Ioffe Physical Technical Institute, 194021 Saint-Petersburg, Russia

Abstract. We present the state–of–the art of large scale computer modeling of electron and hole polarons in advanced ABO_3 perovskites performed by means of semi–empirical quantum chemical (INDO) method. Our calculations confirm existence of the self–trapped electron polarons in $KNbO_3$, $KTaO_3$, $BaTiO_3$, and $PbTiO_3$ crystals. The self–trapped electron is mostly localized on B-type ion due to a combination of breathing and Jahn–Teller modes of nearest 6 O ion displacements. The relevant lattice relaxation energies are typically 0.2–0.3 eV, whereas the optical absorption lies at 0.7–0.8 eV, respectively. The optical absorption energies for the electron polaron bound to cation impurity in $KNbO_3$ (0.88 eV) and hole polaron bound to K vacancy ($\approx$ 1 eV) are also in a good agreement with the relevant experimental data.

INTRODUCTION

Ternary ABO_3 ferroelectric perovskites have numerous technological applications [1]. In particular, $KNbO_3$ crystals are widely used for laser frequency doubling, and doped $KTaO_3$ is prospective for electrically controlled holographic and graded pyroelectric devices. Their properties are influenced by point defects, primarily by vacancies. Transient optical absorption around 1 eV has been associated recently [2], in analogy with other perovskites, with a hole polaron (a hole bound to some defect), whereas IR absorption (around 0.8 eV) with an electron polarons [3]. The ESR study of $KNbO_3$ doped with Ti^{4+} provides strong evidence that holes can be trapped by negatively charged defects [4]. For example, hole polarons were found in $BaTiO_3$ doped with Na or K alkali ions replacing for Ba and thus forming a negatively charged site attracting a hole [5]. Primary candidates for such defects are cation vacancies. In irradiated MgO they are known to trap one or two holes giving rise to the V^- and V^o centers [6] which are in fact bound hole polaron and bipolaron, respectively. For a long time it was believed that the electron self–trapping is not energetically favourable in ionic solids due to the large energy loss necessary for an electron localization on a single cation, which is the first stage of the trapping process, and is not compensated by the energy gain due to crystal polarization, at the second stage of the self–trapping. However in 1993 self–trapped electrons (STE) were reported in $PbCl_2$ crystals [7], and in 1994 another ESR evidence appeared [8] for the electron self–trapping in $LiNbO_3$ perovskite crystals, accompanied by the IR absorption band around 1 eV. Lastly, very recently, experimental arguments for the existence of the STE in $KNbO_3$ were discussed in Ref. [3].

INDO METHOD

The modifed INDO approximation[9, 10, 11, 12] is a simplified version of the Hartree–Fock method which permits large–scale modeling of atomic and electronic structure of materials; it has been successfully applied to defect calculations in many oxides [13, 14], including perovskites [15, 16, 17, 18, 19, 20, 21]. With decreasing temperature, $KNbO_3$ undergoes a series of phase transitions from a paraelectric cubic to ferroelectric tetragonal, orthorhombic, and rhombohedral phases. The displacement of Nb atoms along <100>, <110> and <111> corresponds to distortions of these symmetries and thus models the transitions to the three relevant ferroelectric phases. Recently [17] the $2 \times 2 \times 2$ supercell model has been used in INDO calculations of the $KNbO_3$ crystal energy as a function of Nb displacements along the three principial directions. Consistent with the experimental data [22], the <111> displacements provided the lowest energy minimum, and <110> the next lowest. All this is in good agreement with the first principles FP–LMTO results. The INDO parametrization [17] was further checked by calculations of the atomic positions in the orthorhombic and rhombohedral phases, and results are in a good agreement with neutron diffraction data [22]. Calculated phonon frequencies in the cubic and rhombohedral phases also agree well with the experimental results. The INDO calculations indicate considerable covalency of the chemical bonding in a pure $KNbO_3$: the effective (static) atomic charges are +0.55 e for K, +2.02 e and -0.85 e for O. The relevant INDO parametrization for $KTaO_3$ is described in Ref. [18]. The effective charges on atoms in $KTaO_3$ (+0.62 e for K, +2.23 e for Ta and -0.95 e for O) show slightly higher ionicity in $KTaO_3$ as compared with $KNbO_3$. In this paper, we used $2 \times 2 \times 2$ times extended supercells of 40 atoms atoms for calculations of free electron polarons in $KNbO_3$ and $KTaO_3$ and even larger – $3 \times 3 \times 3$ times extended supercells, containing 135 atoms, for calculations of free electron polarons in $BaTiO_3$, $PbTiO_3$ and Mg–bound electron polaron in $KNbO_3$.

MAIN RESULTS

Electronic polarons in ABO_3 perovskites

We have modeled the free *electron* polarons in ABO_3 perovskites, starting with $KNbO_3$. We allowed six nearest oxygen atoms in the O_6 octahedron around a central Nb atom in $KNbO_3$ to relax, in order to find the energy minimum of the system. All other Nb and K atoms, as well as the rest of O atoms, were kept fixed at their perfect lattice sites. According to our INDO calculations, initially the ground state is three-fold degenerate t_{2g} (at Γ point of the Brillouin Zone). This degeneracy is lifted as a result of the combination of the breathing mode and the Jahn–Teller effect: 1.4 % a_o (a_o is the lattice constant) outward displacement of the four nearest equatorial O atoms (with the energy gain of 0.12 eV) an an *inwards* 1 % a_o relaxation of the two oxygens along the z direction (an additional gain of 0.09 eV). That is, the total lattice energy gain is 0.21 eV. As a result, a considerable (0.5 e) electron density is localized on the central Nb atom producing three closely spaced energy levels in the band gap. They consist mainly of the xy, xz and yz atomic orbitals (split t_{2g} energy level in an isolated ion, where xz and yz are degenerate and lie above the xy level), whereas two further empty levels are located close to the conduction band bottom. The electron polaron absorption energy was estimated by means of ΔSCF to be 0.78 eV. The relevant absorption process corresponds to an electron transfer to the nearest Nb atom. This absorption band has been indeed observed under picosecond laser exciton [3]. In similar calculations for $KTaO_3$ we obtained the optical absorption of 0.75 eV, with the following ground state relaxation: 1.7 % a_o for four O equatorial atoms and 1.2 % a_o inward displacement of two other O atoms along the z axis. This results in the total lattice relaxation energy of 0.27 eV. This is larger than in the $KNbO_3$ case because of higher $KTaO_3$ ionicity. Our quantum chemical INDO calculations argue for an existence of the self–trapped *electrons* also in $BaTiO_3$, and $PbTiO_3$, associated with the net lattice relaxation energy of 0.24 eV and 0.22 eV, respectively. An electron in the ground state occupies t_{2g} orbital of a single Ti^{3+}. Relaxation of other atoms, more distant from Ti^{4+} than 6 nearest O atoms, gives only a very small ($\approx$ 0.02 eV) additional contribution to the total energy gain in $BaTiO_3$ [19].

Bound electron polaron

Along with the self–trapped (intrinsic) electron polaron in $KNbO_3$ discussed above, we have also modeled the electron polaron (Nb^{+4}) bound to the *nearest* neighbor Mg^{2+} impurity. As initial guess, we used the atomic configuration for the self–trapped electron polaron described above. However, the Coulomb field of Mg impurity results in a considerable additional relaxation of three O atoms nearest to both Nb and Mg ions. The optimized displacement of these three O atoms directed towards the Mg atom along the relevant cubic faces is 1.84 % a_o and gives an additional 0.15 eV energy gain additionally to already existing 0.21 eV relaxation energy. It means that a total relaxation energy turns out to be 0.36 eV, i.e. the bound electron polaron is more stable than the free polaron. As a result of asymmetric O relaxation, similarly to a free–electron polaron, around 0.5 e is localized on the central Nb atom. In the bound electron polaron the degeneracy of Nb 4d atomic orbitals (having t_{2g} symmetry and consisting mainly of Nb $4d_{xy}$, d_{xz} and d_{yz} atomic orbitals, where an electron polaron is localized) is lifted. According to our calculations, the absorption energy of bound electron polaron exceeds by 0.1 eV that for the free electron polaron and equals 0.88 eV (Table 1).

Bound hole polarons

We modeled hole polarons bound to cation vacancies. In particular, our $KNbO_3$ calculations show that there are two energetically favourable atomic configurations in which a hole is well localized nearby K vacancy: one–site and two site (molecular) polarons

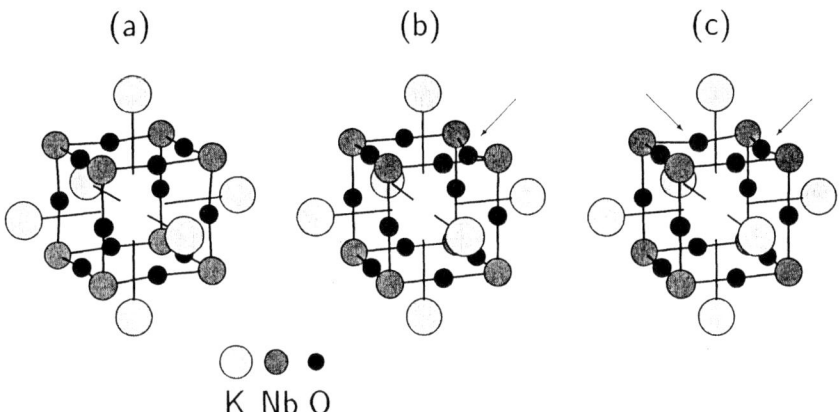

FIGURE 1. Uniform relaxation of O atoms around the K vacancy in $KNbO_3$ (a) and the relaxation corresponding to the formation of one–site (b) and two–site (c) bound hole polarons. Arrows in (b) and (c) indicate the displaced O atoms.

(Fig. 1). In the former case, a single O^- ion is displaced in our INDO calculations *towards* the K vacancy by 3 % a_o. Simultaneously, 11 other nearest oxygens surrounding the K vacancy are slightly displaced outwards from the vacancy.

In contrast, in the two–site (molecular) configuration, a hole is shared by the *two* O atoms which approach each other by 3.5 % and their center–of–mass shifts towards a vacancy by 2.2 % a_o. The relevant lattice relaxation energies are close, 0.4 eV and 0.53 eV, respectively. Despite the fact that the two–site configuration of a polaron is lower in energy, proper incorporation of the electron correlation effects could reverse this delicate energy balance. In its turn, ab initio FP LMTO calculations give close atomic displacements but predict smaller relaxation energies, 0.12 eV and 0.18 eV, respectively [16]. In spite of general observation of a considerable degree of covalency in $KNbO_3$ and contrary to a delocalized character of the F center state, the one–site polaron state remains well localized at a single displaced O atom, with only a small contribution from atomic orbitals of other O ions but not from those of the K or Nb ions. In agreement with a general theory for the small–radius polarons in ionic solids, the optical absorption corresponds to a hole transfer to the state delocalized over nearest oxygens. The calculated absorption energies for one–site and two–site polarons are close, both around 1 eV which is considerably smaller than the experimental value for a hole polaron trapped near Ti [4]. This shows that the optical absorption energy of small bound polarons could be strongly dependent on the defect involved.

CONCLUSIONS

Our INDO calculations confirm a possibility of the self–trapped electron polaron formation in $KNbO_3$, $KTaO_3$, $BaTiO_3$, and $PbTiO_3$ crystals. The self–trapped electron

TABLE 1. Optical absorption energies of electron and bound one–site (1-s) and two–site (2-s) hole polarons in ABO_3 perovskites as calculated by means of the INDO method. Results for lattice relaxation energies are compared with *ab initio* FP–LMTO calculations.

Crystal	Type of polarons	Absorption energy (eV)	Relaxation energy (eV) INDO LMTO[16]	
$BaTiO_3$	Electron polaron	0.69	0.24	–
$PbTiO_3$	Electron polaron	0.73	0.22	–
$KNbO_3$	Electron polaron	0.78	0.21	–
	Bound electron-polaron	0.88	0.36	–
	Bound 1-s hole polaron	0.90	0.40	0.12
	Bound 2-s hole polaron	0.95	0.53	0.18
$KTaO_3$	Electron polaron	0.75	0.27	–

is mostly localized on B–type ion due to a combination of breathing and Jahn–Teller modes of nearest six O ions displacements. The lattice relaxation and the optical absorption energies are typically ~ 0.2 eV and ~ 0.8 eV, respectively. The calculated optical absorption (0.78 eV) for $KNbO_3$ is close to the experimental observation [3].

As follows from results of our calculations, both one–center and two–center configurations of bound hole polarons in $KNbO_3$ are energetically favorable and close in energy (with slight preference for the two–center configuration). The calculated bound hole polaron absorption (≈ 1 eV) is close to the observed short–lived absorption band energy [2]; hence this band could indeed arise due to a hole polaron bound to a cation vacancy. We are not aware of any *ab initio* calculations of free polarons in ABO_3 perovskites, which could be of great interest.

ACKNOWLEDGEMENTS

Authors are indebted to N.E. Christensen, J. Maier and A. Postnikov for many stimulating discussions. This study is supported in part by DFG, by NATO (grant PST.CLG.977561), and by Excellence Centre for Advanced Materials Research and Technology (Riga, Latvia).

REFERENCES

1. H.-J. Donnerberg, *Atomic Simulations of Electro–Optical and Magnetico–Optical Materials*, Springer Tracks in Modern Physics, Vol. **151**, Berlin, 1999.
2. L. Grigorjeva, D. Millers, E.A. Kotomin and E.S. Polzik, Solid State Commun. **104**, 327 (1997).
3. Q. Yong, K.B. Ucer, R.T. Williams, L. Grigorjeva, D. Millers and V. Pankratov, Nucl. Instr. Methods B **191**, 98 (2002).
4. E. Possenriede, O.F. Schirmer, H.-J. Donnerberg, J. Phys.: Cond. Matt. **1**, 7267 (1989).
5. T. Varnhorst, O.F. Schirmer, H. Krose, R. Scharsfswerdt and T.W. Kool, Phys. Rev. B **53**, 116 (1996).
6. Y. Chen and M.M. Abraham, J. Phys. Chem. Sol. **51**, 747 (1990).
7. S.V. Nistor, E. Goovaerts, D. Schoemaker, Phys. Rev. B **48**, 9575 (1993).
8. B. Faust, H. Müller and O.F. Schirmer, Ferroelectrics **153**, 297 (1994).
9. J.A. Pople and D.L. Beveridge, *Approximate Molecular Orbital Theory* (McGraw–Hill, New York, 1970).

10. A.L. Shluger, Theoret. Chim. Acta **66**, 355 (1985).
11. E. Stefanovich, E. Shidlovskaya, A.L. Shluger, and M. Zakharov, Phys. Stat. Sol. B **160**, 529 (1990).
12. A.L. Shluger and E. Stefanovich, Phys. Rev. B **42**, 9664 (1990).
13. E. Stefanovich, A.L. Shluger, and C.R.A. Catlow, Phys. Rev. B **49**, 11560 (1994).
14. E.A. Kotomin, A. Stashans, L.N. Kantorovich, A.I. Livshitz, A.I. Popov, I. Tale, and J.-L. Calais, Phys. Rev. **51**, 8770 (1995).
15. R.I. Eglitis, N.E. Christensen, E.A. Kotomin, A.V. Postnikov, and G. Borstel, Phys. Rev. B **56**, 8599 (1997).
16. E.A. Kotomin, R.I. Eglitis, A.V. Postnikov, G. Borstel, and N.E. Christensen, Phys. Rev. B **60**, 1 (1999).
17. R.I. Eglitis, A.V. Postnikov, and G. Borstel, Phys. Rev. B **54**, 2421 (1996).
18. R.I. Eglitis, A.V. Postnikov, and G. Borstel, Phys. Rev. B **55**, 12976 (1997).
19. R.I. Eglitis, E.A. Kotomin, and G. Borstel, J. Phys.: Cond. Matter **14**, 3735 (2002).
20. R.I. Eglitis, E.A. Kotomin, V.A. Trepakov, S.E. Kapphan, and G. Borstel, J. Phys.: Condens. Matter **14**, L647 (2002).
21. R.I. Eglitis, E.A. Kotomin, G. Borstel and S. Dorfman, J. Phys.: Cond. Matter **10**, 6271 (1998).
22. A.W. Hewat, J. Phys. C **6**, 2559 (1973).

Calculations of Perovskite Polar Surface Structures

E. Heifets[a*], R. I. Eglitis[b], E. A. Kotomin[c,d], W.A. Goddard III[a], and G. Borstel[b]

[a] *Materials and Process Simulation Center, Beckman Institute (139-74), California Institute of Technology, MS 139-74, Pasadena CA 91125, USA*
[b] *Universität Osnabrück, Fachbereich Physik, D-49069 Osnabrück, Germany*
[c] *Institute for Solid State Physics, University of Latvia, 8 Kengaraga, Riga LV-1063, Latvia*
[d] *Max Planck Insitut für Festkörperforschung, Heisenbergstr., 1, D-70569 Stuttgart, Germany*

Abstract. Results of calculations for the (110) polar surfaces of three ABO_3 perovskites – STO, BTO and LMO – are discussed. These are based on *ab initio* Hartree-Fock method and classical Shell Model. Both methods agree well on both surface energies and on near-surface atomic displacements. A novel model of the "zig-zag" surface termination is suggested and analyzed. Considerable increase of the Ti–O chemical bond covalency nearby the surface is predicted for STO.

I INTRODUCTION

Thin films of ABO_3 perovskite ferroelectrics are important for many technological applications, including catalysis, microelectronics, substrates for growth of high T_c superconductors, where surface structure and its quality are of primary importance [1,2]. Several *ab initio* quantum mechanical [3-9] and classical Shell Model (SM) [10,11] theoretical studies were published recently for the (100) surface of $BaTiO_3$ and $SrTiO_3$ crystals (hereafter BTO and STO). In order to study dependence of the surface relaxation properties on exchange-correlation functionals and localized/plane wave basis sets used in calculations, we performed recently a detailed comparative study based on a number of different quantum mechanical techniques [12-14]. The main conclusion was drawn there that the Hartree-Fock (HF), Density Functional Theory (DFT), and even SM calculations give quite similar results for the atomic structure relaxation and surface energies.

We performed also SM calculations of the atomic relaxation for the polar (110) surfaces of STO and BTO [11]. To our knowledge, only semi-empirical quantum mechanical calculations [15] exist so far for such perovskite surfaces. In this paper, we present a novel, "zig-zag" model for the polar (110) surface termination, and

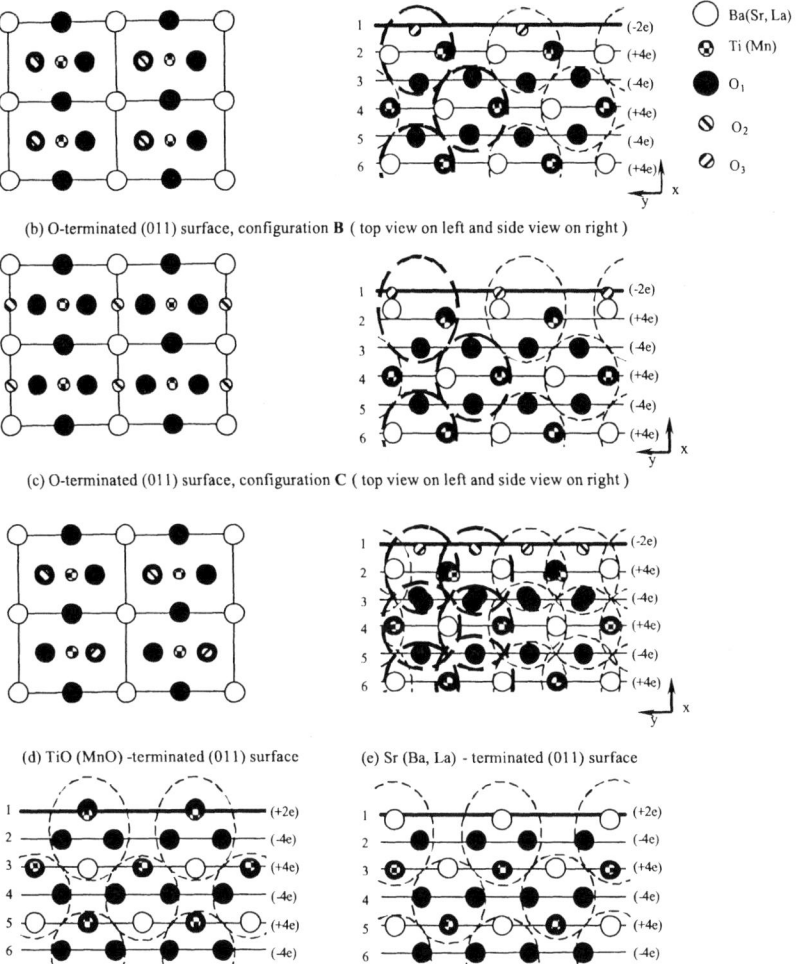

FIGURE 1. The top and side view of the (110) perovskite surfaces. (a), (b), (c) are three possible configurations of O-terminated surface, (d) and (e) same for TiO- and Sr (Ba) terminations, see details in the text.

perform calculations of the relaxed atomic structure of the STO, BTO and LaMnO$_3$ (LMO) (110) surfaces, combining the *ab initio* HF and classical SM methods.

II METHODS AND SURFACE MODELS

In this study, we restrict ourselves to simulations of ABO$_3$ perovskites in the cubic crystalline phase, stable at high temperatures. Description of SM and its parameterization is available in Ref. [11]. Use of this model permits us to find the atomic relaxation for several hundreds of atoms, surface energies, along with the surface polarization, characterized by dipole moments perpendicular and parallel to the surface. This information is of great importance for analysis of dielectric properties of thin ferroelectric films. We allow atoms in a given number of near-surface planes (varied from 2 to 16) to relax to the minimum of total energy, and then analyze, how the major properties are affected by a number of relaxed planes. This is important since in time-consuming ab initio calculations only 2-3 near-surface planes are typically allowed to relax.

In HF calculations for STO, performed to check accuracy of the SM calculations, we use the CRYSTAL–98 computer code (see [16] and references therein for description of all mentioned techniques), in which both (HF/DFT) types of calculations are implemented on equal grounds. Unlike previous plane–wave calculations, this code uses the localized Gaussian-type basis set. In our simulations we applied the basis set recommended for SrTiO$_3$ [16]. Another advantage of the CRYSTAL–98 code is its treatment of purely 2D slabs, without an artificial periodicity in the direction perpendicular to the surface, commonly employed in all previous surface–band structure calculations (e. g., [3,9]). In HF calculations, along with the atomic displacements in several planes near the surface, we calculate effective (Mulliken) atomic charges, bond populations between nearest atoms, characterizing the covalency effects, and dipole, quadrupole moments characterizing atomic polarization and deformation. In particular, the dipole moments p_z and p_y characterize atomic deformation and polarization along the z axis and the y axis perpendicular and parallel to the surface, respectively.

For optimization of atomic coordinates through minimization of the total energy per unit cell, we use our own computer code that implements the Conjugated Gradients optimization technique with numerical computation of derivatives. Using this code, we optimized the atomic positions in three top layers of a STO slab consisting of seven planes.

The problem of the (110) polar surface modeling is that it consists of charged planes. This is why, if the (110) surface were to be modeled exactly as one would expect after crystal cleavage, it would have in *infinite* dipole moment perpendicular to the surface, which makes such the surface unstable [17]. To avoid this problem, in our calculations we removed half the O atoms from from the O-terminated surface, the Sr(Ba) atoms from the Ti–terminated surface, and both the Ti and O atoms from the Sr (Ba) – terminated surface. As a result, we obtain the surface with

TABLE 1. Atomic relaxation of three top layers (in percent of the lattice constant) for four terminations, calulated by means of the *ab initio* HF and Shell Model [11].

Ti–O terminated		SM		HF	
Layer	Atom	δz	δy	δz	δy
1	Ti	-5.99		-6.49	
1	O	8.48		6.85	
2	O	-1.72		-1.47	
3	O	-4.10		-3.85	
3	Ti	2.14		2.20	
3	Sr	-6.96		-5.78	
O–terminated	A-type				
1	O	-14.2	-8.54	-10.41	-10.53
2	Ti	-2.37	-8.27	-1.36	-7.71
2	Sr	4.10	-10.79	2.20	-7.30
2	O	5.71	8.20	6.65	6.15
3	O	-11.06	-11.01	-7.02	-7.46
O–terminated	B-type				
1	O	-2.78		-3.95	
2	Ti	-5.14		-4.26	
2	Sr	30.32		22.67	
2	O	9.68		8.23	
3	O	-2.41		-1.68	
O–terminated	C-type				
1	O	-13.76	-9.08		
2	Ti	-4.87	-5.52		
2	Sr	4.31	0.0		
2	O	1.21	0.0		
3	O	-9.60	7.54		
Sr–terminated					
1	Sr	-19.07		-17.38	
2	O	3.18		2.72	
3	Sr	4.67		3.95	
3	O	-0.25		-0.21	
3	Ti	-0.89		-0.86	

charged planes but a zero dipole moment (before atomic relaxation). The relevant surface cells are built from neutral five–atom elements from three successive planes which are shown as encircled dashed ellipses in Fig.1.

The initial atomic configuration for the O–terminated surface, where every second surface O atom is removed and others occupy the same sites as in the bulk structure, we call asymmetric (A), Fig. 1.a. Since such a removal of half of O atoms disturb the balance of interatomic forces along the surface, we also studied another, symmetric initial surface configuration (B) in which the O_2 atom is placed in the *middle* of the distance between two equivalent O atoms in the bulk (Fig. 1.b). The A-type surface reveals considerable atomic displacements not only perpendicular to the surface, but also parallel to the surface. Preliminary results for the A-, B–cases were discussed in Ref. [11]. In this paper, we study one more configuration (C) which corresponds to the 2×1 surface reconstruction where O atoms are removed in pairs of nearest surface cells in a "zig–zag" way (O_2 and O_3 in Fig. 1.c). In this case, there is no artificial dipole moment parallel to the surface, in contrast to the case A.

The effective charges for Ti and O ions, both in the bulk and on the (100) surface, calculated by means of the HF and DFT methods [13], are much smaller than formal ionic charges (4 e , -2 e, respectively). This arises due to partly covalent nature of the Ti-O chemical bond. In contrast, Sr charge remains close to the formal charge, +2 e. The Ti–O chemical bond covalency is confirmed by calculated bond populations, which vary from 0.05 e (DFT–LDA) to 0.11 e (HF), dependent on the particular method. Obviously, there is no chemical bonding between any other types of atoms, e.g. Sr–O or O–O.

Our atomic displacements in the (100) outermost $SrTiO_3$ planes, obtained by means of various *ab initio* methods, were analyzed recently [13,14].

In all calculations of the (100) surface energy, that for the SrO termination is only slightly smaller than for the TiO_2 termination. Thus, both (100) surfaces can co-exist, in agreement with the experimental observation [13].

III MAIN RESULTS

A Shell Model

Atomic relaxation of the first three top layers are given in Table 1. SM calculations predict large, ≈ 14 % a_0 *rumpling* for the TiO–terminated surface (the distance along the z axis between O and Ti atoms displaced from the first plane in opposite directions). Atomic displacements in the third plane are still considerable, unlike the (100) case. For the Sr-termination, top Sr atoms are displaced inwards, by ≈ 19 %, whereas the O atoms in the second plane go outwards, by ≈ 3 %.

The top O atom on the O–terminated symmetric surface (A) is strongly displaced inwards, by ≈ 14 %, whereas the Ti atom in the second plane is also displaced inwards, but only by ≈ 2 %. Along with the displacements along the z axis perpendicular to the surface, all atoms here reveal also considerable displacements parallel to the surface. This results in the dipole moments p_y, to be duscussed below.

In contrast, for the symmetric termination B, atoms are displaced only along the z axis and reveal much smaller displacements (e.g., the top O atoms go inward, by $\approx$ 3 %.) However, Sr atoms in the second plane are strongly ($\approx$ 30 %) displaced outwards the surface. Similar effect was observed by us for the (100) surface [11]. The (110) surface polarization, characterized by the relaxation–induced dipole moments (per surface unit cells) p_z and p_y perpendicular and parallel to the surface, respectively, is shown in Fig. 2. For the asymmetric termination (A) the surface polarization p_z oscillates around 1.1 e Å, with an increase of a number of of relaxed near– surface planes (varied between 2 and 16). This is accompanied by a considerable dipole moment p_y parallel to the surface. In contrast, the p_z dipole moment for the B-type termination rapidly saturate at 1.13 e Å whereas p_y strives for zero, with an increase of a number of relaxed layers. Lastly, for the "zig-zag" termination C, p_z oscillates around 1.1 e Å similarly to the asymmetric case (A) but without any dipole moment parallel to the surface. In other words, surface relaxation of cubic perovskite structure leads to the considerable polarization perpendicular to the surface which results from near-surface relaxation. This could considerably affect the dielectric properties of thin films.

Table 2 shows cosiderable difference for surface energies for STO, BTO and LMO (110) surfaces, obtained for two and 16 relaxed layers. (Our calculations show that the (110) surface energy saturates at about 6-8 relaxed layers only, whereas in *ab initio* calculations only 1-2 layers are typically relaxed.) Unlike the (100) surface, different (110) surface terminations strongly differ in energies. For all three perovskites, the novel, "zig-zag" termination (C) is lowest in surface energy.

B Hartree–Fock Calculations

The HF–calculated atomic relaxations for STO, shown in Table 1, confirm results of much simpler SM calculations. The agreement for all four termination is remarkable, indeed. This demonstrates that semi–empirical classical calculations with a proper parameterization could serve as a very useful tool for modeling perovskite thin films. Table 2 demonstrates also a good agreement between the HF calculated surface energies and those obtained by means of the SM (comparing in both cases results for two relaxed planes.)

We calculated also the effective atomic charges, dipoles and quadrupole moments for atoms near the surface. For the TiO termination, the Ti atom charge is reduced by 0.14 e, as compared to that in the bulk, due to the additional electron charge transfer from O atoms. The effective charge for O atoms turns out to be more positive, by 0.11 e. The Sr effective charge on the Sr-terinated surface is also reduced by 0.13 e, as for the TiO terminated (110) surface, and metal-terminated (100) surfaces. Changes in atomic charges in deeper layers become small. On

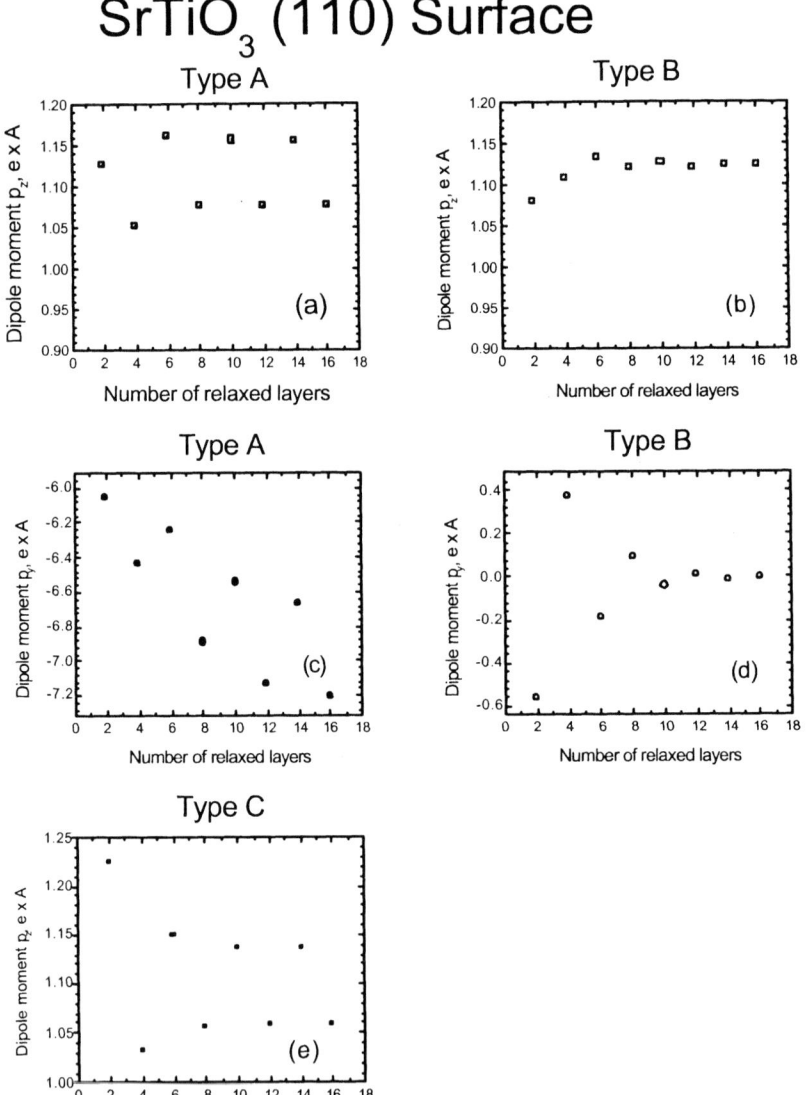

FIGURE 2. Surface polarization and dipole moments perpedicular and parallel to the STO (110) surface with different terminations.

TABLE 2. Surface energies for four different O (110) terminations shown in Fig.1, as calculated for STO, BTO and LMO perovskites, using the Shell Model (SM) [11] and *ab initio* HF method (for STO only). SM-2, SM-16 refer to the number of near-surface planes allowed to relax.

Type	HF	SM-2	SM-16
STO–A-type	1.40	1.54	0.92
B	3.08	3.13	3.31
C		1.63	0.76
TiO	2.10	2.21	2.36
Sr	2.97	3.04	3.37
BTO–A-type		1.58	1.83
B		4.66	4.84
C		1.84	1.82
TiO		2.11	2.36
Ba		3.79	4.16
LMO–A-type		2.59	2.06
B		4.11	4.06
C		2.81	1.95

the TiO– and Sr–terminated (110) surfaces both Ti and Sr atoms reveal negative dipole moments, directed outwards the surface which means a contraction of the near–surface cations.

The interatomic bond populations for three terminations are given in Table 3. The major effect observed here is a strong increase of the Ti–O chemical bonding near the surface as compared to (already large) bonding in the bulk (112 me). The Ti–O bond population for the O–terminated A-type surface is as large as 294 me, i.e. by a factor larger than 2 larger than in the bulk. (This factor for the (100) surface was 1.5.) The Ti–O bond population reaches practically the bulk value for atoms in a third plane. An increased Ti–O bond population near the (110) surface obviously does not arise from the surface relaxation. As shows Table 3, for the TiO termination and *unrelaxed* surface P[Ti(I)–O(II)]=182 me, which increases up to 240 me after surface relaxation. Second, for the same interatomic distance, the Ti–O bond populations are larger in the direction perpendicular to the surface (182 me) than on-plane (126 me).

IV CONCLUSIONS

Our ab initio calculations indicate a considerable increase of the Ti–O bond covalency near the (110) surface, much larger than that for the (100) surface. This should have impact on the electronic structure of surface defects (e.g., F centers), as well as affect an adsorption and surface diffusion of atoms and small molecules

TABLE 3. The A-B bond populations, P (in milli e= me) and the relevant interatomic distances R (in Å) for three different O (110) terminations in STO. I to IV are number of planes enumerated from the surface. The nearest neighbour Ti–O distance in the unrelaxed lattice is 1.945 Å. Numbers in brackets are bond populations for unrelaxed lattice.

Atom A	Atom B	P	R
Ti–O terminated			
Ti(I)	O(I)	176 (126)	2.01
	O(II)	240 (182)	1.81
O(II)	Ti(III)	140 (130)	1.85
	Sr(III)	-10	2.84
	O(III)	-22	2.80
Ti(III)	Sr(III)	0	3.38
	O(III)	126 (136)	1.96
Sr(III)	O(III)	-22	2.75
	O(IV)	-24	2.64
Ti(III)	O(IV)	108 (112)	2.00
O(III)	O(IV)	-24	2.68
O-terminared	A-type		
O(I)	Sr(I)	-28	2.47
	Ti(II)	294	1.80
	O(II)	-26	2.90
Sr(II)	O(II)	-30	2.23
	Ti(II)	0	3.36
Ti(II)	O(II)	90	2.04
	O(III)	104	2.10
O(II)	O(III)	-28	2.85
Sr(II)	O(III)	-6	2.94
O(III)	O(IV)	-20	2.48
	Ti(IV)	110	2.00
	Sr(IV)	-14	2.48
O-terminared	B-type		
O(I)	Sr(I)	-30	1.97
	Ti(II)	16	3.08
	O(II)	-4	3.49
Sr(II)	O(II)	-20	2.81
	Ti(II)	0	3.53
Ti(II)	O(II)	130	2.00
	O(III)	204	1.87
O(II)	O(III)	-18	2.96
Sr(II)	O(III)	4	3.33
O(III)	O(IV)	-22	2.72
	Ti(IV)	114	1.90
	Sr(IV)	-22	2.72

relevant for catalysis. Atomic displacements calculated by means of classical SM are in surprisingly good agreement with the *ab initio* HF calculations for STO. The lowest surface energies in all three perovskites studied are found for the novel "zig-zag" 2 × 1 reconstructed surface termination. This surface termination reveals no dipole moment parallel to the surface, but considerable dipole moment perpendicular to the surface, which certainly can affect the dielectric properties of thin perovskite films.

V ACKNOWLEDGEMENTS

This study was partly supported by DFG (G. Borstel and R. Eglitis) and European Center of Excellence in Advanced Material Research and Technology in Riga, Latvia (contract No. ICA–I–CT–2000–7007 to EK).

*Corresponding author, e-mail: heifets@wag.caltech.edu

REFERENCES

1. J. F. Scott, *Ferroelectric Memories* (Springer, Berlin, 2000).
2. M. E. Lines and A. M. Glass, *Principles and Applications of Ferroelectrics and Related Materials*, Clarendon, Oxford, 1977.
3. J. Padilla and D. Vanderbilt, Surf. Sci. **418**, 64 (1998).
4. J. Padilla and D. Vanderbilt, Phys. Rev. B **56**, 1625 (1997).
5. B. Meyer, J. Padilla and D. Vanderbilt, Faraday Discussions **114**, 395 (1999).
6. F. Cora, and C. R. A. Catlow, Farady Discussions **114**, 421 (1999).
7. R. E. Cohen, Ferroelectrics **194**, 323 (1997).
8. L. Fu, E. Yashenko, L. Resca, and R. Resta, Phys. Rev. B **60**, 2697 (1999).
9. C. Cheng, K. Kunc, and M. H. Lee, Phys. Rev. B **62**, 10409 (2000).
10. S. Tinte, and M. G. Stachiotti, *AIP Conf. Proc.* (ed. R. E. Cohen) **535**, 273 (2000).
11. E. Heifets, E. A. Kotomin, and J. Maier, Surf. Sci. **462**, 19 (2000).
12. E. A. Kotomin, R. I. Eglitis, J. Maier, and E. Heifets, Thin Solid Films **400**, 76 (2001).
13. E. Heifets, R. I. Eglitis, E. A. Kotomin, J. Maier, and G. Borstel, Phys. Rev. B **64**, 235417 (2001).
14. E. Heifets, R. I. Eglitis, E. A. Kotomin, J. Maier, and G. Borstel, Surf. Sci. **513**, 211 (2002).
15. A. Pojani, F. Finocchi, and C. Noguerra, Surf. Sci. **442**, 179 (1999).
16. V. R. Saunders, R. Dovesi, C. Roetti, M. Causa, N. M. Harrison, R. Orlando, and C. M. Zicovich-Wilson, *Crystal–98 User Manual* (University of Torino, Italy, 1999).
17. P. W. Tasker, J. Phys. C: Solid State Phys. **12**, 4977 (1979).

Extending first principles modeling with crystal chemistry: a bond-valence based classical potential

Valentino R. Cooper*, Ilya Grinberg* and Andrew M. Rappe*

Department of Chemistry and Laboratory for Research on the Structure of Matter, University of Pennsylvania, Philadelphia, PA 19104-6323

Abstract. We demonstrate how a phenomenological model which reproduces first-principles density functional theory (DFT) calculations can be constructed to study complex ferroelectric oxides. This model is derived from a well-known crystal chemistry approach, in which Brown's Rules of Valence[1, 2] are used to determine the configurational energy of the system. Our previous work has shown that this model can be used to explain the atomic interactions in the $Pb(Zr_{1-x}Ti_x)O_3$ (PZT) solid solution. We extend this model by parameterizing to a distorted DFT structure as well as a structural minimum. Finally, we shall comment on the applicability of this model for finite-temperature and composition-dependent studies.

INTRODUCTION

A wide variety of complex oxides are used as ferroelectric materials in modern technology. These materials, usually in the perovskite form, have been found favorable for many applications, such as ultrasound machines, cell phones, radios and even Naval SONAR devices. In order to improve upon or propose new materials for use in these applications, it is necessary to understand the inter-atomic interactions in these systems. Density functional theory (DFT)[3] and other *ab initio* techniques have been proven to be effective in determining the local structural properties of many of these materials. However, like many other probes, DFT has its limitations. Due to the computational demands of DFT, it is not possible to use DFT to model large supercells. This makes it difficult to study disordered materials or materials with extensive long range correlated fluctuations. This also inhibits the ability to use DFT to observe phenomena such as domain wall shifting, ion transport, and doping and vacancy effects. In addition, DFT is a zero temperature probe making it difficult to extract finite temperature properties from these calculations.

In this regard, extensive research has been performed in developing DFT based models for studying finite temperature properties of large ferroelectric systems [4, 5, 6, 7]. It has been shown that such methods can be used to perform molecular dynamics and Monte Carlo simulations to predict temperature dependent phase transitions in oxides and to calculate phonon mode spectra. However, it is not always easy to use these methods to effectively study these systems, as they often have a large number of parameters which require fine tuning before they can be used in practice.

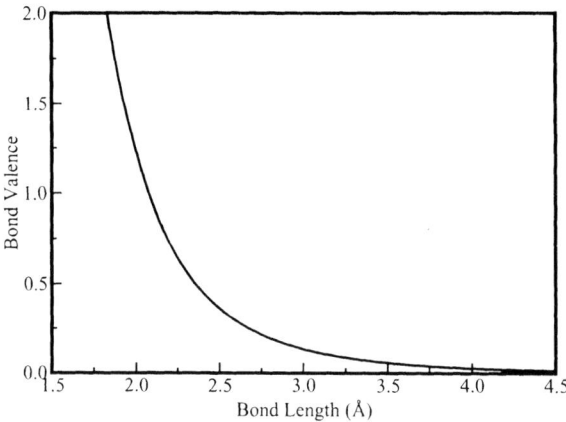

FIGURE 1. This plot depicts the bond-valence relation for a diatomic molecule. Here we see that the partial valence (or bond strength) contribution for each atom will increase with an inverse power law as the two atoms move closer to each other.

THE BOND VALENCE THEORY

For years, crystal chemists have used the bond-valence theory to assess the validity of various chemical structures. The bond-valence theory relates the bond strength (or valence) of an ionic pair to the inter-atomic distance (Figure 1). When the ions are sufficiently separated, there is no bonding interaction between them. As they move closer to each other there is an inverse power relation between the bond-strength and the bond distance (Equation 1).

$$s_{ij} = \left(\frac{R_{ij}}{R_0}\right)^{-N_{ij}} \quad (1)$$

Here R_{ij} is the distance between the two ions, R_0 is the length of a bond that would give a valence of 1, and N determines the rate of decay of the bonding interaction between the ith and jth ions. In a crystal structure, each ion will make bonding interactions such that the sum of all its bond-valences is equal to its nominal valence.

Figure 2 shows a valence map for a single Ti ion in the xy plane of on oxygen octahedron. If the Ti ion moves too close to a single oxygen ion there will be a dramatic increase in the ion's valence. In this configuration, the Ti ion will be over-bonded. Similarly, if the Ti ion remains in the center of the O octahedron, it will be under-bonded and the Ti ion will have to off-center to obtain its nominal valence of 4. Brown et al.[1, 2] have shown that the bond-valence theory corresponds remarkably with experimental data.

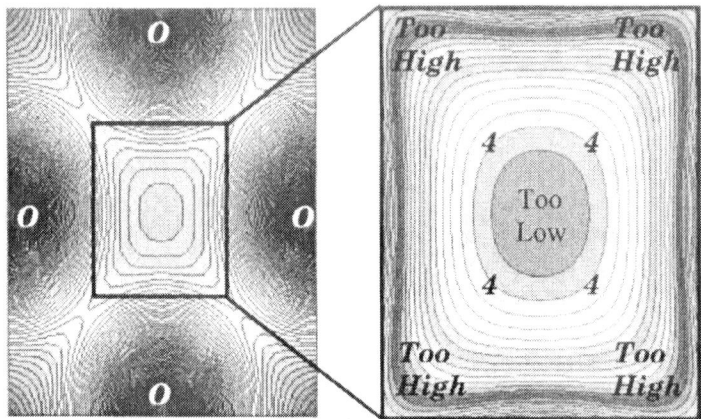

FIGURE 2. Valence of a single Ti ion as it moves in the x-y plane of its oxygen octahedron. As the Ti ion moves close to a single O ion there is an increase in the valence of the Ti ion and the Ti ion is said to be over-bonded. Conversely, if the Ti ion remains in the center of its oxygen octahedra it will be under-bonded as its valence would be significantly less than its nominal valence of 4.

PRACTICAL APPLICATIONS OF THE BOND VALENCE THEORY

Bond-valence arguments are used by X-ray crystallographers to eliminate chemically unreasonable structures and to give more credence to proposed structure[8, 9, 10]. In these approaches, the bond-valence analogy is further applied to identifying the valence states of the elements involved in the oxides studied.

Recently, theorists have begun to employ methods based on the bond-valence concept to model various chemical systems. Eck and co-workers have shown how an empirical potential based on the bond-valence concept, and parameterized from experimental data, can be used to effectively simulate solid systems[11]. They have extended this concept to molecular dynamics simulations (MD), extracting finite-temperature, physical and chemical information (such as structures, energetics, mobilities and reactivities) for an array of compounds[12]. These compounds include ionic/covalent solids, ternary compounds, metals and alloys.

Adams and Swenson have used a combination of bond-valence concepts and Reverse Monte Carlo (RMC) methods to investigate the migration pathways in Ag-based superionic glasses[13, 14]. They employ a bond valence potential (parameterized from experimental data) to evaluate the energy of structures produced by RMC techniques performed on experimental data.

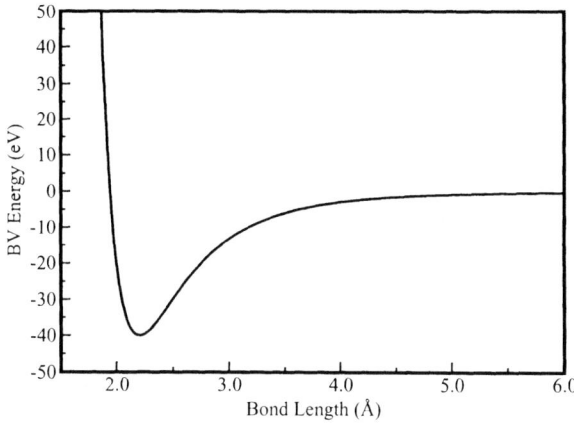

FIGURE 3. This plot shows the bond-valence energy dependence on distance for a diatomic molecule, as defined in our BV model. The bond-valence concept has been modified to assign an energy cost to each atom's inability to obtain its desired valence.

OUR MODIFIED BOND VALENCE MODEL

Bond-valence arguments can correctly predict which crystal structures are favored. However, we find that this concept alone cannot model structural distortions in complex ionic crystals. Therefore, we proceed to construct a potential which reflects each atom's desire to fulfill its bond valence. We augment this potential with other physically-motivated terms, so that the resulting potential is versatile enough to accurately reproduce DFT data for a wide variety of structures.

Our method, similar to the above mentioned studies, extends the bond-valence concept by assigning an energy cost to an atom's inability to achieve its desired valence. Our Bond-Valence (BV) model proposes the idea that in a relaxed structure each atom wants to achieve its desired valence by surrounding itself with the appropriate number of bonds. From this the energy due to the bond-valence interactions can be computed:

$$E_{BV} = \sum_i A_i \left[|V_i - V_i^0|^{\alpha_i} - V_i^{\alpha_i} \right], \qquad (2)$$

Here V_i is the actual valence of the ith ion and V_i^0 is the nominal valence of that ion. A_i and α_i affect the total energy and the forces of the ion and are necessary for matching the BV forces and energies to density functional energies. Figure 3 shows the form of the bond valence potential.

One shortcoming of the bond-valence theory is that it does not include long-range electrostatic interactions, which play an important role in ionic systems such as the ferroelectric perovskites. To account for this we use an Ewald summation (E_{ewald}) to calculate the long range, Coulombic interactions. An additional short-range repulsive term (E_{rep}) is included (Equation 3) to prevent unphysically short distances which arise when only the Ewald and BV terms are used.

$$E_{\text{rep}} = \sum_{ij} A_{ij} \left[e^{-B_{ij} R_{ij}} \right] \tag{3}$$

The BV Model in Practice

Previously, we have shown that our BV model can be used to describe the local structure of the $Pb(Zr_{1-x}Ti_x)O_3$ (PZT) solid solution[15]. Figures 4 and 5 show the displacement patterns obtained from DFT and BV model calculations on a 4×2×1 structure. The model gives good agreement with DFT minimum energy structures on small 4×2×1 supercells. To compare these results to experimental data, we compute the pair distribution function (PDF) of the system. The PDF reveals the inter-atomic distances within a material. It is constructed from the Fourier transform of data obtained from pulsed neutron scattering experiments. The peak widths in an experimental PDF are related to both the thermal motion of the ions and the ionic disorder in the material. Figure 6 depicts the experimental PDF for the 50/50 monoclinic phase of PZT[16] and the PDFs obtained for a 40-atom DFT and a 320-atom BV model 50/50 monoclinic supercell calculation. The experimental results were obtained at 10K, thus limiting the effects of thermal motion within the material. The widths of the peaks show the extensive structural disorder within PZT.

The narrow peaks of the DFT PDF show that the DFT supercell is too small and ordered to explain the experimental PDF. In particular, the extra peaks at 4 Å and 8 Å are a result of the periodicity of the unit cell. The BV model has the advantage of being able to model larger, more disordered supercells. This can be seen through the excellent agreement of the 50/50 monoclinic experimental PDF with the PDF obtained from the 320

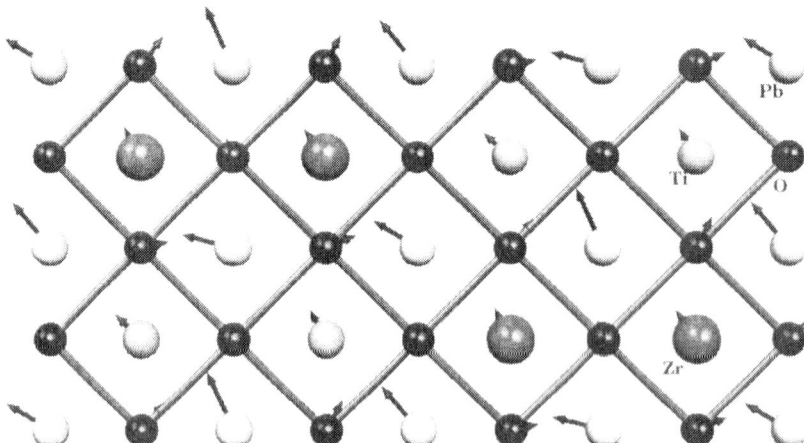

FIGURE 4. Projection of the 4×2×1 50/50 supercell DFT PZT structure onto the x-y plane. The oxygen octahedra are depicted by diamonds, and the distortions from the ideal cubic perovskite positions are shown by arrows. Pb atoms are 1/2 unit cell above the plane and apical O atoms are omitted.

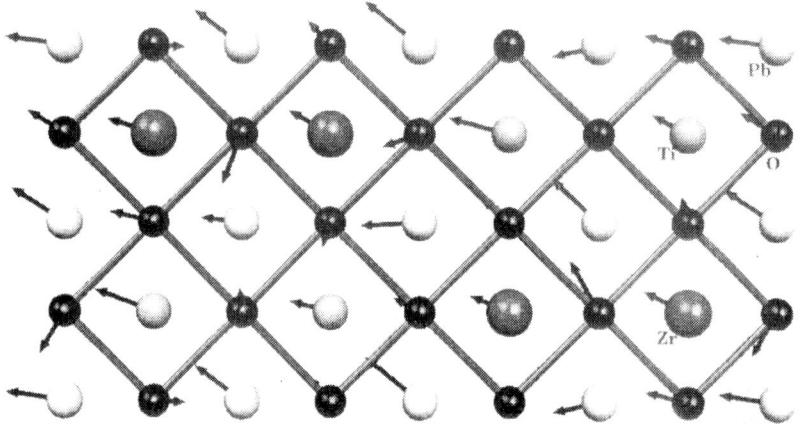

FIGURE 5. BV distortion pattern for structure shown in figure 4

atom BV supercell. Similar agreement was observed for both the 40/60 tetragonal and 60/40 rhombohedral phases. The fact that the tetragonal and rhombohedral supercells were all parameterized from the $4\times 2\times 1$ monoclinic PZT data points to the transferability of the BV model.

The BV model can be further applied as an analytical tool to explain the off-centering of the ions in PZT by comparing the valences and bond distances of Ti and Zr ions when structurally relaxed versus that of the ideal perovskite cube. Table 1 shows the partial valence and the nearest oxygen neighbor bond lengths for Ti in both the relaxed

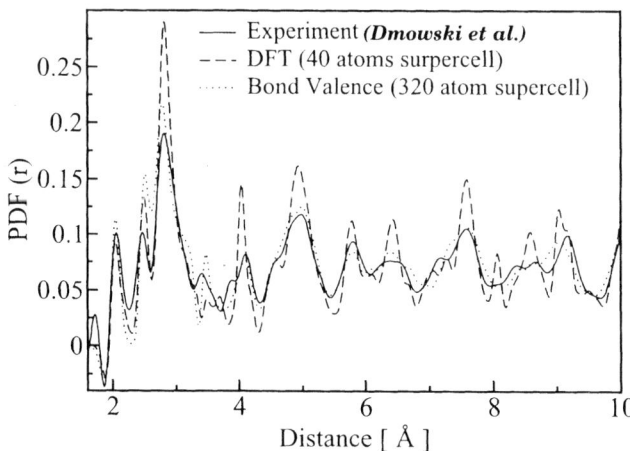

FIGURE 6. PDFs for 50/50 monoclinic PZT. Data are from experiment[16], DFT, and the BV model. Similar agreement with experiment was obtained with the BV model for the tetragonal and rhombohedral phases of PZT.

TABLE 1. Ti ion valence in ideal and DFT relaxed structures

Ti-O Bond Length	Ti Partial* Valence	Ti-O Bond Length	Ti Partial† Valence
2.015	0.631	1.822	1.071
2.015	0.631	2.358	0.279
2.015	0.631	1.939	0.775
2.015	0.631	2.027	0.614
2.070	0.549	1.999	0.660
2.070	0.549	2.067	0.555
Total	3.622	Total	3.954

* ideal perovskite structure
† DFT relaxed structure

and ideal perovskite structures for a $4\times2\times1$ DFT supercell. If the Ti ion remains in the center of the perfect perovskite structure it will be under-bonded with a valence of 3.622. To compensate for this under-bonding, the Ti ion off-centers to make one short Ti-O bond and one long bond, keeping the other bond distances relatively unchanged, and producing a more favorable Ti valence of 3.954.

Table 2 shows similar data for the interaction of the Zr ion with its oxygen octahedron. In this scenario, the ideal perovskite arrangement of this ion results in the Zr ion being over-bonded, with a total valence of 4.499. This is because the Zr ion is larger than the Ti ion and will interact more with its neighboring oxygen ions. In order to obtain its nominal valence, the Zr octahedron expands significantly, accompanied by a small Zr off-centering and octahedral rotation. This create three shorter Zr-O bonds and three longer ones, reducing the Zr valence to 3.996.

TABLE 2. Zr ion valence in ideal and DFT relaxed structures

Zr-O Bond Length	Zr Partial* Valence	Zr-O Bond Length	Zr Partial† Valence
2.015	0.789	1.984	0.865
2.015	0.789	2.214	0.449
2.015	0.789	2.028	0.759
2.015	0.671	2.086	0.641
2.070	0.671	2.058	0.696
2.070	0.789	2.118	0.586
Total	4.499	Total	3.996

* ideal perovskite structure
† DFT relaxed structure

SECOND GENERATION BOND VALENCE MODEL

Our first generation BV model (described above) gives excellent agreement with minimum energy DFT structures. It is able to reproduce the DFT displacement patterns in the relaxed structures, making it suitable for explaining minimum energy structure phenomena such as low temperature PDFs. However, in order to extend the use of the BV model to investigate finite temperature properties, it is necessary to parameterize our BV model using forces and energies from a database of DFT structures, including structures far from equilibrium.

Before BV parameters can be optimized, two things must be considered. First, a suitable database of structures must be constructed using DFT, from which to match forces and energies. Second, an efficient minimization routine must be chosen to automate the fine tuning of the BV parameters.

Constructing a Database of DFT Structures

We have constructed a database of 1000+ DFT PZT structures for the parameterization of the BV model. This database consists of 30 minimum energy structures, numerous structures obtained during the ionic relaxation steps used to determine the minimum energy structures, and structures resulting from further calculations in which a single atom was moved up to 0.8 Å in each direction from its DFT relaxed structure positions. The database contains a variety of cation arrangements and appropriately contains structures with forces up to 2 eV/Å on a single ion in a particular direction. High force structures are necessary to insure the accurate modeling of high-energy modes that may arise during molecular dynamics simulations.

Simulated Annealing: An Optimization Routine

In an attempt to optimize the BV parameters to produce the best quantitative agreement with DFT data, a simulated annealing[17] method was employed. In our implementation of simulated annealing, each of the BV parameters is randomly adjusted to make a trial step. Next, a penalty function is evaluated, based solely on the difference in the forces on the atoms obtained from the BV and the DFT calculations. Finally, the Metropolis Monte Carlo[18] algorithm is used to either accept or reject the proposed move. These steps are repeated for a selected number of moves after which the temperature is decreased.

This minimization routine is analogous to the cooling of a liquid. At high "temperatures" the parameters being minimized are allowed to change freely with respect to each other. As the temperature is reduced slowly, each parameter will adjust so as to settle into its preferred minimum configuration. This allows the routine to find a global minimum.

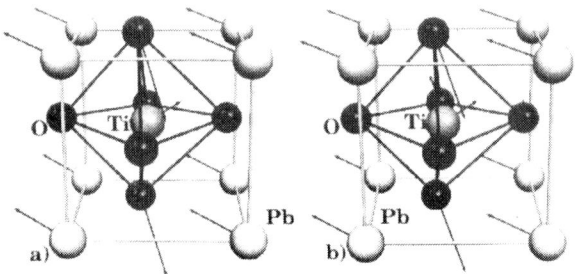

FIGURE 7. Optimized BV parameters are able to accurately reproduce DFT(a) forces (portrayed as arrows) on a high-energy test structure. The parameters obtained from the BV optimization routine give BV forces (b) that agree with DFT forces within an average deviation of 0.032 eV/Å per coordinate.

Preliminary Results: Lead Titanate (PbTiO$_3$)

Lead Titanate (PT) was used as a model system to test the effectiveness of the BV parameter optimization routine. PT has been extensively studied both theoretically and experimentally. It has a simple tetragonally distorted five-atom unit cell. Comparison of BV PT results with previous studies will give an indication as to the accuracy of both the fitting routine and the BV model for use in MD simulations.

For our initial investigations, BV parameters were optimized with two DFT structures. The first DFT structure was a high-energy structure, with ionic positions close to the ideal perovskite positions. The second structure was obtained from the ionic relaxation of the high-energy structure.

Preliminary results show that the current minimization routine is able to optimize parameters to an average deviation from DFT of 0.032 eV/Å per coordinate. Figure 7 depicts the forces on each ion in the high-energy structure obtained using DFT and the optimized BV parameters. Table 3 shows the actual forces on each atom in the high energy structure. Considering both Figure 7 and Table 3, we see that there is excellent agreement between the BV forces and the DFT forces. Furthermore, our DFT calculations report a cohesive energy of 38.45 eV/cell for the test PT structure. The BV structure is in good agreement with these results producing a cohesive energy of 39.62 eV/cell. These results give us confidence that the optimized BV parameters will be able to accurately reproduce DFT results and can be further applied to MD simulations to observe finite-temperature properties of complex oxide materials.

CONCLUSION

BV arguments have been used extensively by X-ray crystallographers to characterize the structure of many materials. Theoretical investigations have shown how these concepts can be used to construct a potential which can be used to explain experimental data such as PDFs of three different phases of PZT. While experimentally determined BV parameters give excellent qualitative agreement with accurate DFT calculations, fine tuning the

TABLE 3. DFT and Optimized Bond Valence Forces on atoms for a High Energy Structure of $PbTiO_3$

Species	x (eV/Å)	y (eV/Å)	z (eV/Å)
Pb(DFT)	-0.97	0.25	-0.59
Pb(BV)	-0.97	0.32	-0.58
Ti(DFT)	0.59	0.29	-0.38
Ti(BV)	0.57	0.28	-0.39
O_1(DFT)	-0.17	0.74	0.93
O_1(BV)	-0.11	0.73	0.88
O_2(DFT)	0.54	-1.49	0.22
O_2(BV)	0.48	-1.54	0.27
O_3(DFT)	0.01	0.21	-0.18
O_3(BV)	-0.02	0.17	-0.19

BV parameters and adding Ewald and repulsive energy terms are essential for extending the model to calculate finite-temperature properties of larger DFT-inaccessible systems. We have employed SA techniques to optimize BV parameters to match DFT forces. Our preliminary investigations have shown that for the sample PT case reasonable agreement (average deviation of 0.032 eV/Å per coordinate) can be achieved. Furthermore, optimization of the BV parameters gives cohesive energies in agreement with DFT.

Future work on this empirical model will seek to investigate the size of the DFT database required to give BV parameters that give accurate forces for any structure. Also, more efficient minimization routines will be examined for the fine tuning of these parameters. An optimized BV model can be further adapted to MD simulations to investigate temperature-dependent properties for PT and other complex-oxide systems.

ACKNOWLEDGMENTS

This work was supported by the Office of Naval Research under grant number N-000014-00-1-0372 and through the Center for Piezoelectrics by Design. Computational support was provided by the High-Performance Computing Modernization Office of the Department of Defense and the Center for Piezoelectrics by Design. AMR would also like to thank the Camille and Henry Dreyfus Foundation for support.

REFERENCES

1. Brese, N., and O'Keefe, M., *Acta. Crystallogr.*, **47**, 192–197 (1991).
2. Brown, I. D., *Structure and Bonding in Crystals II*, Academic, New York, 1981, pp. 1–30.
3. Kohn, W., and Sham, L. J., *Physical Review*, **140**, A1133–A1138 (1965).
4. Ghosez, P., Cocakayne, E., Waghmare, U., and Rabe, K. M., *Physical Review B*, **60**, 836–843 (1999).
5. Huff, N. T., Demiralp, E., Cagin, T., and Goddard III, W. A., *Journal of Non-Crystalline Solids*, **253**, 133–142 (1999).

6. Bellaiche, L., and Vanderbilt, D., *Physical Review Letters*, **81**, 1318–1321 (1998).
7. Sepilarsky, M., Phillpot, S. R., Wolf, D., Stachiotti, M. G., and Migoni, R. L., *Applied Physical Letters*, **76**, 3986–3988 (2000).
8. Mercurio, D., Champarnaud, J. C., Gouby, I., and Frit, B., *European Journal of Solid State and Inorganic Chemistry*, **35**, 49–65 (1998).
9. Cooper, M. A., Hawthorne, F. C., and Grew, E. S., *Canadian Mineralogist*, **36**, 1305–1310 (1998).
10. Sawa, H., Ninomiya, E., Ohama, T., Nakao, H., Ohwada, K., Murakami, Y., Fujii, Y., Noda, Y., Isobe, M., and Ueda, Y., *Journal of the Physical Society of Japan*, **71**, 385–388 (2002).
11. Eck, B., and Dronskowski, R., *Journal of Alloys and Compounds*, **338**, 136–141 (2002).
12. Eck, B., Kurtulus, Y., Offermans, W., and Dronskowski, R., *Journal of Alloys and Compounds*, **338**, 142–152 (2002).
13. Adams, S., and Swenson, J., *Physical Review Letters*, **84**, 4144–4147 (2000).
14. Adams, S., and Swenson, J., *Physical Review B*, **63**, 0524201-1–054201-11 (2001).
15. Grinberg, I., Cooper, V. R., and Rappe, A. M., *Nature*, **419**, 909–911 (2002).
16. Dmowski, W., Egami, T., Farber, L., and Davies, P. K., "Structure of Pb(Zr, Ti)O_3 near the morphotropic phase boundary," in *Fundamental Physics of Ferroelectrics - Eleventh Williamsburg Ferroelectrics Workshop*, edited by R. E. Cohen, AIP, Woodbury, New York, 2001.
17. Press, W. H., Teukolsky, S. A., Vetterling, W. T., and Flannery, B. P., *Numerical Recipes in C: The Art of Scientific Computing*, Cambridge University Press, New York, 1997, pp. 444–451.
18. Metropolis, N., Rosenbluth, A., Rosenbluth, M., Teller, A., and Teller, M., *Journal of Chemical Physics*, **21**, 1087–1092 (1953).

Large-scale quantum chemical modeling of the phase transitions in KTN solid solutions

R. I. Eglitis[1], D. Fuks[2], S. Dorfman[3], E. A. Kotomin[4,5], G. Borstel[1], and V. A. Trepakov[6,7]

[1] *Fachbereich Physik, Universität Osnabrück, D-49069 Osnabrück, Germany*
[2] *Materials Engineering Department, Ben-Gurion University of Negev, IL-84105 Beer Sheva, Israel*
[3] *Department of Physics, Technion–Israel Institute of Technology, IL-32000 Haifa, Israel*
[4] *MPI für Festkörperforschung, Heisenbergstrasse 1, D-70569 Stuttgart, Germany*
[5] *Institute for Solid State Physics, 8 Kengaraga str., Riga LV-1063, Latvia*
[6] *A.F. Ioffe Physical Technical Institute, 194021 Saint-Petersburg, Russia*
[7] *Institute of Physics AS CR, Prague 18221, Czech Republic*

Abstract. The large–scale modeling of the atomic and electronic structure of $KNb_xTa_{1-x}O_3$ (KTN) perovskite solid solutions is performed using the Intermediate Neglect of the Differential Overlap (INDO) method based on the Hartree-Fock formalism. It is found that periodic Nb impurities in $KTaO_3$ reveal coherent off–center displacements already at the smallest calculated concentration, $x=0.125$. The calculated magnitude of $<111>$ Nb off-center displacement is 0.27 a.u., which is close to the XAFS observation at 70 K and $x=0.09$. In contrast, Ta impurities in $KNbO_3$ remain on–center, due to higher ionicity of Ta, as compared to Nb. Using the calculated energy gain caused by the off–center displacements of Nb atoms, the non–empirical Ginzburg–Landau–Devonshire functional with concentration–dependent coefficients is constructed. Analysis of INDO results for several Nb concentrations in KTN allows calculate the lowest critical Nb concentration, $x_{cr}=0.025$, corresponding to the quantum displacive limit for the paraelectric–ferroelectric transition at 0 K. This value is only slightly higher than the experimental one.

I INTRODUCTION

ABO_3-type perovskite solid solutions, e.g. potassium tantalate–niobate crystals $KTa_{1-x}Nb_xO_3$, conventionally called KTN, are of great interest for many advanced non-linear optic and holographic applications [1]. The parent pure $KTaO_3$ (KTO) is an incipient ferroelectric, and very likely, a *quantum paraelectric*, whose low-temperature ferroelectric (FE) phase transition (PT) is suppressed by quantum vibrations [2–5]. On the other hand, pure KNbO$_3$ is a ferroelectric; exhibiting the transitions from a cubic paraelectric to three succesive ferroelectric states; cubic–tetragonal–orthorhombic–rhombohedral [6]. All these are of the first order (PT I), being accompanied by a thermal hysteresis of 5 K (at 708 K), 15 K (498 K) and 25 K (263 K), respectively [7]. Because of practical coinsidence of ionic radii for Ta and Nb, the same symmetry of paraelectric phase and a small difference in lattice parameters, KNO and KTO form a continuous series of solid solutions over the entire composition range [8–10]. Nb doping of KTO leads to its ferroelectric instability at the critical concentration $x_{cr} = 0.008$, at which $T_C = 0$ K and characteristic critical exponents indicate the quantum displacive limit [11–16]. At Nb concentrations higher than critical, a ferroelectric transition of the second order (PT II) arises. The temperature dependence for the cubic-rhombohedral transition $T_C(x) \sim (x - x_{cr})^{\frac{1}{\zeta}}$, where the critical exponent $\zeta \approx 2$ for $x \leq 0.028$ [16], and $x = 1$ at higher Nb concentrations, demonstrates the transition from a quantum to a classical limit [15]. At $x_{mc}=0.04$ all three characteristic for KNbO$_3$ PTs, cubic–tetragonal–orthorhombic–rhombohedral, merge at a multi–critical point, $T_{mc} \sim 40$ K [16] (according to [17], $T_{mc} \sim 33$ K at $x_{mc} \sim 0.02$) and then, transition temperature increases with the Nb concentration, up to ~ 700 K, where cubic-tetragonal FE PT occurs in KNbO$_3$ [12,13]. The study of spontaneous polarization P_S, determined from the measurements of the hysteresis loops [16], has shown that the edge characteristic of P_S appears in the polarization vs. field diagram only at sufficiently low, mHz-scale frequencies, due to a finite speed of domain wall motion, and P_S increases with Nb concentration (at least, for $x \leq 0.12$). Replacement of Nb ions by Ta ions changes the nature of the FE transition from first-order to the second-order. That is, not only the transition temperature itself, but also its behaviour and main characteristics of PTs in KTO crucially depend on the presence and concentration of Nb impurities. The knowledge of their microstructure in KTN is important for the correct explanation of numerous experimental data, but this remains still a subject of controversy. Indeed, in spite of the fact that octahedrally-coordinated Nb^{5+} and Ta^{5+} ions have nearly identical radii, the model of multi–well < 111 > Nb^{5+} off-centers, existing already far above T_C, is often used in interpreting experiments (see, e.g., [17–21]), thus providing explanation to the polar microregions and glass–like behaviour observed in the Raman scattering and other experiments. This model has been developed in Ref. [22] and is supported by both x-ray absorption fine structure (XAFS) experiments [23] and theoretical calculations [24–30]. In particular, the Nb off-center displacements below T_C are found in XAFS experiments to be 0.15 Å(i.e. 0.27 a.u.) at 70 K, and they change by less than 20 percent as the temperature increases up to the room temperature [23].

II INDO METHOD

We have used the semi-empirical, quantum chemical method of the Intermediate Neglect of the Differential Overlap (INDO) [31]. The modification of the standard INDO method for ionic solids is described in detail in Refs. [32-34]. This method is based on the Hartree-Fock formalism and allows self-consistent calculations of the atomic and electronic structure of pure and defective crystals. In the last decade the INDO method has been used to study bulk solids and defects in many oxides and perovskites [32-41]. This method has been earlier applied to the study of phase transitions and frozen phonons in pure $KNbO_3$ [40], pure and Li-doped $KTaO_3$ [41], point defects in $KNbO_3$ and $BaTiO_3$ [38,39,42], and perovskite solid solutions $KNb_xTa_{1-x}O_3$ [43,44]. In the present calculations we use periodic, the so-called *large unit cell* (LUC) model. In this model the electronic structure calculations are performed for an extended unit cell at the wave vector $\mathbf{k} = \mathbf{0}$ in the narrowed Brillouin zone (BZ) which is equivalent to the band-structure calculations at several special points of the normal BZ, transforming to the narrow BZ center after the corresponding extension of the primitive unit cell. In the ABO_3 crystals under study, the unit cell contains five atoms and the 2×2×2 extended LUC used in our calculations consists of 40 atoms.

The detailed analysis of the development of the INDO parametrization for pure $KNbO_3$ and $KTaO_3$ is given in Refs. [40,41]. The INDO method reproduced very well both available experimental data and results of ab initio LDA-type calculations. In particular, this method reproduces the effect of a ferroelectric instability in $KNbO_3$ due to off-center displacement of Nb atoms from the regular lattice sites, as well as the relative magnitudes of the relevant energy gains for the [100], [110] and [111] Nb displacements, which are consistent with the order of the stability of the tetragonal, orthorombic and rhombohedral ferroelectric phases, respectively, with decreasing temperature. This is a very non-trivial achievement since the typical energy gain due to the Nb off-center displacement is of the order of several milli Rydbergs (mRy) per unit cell.

The calculated frequencies of the transverse-optic (TO) phonons at the Γ point in the BZ of cubic and rhombohedral $KNbO_3$ and the atomic coordinates in the minimum energy configuration for the orthorhombic and rhombohedral phases of $KNbO_3$ are also in good agreement with experiments thus indicating that a highly successful INDO parametrization has been achieved. The frozen-phonon calculations for T_{1u} and T_{2u} modes of cubic $KTaO_3$ are also in good agreement with experiment. An appreciable covalency of the chemical bonding is seen from the calculated (static) effective charges on atoms (calculated using the Löwdin population analysis): 0.62 e for K, 2.23 e for Ta and -0.95 e for O in $KTaO_3$, whic are far from those expected in the purely ionic model (+1 e, +5 e and -2 e, respectively) often used. These charges show slightly higher ionicity in $KTaO_3$ as compared with the relevant effective charges calculated for $KNbO_3$: 0.54 e for K, 2.02 e for Nb and -0.85 e for O.

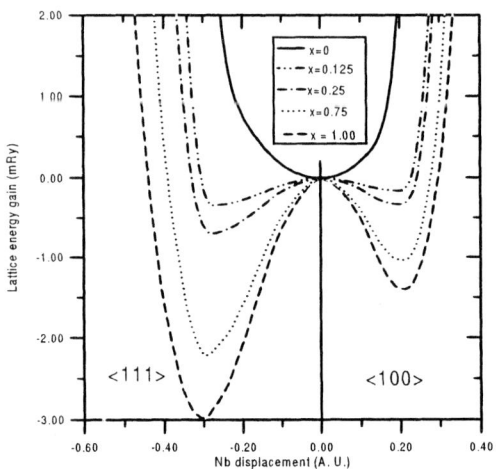

FIGURE 1. The INDO lattice energy gain (per unit cell) vs Nb off–center displacement in KTN for fractional concentrations: $x = 0$ (Ta displacement), 0.125, 0.25, 0.75 and 1 (see a legend).

III INDO RESULTS

Figure 1 shows the total energy dependence on impurity off–center displacements (per unit cell of five atoms) for a series of Nb fractional concentrations ($x = 0$, 0.125, 0.25, 0.75, and 1). The immediate conclusion suggests itself from the calculation for a pure $KTaO_3$ ($x = 0$) that Ta atoms definitely reveal no off–center displacement and thus no phase transition occurs in pure KTO, in agreement with experiment. Any attemps to displace the Ta ions from their sites are costly in energy.

In another extreme case of a pure $KNbO_3$ ($x = 1$) Nb atoms demonstrate that their sinhronous (coherent) off–center displacements along the $<100>$ and $<111>$ axes are energetically favourable. For a given Nb concentration the energy gain is larger for the $<111>$ direction, again in agreement with experiment.

Table 1 shows the relevant calculated atomic displacements (with respect to the K atoms) and experimental data. For $<100>$ and $<111>$ directions the Nb displacements are larger that those from the experimental data by 39 % and 65 %, respectively, whereas the O displacements are slightly smaller than experimental. As a result, the *relative* Nb–O displacements are close to those experimentally observed (0.340 a.u. and 0.412 a.u. for the $<100>$ and $<111>$ directions, respectively.) After Nb atom is relaxed towards the nearest O atom along the $<100>$ axis, their effective charges reduce additionally by $\approx 0.1\ e$ due to an increased chemical bonding, and become $1.92\ e$ and $-0.75\ e$, respectively.

In KTN calculations with $0 \leq x \leq 1$ we fixed Ta and O atoms in their lattice sites by the two reasons: (i) our previous calculations demonstrate that Ta atom prefer to stay on–center, (ii) a complete relaxation of all O atoms in KTN without any

TABLE 1. INDO-calculated Nb and O ion displacements (in a.u.) in pure KNbO$_3$ with respect to the K atoms and lattice energy gain (in brackets, in mRy). The energy gain corresponds to the energy contribution from all three O atoms in the unit cell. The $<100>$ and $<111>$ displacements correspond to the tetragonal and rhombohedral phase transitions, respectively.

Displacement $<100>$	40 atoms	Exp. [45]
Nb	0.213 (-1.40)	0.145
O	-0.179 (-3.80)	-0.195
Displacement $<111>$	40 atoms	Exp. [45]
Nb	0.300 (-2.99)	0.172
O	-0.160(-3.69)	-0.240

TABLE 2. Energy gain (per unit cell) due to Nb impurity off-center displacements in KTaO$_3$. Calculations have been performed for 2×2×2 extended supercell containing 40 atoms for different Nb concentrations.

Nb concentration in at. frac.	$<100>$ displacement δ, a.u. ΔE(mRy)	$<111>$ displacement δ, a.u. ΔE(mRy)
x=0.125	0.1959 -0.1616	0.2759 -0.3445
x=0.25	0.1984 -0.3321	0.2795 -0.708
x=0.75	0.2080 -1.0412	0.293 -2.22
x=1.00	0.2130 -1.4000	0.300 -2.99

symmetry is very time-consuming. Figure 1 and Table 2 show clearly that the Nb atom displacements only slightly depend on their concentration. The displacements along the $<111>$ axis are larger than those along $<100>$ and agree very well with XAFS experimental estimate of 0.27 a.u. [23].

The lattice energy gain is essentially sub-linear function of the Nb concentration and is larger for the $<111>$ direction, in an agreement with experiment. The only previous *ab initio* LMTO calculations [46] predict off-center Nb displacements at high concentrations, $x=0.22$, in contradiction with experiments, suggesting $x_{cr}=0.008$.

IV DEVELOPMENT OF THE MICROSCOPIC FUNCTIONAL FOR THE EXCESS ENERGY

The described above modeling of KTN solid solutions for tetrahedral and rhombohedral phases for a set of Nb concentrations creates a basis for the development of the microscopic Landau-Ginzburg-Devonshire excess energy functional. It was suggested in 60ies that the phase transition from the paraelectric to ferroelectric

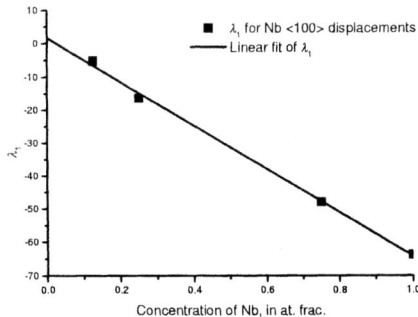

FIGURE 2. Extrapolated plots for λ_1 cofficients of the Φ^6–type functional for the displacements of Nb atoms in $<100>$ direction. The extrapolations were done on the whole interval of concentrations.

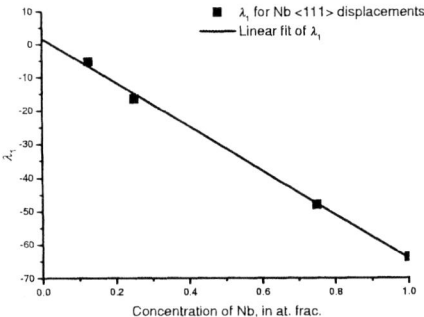

FIGURE 3. Extrapolated plots for λ_1 coefficients of the Φ^6–type functional for the displacements of Nb atoms in $<111>$ direction. The extrapolations were done on the whole interval of concentrations.

phase in KTN-type solid solutions could be described in terms of the off-site displacements of the impurity atoms (see for example Ref. [47] and references therein). This simple model was further developed in Refs. [29,48,49].

Our present simulations allow to return to the original model and to estimate its abilities, using the parameters obtained from atomistic quantum chemical calculations. We used the Nb displacement, δ, as the *order parameter* in our functional. In ABO$_3$ perovskites the functional for the excess energy could be presented in the following way [50,51]

$$F = \lambda_1 \delta^2 + \lambda_2 \delta^4 + \lambda_3 \delta^6 \tag{1}$$

This functional was used to fit the excess energies (Fig. 1 and Tables 1 and 2). As a result, Table 3 presents a set of λ_1 parameters obtained for the two different KTN phases and several calculated compositions. The dependence of λ_1 coefficient on the Nb concentration in the low concentration region is plotted in Fig. 2 and Fig. 3. On the basis of these plots, the conclusion could be drawn that λ_1 coefficients change their sign at small Nb concentrations, x_o, for both tetragonal and rhombohedral phases of KTN. This happens for $x_{cr} \cong 0.025$. In the virtual crystal approach calculations the tricritical point for KTN were estimated in the interval $0.01 \leq x_{cr} \leq 0.03$. For sure, this is an interesting example of the correlation of the very approximate virtual approach with our direct semiempirical calculations. As is well known from the literature on phase transitions (see, for example Ref. [50]), such the change of the sign of the first coefficient means the change of the type of the phase transition from paraelectric to the ferroelectric phase. For $\lambda_1 > 0$ and $\lambda_2 < 0$, Eq. (1) describes the PT I, with a tri-critical point at $\lambda_1 = 0$.

TABLE 3. The coefficient λ_1 of the Ginzburg–Landau–Devonshire functional, Eq. (1), for different Nb concentrations. λ_1 is in mRy/(a.u.)^2.

Nb concentration in at. frac.	$<100>$ displacement	$<111>$ displacement
x=0.125	-5.1471	-5.22186
x=0.25	-16.3348	-13.5734
x=0.75	-47.9672	-48.914
x=1.00	-63.719	-67.5694

From the results of our fitting, it follows that λ_1 may be approximated by the linear dependence on Nb concentration, $\lambda_1 = \lambda_1^{(1)}(x - x_{cr})$, where x_{cr} is the Nb concentration corresponding to the stability limit of the paraelectric phase.

V CONCLUSIONS

Our quantum chemical calculations have clearly demonstrated that Nb impurities in KTaO$_3$ reveal coherent off center displacements, with the magnitudes close to those found in XAFS experiments [23]. We observe well-pronounced collective effect due to which the lattice energy gain and related polarisation increse sub-linearly with the Nb concentration. Unlike Nb, Ta impurities in $KNbO_3$ reveal no off–center displacements, very likely due to higher ionicity of Ta compared to Nb atoms.

Combining the Landau–Ginzburg–Devonshire functional for the excess energy with the INDO calculations, we have calculated the tricritical concentration for the KTN phase diagram to be as small as $x_{cr}=0.025$. We also predict that the phase transitions in KTN at lower Nb concentrations change its character. Note, that we assumed the off-center displacement of the Nb atom serves a role of the critical order parameter in KTN phase transitions. It should be noted that the tricritical point in the experimental measurements for the cubic-trigonal phase transition for Nb concentration is about 0.8 at %. At T=40K multicritical point appears at Nb concentration equal to 4 at % [16] (2 at % [17]). In some sense we are estimating this multicritical point because the values of the concentrations are very close for $<100>$ and $<111>$ displacements of the Nb atom. We would like to emphasize in conclusion that, to our knowledge, this is the first attempt to derive microscopically the coefficients of the Ginzburg–Landau–Devonshire functional. Our results could be improved by calculations of the lattice energy gain (per unit cell) vs Nb off–center displacement in KTN for more extended set of Nb concentrations. These calculations could be especially important in the low concentration limit.

VI ACKNOWLEDGEMENTS

Authors are indebted to J. Maier and V. Vikhnin for many stimulating discussions. This study is supported in part by DFG, by NATO (grant PST.CLG.977561), by Excellence Centre for Advanced Materials Research and Technology (Riga, Latvia), RFBR Grants 01-02-17799, 01-02-16875, LN00A015 of the MSMT, Czech Republic, PR program of the Presidium of RAS *Low- dimensional quantum structures* and Scientific program of RAS St.-Petersburg Scientific Center. This research was also supported by the Niedersachsen Ministry of Science and Art, and by the KAMEA program (Israel).

REFERENCES

1. P. Günter and J.-P. Huignard (eds.) *Photorefractive Materials and Their Applications, Topics in Applied Physics*, **61,62** (Springer, Berlin, 1988).
2. J.H. Barrett, Phys. Rev. **86**, 118 (1952).
3. H. Brukard and K.A. Müller, Helv. Phys. Acta **49**, 725 (1976).
4. V.G. Vaks, *Introduction into Microscopic Theory of Ferroelectrics* (Moscow, Nauka, 1977, in Russian).
5. H. Voght, Phys. Rev. B **51**, 8046 (1995).
6. B.T. Matthias and J.P. Remejka, Phys. Rev. **82**, 727 (1951).
7. R.M. Cotts and W.D. Knight, Phys. Rev. **96**, 1285 (1954).
8. S. Reisman, S. Triebwasser, and J. Holtzberg, J. Am. Chem. Soc. **77**, 4228 (1955).
9. A. Reisman and E. Banks, J. Am. Chem. Soc. **80**, 1877 (1958).

10. S. Triebwasser, Phys. Rev. **114**, 63 (1959).
11. U.T. Hochli, H.E. Weibel, and L.A. Boatner, Phys. Rev. Lett. **39**, 1158 (1977).
12. L.A. Prater, L.L. Chase, and L.A. Boatner, Phys. Rev. B **23**, 221 (1981).
13. T. Schneider, H. Beck, and E. Stoll, Phys. Rev. B **13**, 1123 (1976).
14. R. Oppermann and H. Thomas, Z. Phys. b **22**, 387 (1975).
15. R. Morf, T. Schneider, and E. Stoll, Phys. Rev. B **16**, 462 (1977).
16. L.A. Boatner, U.T. Hochli, and H. Weibel, Helv. Phys. Acta **50**, 620 (1977).
17. D. Sommer, D. Friese, W. Kleeman, and D. Rytz, Ferroelectrics **124**, 231 (1991).
18. D. Sommer, D. Friese, W. Kleeman, and D. Rytz, Ferroelectrics **124**, 137 (1990).
19. G.A. Samara, Phys. Rev. Lett. **53**, 298 (1984).
20. G.A. Samara, Jpn. J. Appl. Phys., **24**, Suppl. 34-2, 80 (1984).
21. E. Bouziane, M. Fontana, and W. Kleeman, J. Phys.: Condens. Matter. **6**, 1965 (1994).
22. Y. Yacoby, in Proc. Int. Conf. on Lattice Dynamics, Paris, France (1977), 453.
23. O. Hanske-Petitpierre, Y. Yacoby, J. Mustre de Leon, and E.A. Stern, J.J. Rehr, Phys. Rev. B **44**, 6700 (1991).
24. A. Postnikov, T. Neumann, G. Borstel, and M. Methfessel, Phys. Rev. B **48**, 5910 (1993).
25. M.G. Stachiotti, M. Sepliarsky, R.L. Migoni, and C.O. Rodriguez, AIP Conf. Proc. **436**, 274 (1998).
26. R.I. Eglitis, E.A. Kotomin, and G. Borstel, J. Phys.: Condens. Matter **12**, L431 (2000).
27. R.I. Eglitis, E.A. Kotomin, and G. Borstel, Comp. Mater. Sci. **21**, 530 (2001).
28. Y. Girshberg and Y. Yacoby, J. Phys.: Condens. Matter **13**, 8817 (2001).
29. H. Abe, K. Harada, R.J. Matsuo, H. Uwe, and K. Ohshima, J. Phys. Cond.: Matter **13** 3257 (2001).
30. S. Prosandeev, E. Cockayne, and B. Burton, AIP Conf. Proc. **626**, 169 (2002).
31. J.A. Pople and D.L. Beveridge, *Approximate Molecular Orbital Theory* (McGraw-Hill, New York, 1970).
32. A.L. Shluger, Theoret. Chim. Acta **66**, 355 (1985).
33. E. Stefanovich, E. Shidlovskaya, A.L. Shluger, and M. Zakharov, Phys. Stat. Sol. B **160**, 529 (1990).
34. A.L. Shluger and E. Stefanovich, Phys. Rev. B **42**, 9664 (1990).
35. V.S. Vikhnin, R.I. Eglitis, S.E. Kapphan, E.A. Kotomin, and G. Borstel, Europhys. Lett. **56**, 702 (2001).
36. E. Stefanovich, A.L. Shluger, and C.R.A. Catlow, Phys. Rev. B **49**, 11560 (1994).
37. E.A. Kotomin, A. Stashans, L.N. Kantorovich, A.I. Livshitz, A.I. Popov, I. Tale, and J.-L. Calais, Phys. Rev. **51**, 8770 (1995).
38. R.I. Eglitis, N.E. Christensen, E.A. Kotomin, A.V. Postnikov, and G. Borstel, Phys. Rev. B **56**, 8599 (1997).
39. E.A. Kotomin, R.I. Eglitis, A.V. Postnikov, G. Borstel, and N.E. Christensen, Phys. Rev. B **60**, 1 (1999).
40. R.I. Eglitis, A.V. Postnikov, and G. Borstel, Phys. Rev. B **54**, 2421 (1996).
41. R.I. Eglitis, A.V. Postnikov, and G. Borstel, Phys. Rev. B **55**, 12976 (1997).
42. R.I. Eglitis, E.A. Kotomin, and G. Borstel, J. Phys.: Cond. Matter **14**, 3735 (2002).

43. R.I. Eglitis, E.A. Kotomin, and G. Borstel, Solid State Comm. **108**, 333 (1998).
44. R.I. Eglitis, E.A. Kotomin, G. Borstel and S. Dorfman, J. Phys.: Cond. Matter **10**, 6271 (1998).
45. A.W. Hewat, J. Phys. C **6**, 2559 (1973).
46. A.V. Postnikov, T. Neumann, and G. Borstel, Ferroelectrics **164**, 101 (1995).
47. W. Kleemann, A. Albertini, R. V. Chamberlin, and J. G. Bednorz, Europhys. Lett. **37**, 145 (1997).
48. G.A. Samara, Ferroelectrics **117**, 347 (1987).
49. S. Dorfman, D. Fuks, A. Gordon, A.V. Postnikov, and G. Borstel, Phys. Rev. B **52**, 7135 (1995).
50. M.E. Lines, and A.M. Glass, *Principles and Applications of Ferroelectrics Related Materials*, (Clarendon Press, Oxford, 1977).
51. A. Gordon, and S. Dorfman, Phys. Rev. B **51**, 9306 (1995).

Progress in quantum Monte Carlo calculations of perovskite transition metal oxides

L.K. Wagner, P. Sen and L. Mitas

Department of Physics, North Carolina State University, Raleigh, NC 27695-8202

Abstract. We present a report on recent progress in application of quantum Monte Carlo (QMC) methods to accurate electronic structure calculations of ferroelectric transition metal oxides. There are three major aspects: i) construction of accurate many-body trial functions using orbitals from one-particle approaches such as Hartree-Fock and density functional theory, in particular, the hybrid functionals such as B3LYP; ii) efficient calculations of excitation energies both in variational and diffusion QMC; iii) method for calculating the small energy differences such as subtle features on total energy surfaces related to ferroelectric distortions.

I. INTRODUCTION.

Solids composed of transition metal (TM) atoms and oxygen belong to the most useful and, at the same time, complicated materials. The past decades revealed a plethora of phenomena exhibited by these systems such as variety of magnetic effects, ferroelectricity and related phenomena and high-T_c superconductivity. One of the most common stoichiometries of these solids is A-TM-O_3 where A denotes a divalent alkaline metal. It leads to a well-known cubic perovskite structure with A in the cube corners, oxygen at the face centers and TM in the center of the cube. The rich variety of effects implies a rich physics behind and indeed transition metal oxides and perovskites in particular belong to the most intensively studied solid materials. The current mainstream studies of the electronic structure are based on commonly used approaches such as Hartree-Fock and density functional theory (DFT) since these methods have been in the forefront of the development for a few decades. It is still interesting that due to the demands and difficulties related to transition metals, the first high accuracy DFT calculations of ferroelectric perovskites, for example, are only about a decade old [1, 2, 3, 4, 5, 6].

The DFT approaches for electronic structure were tremendously successful in many respects [7, 8]. They enabled, at a low cost, estimation of basic features such as atomic structures, cohesive energies and other quantities for a large variety of condensed and molecular systems. However, there are still open problems with the accuracy of cohesive energies or band gaps. In addition, for transition metals the DFT method sometimes has more fundamental difficulties and require significant modifications such as LDA+U or LDA+J methods which mimic more accurately the large Coulomb interactions of the d-levels.

As an alternative to DFT methods, a high degree of accuracy is offered by post-Hartree-Fock approaches such as the Configuration Interaction or Coupled Cluster [6]. Unfortunately, the high accuracy of these methods for small systems (10 to 20 correlated electrons) are rather difficult to apply to solids or larger molecular systems. These methods, at least in their canonical formulation, scale rather unfavorably in the number of correlated electrons and that makes calculations of larger systems problematic.

One of the most promising alternatives for accurate electronic structure calculations has emerged in the methodology of quantum Monte Carlo (QMC) [9]. This approach uses the results of one-particle approaches such as sorting out of one-body states as a zeroth-order approximation for the wave function. This is then used as the starting point for many-body treatment. The many-body strategy includes actual, explicit and direct construction of correlated wave functions and their optimization within the variational Monte Carlo (VMC) method. The next step in pursuing the many-body effects is the diffusion Monte Carlo method which is based on a projection of the ground state using the propagator $\exp(-tH)$ in the imaginary time t. The QMC has recently emerged as a new alternative for electronic structure calculations and has led to a number of intersting results such as identification of the lowest energy clusters of carbon and silicon, reaction barriers of challenging reactions, band gaps in solids, excited states in nanocrystals [10, 11, 12, 13, 14, 15, 16, 17, 18].

There are three important issues in applying the QMC methods to complicated systems such as transition metal oxides: First, construction of accurate many-body trial functions using orbitals from one-particle approaches such as Hartree-Fock and density functional theory, in particular, the hybrid functionals such as B3LYP; second, efficient calculations of excitation energies both in variational and diffusion QMC; and lastly, method for calculating the small energy differences such as subtle features on total energy surfaces related to ferroelectric distortions. In this paper we will briefly explain how far the calculations have progressed so far and discuss the results obtained.

VARIATIONAL AND DIFFUSION QUANTUM MONTE CARLO METHODS.

In the variational Monte Carlo (VMC) method any expectation value for a Hamiltonian H of a system with N electrons is evaluated by a Monte Carlo integration. For example, the energy is given by

$$E_V = \frac{\int \Psi^*(R) H \Psi(R) dR}{\int \Psi^*(R) \Psi(R) dR} = \frac{\int |\Psi(R)|^2 [H\Psi(R)/\Psi(R)] dR}{\int |\Psi(R)|^2 dR} \qquad (1)$$

where $\Psi(R)$ denotes the trial or variational function and the integration is over $3N$-dimensional space $R = (r_1, r_2, ..., r_N)$. The integral is estimated as an average of local energies

$$E_V = (1/M) \sum_{m=1}^{M} \frac{H\Psi_T(R_m)}{\Psi_T(R_m)} + \varepsilon \qquad (1a)$$

where M samples of electron configurations $\{R_m\}$ are distributed according to $|\Psi(R)|^2$ and the error ε is proportional to $1/\sqrt{M}$. The Monte Carlo method of integration enables us to to include many-body effects into the wave function in an explicit way. In particular, it is very easy to incorporate the electron-electron cusp, a non-analytic behavior of the wave function at the point where two electrons coalesce [19]. It is well-known that this and other non-analytic features cause a slow convergence of the wave function expansion in space of Slater determinants. The VMC method is robust and relatively fast; however, the variational energy is biased by the choice of the trial function, and the number and quality of variational parameters optimized. The problem of the variational bias is particularly important, for example, when comparing energies of two different systems (like a solid in two different phases) because the quality of trial functions in the two cases might be different.

Fortunately, the diffusion Monte Carlo (DMC) which stochastically solves the Schrödinger equation with a high degree of accuracy, is capable of removing a large part of the variational bias. It is straightforward to show that

$$\lim_{t\to\infty} \exp(-tH)\Psi(R) = <\Psi|\Phi_0> \exp(-tE_0)\Phi_0(R), \qquad (2)$$

where $\Phi_0(R)$ is the ground state of the prescribed symmetry and E_0 is the corresponding eigenvalue. This projection can be carried out by solving the imaginary time Schrödinger equation. In its integral form, which also includes the importance sampling by the trial function $\Psi(R)$, it is given by

$$f(R, t+\tau) = \int G(R, R', \tau) f(R', \tau) dR' \qquad (3)$$

where

$$f(R,t) = \Psi(R)\Phi(R,t) \qquad (4)$$

and

$$G(R, R', \tau) = \frac{\Psi(R)}{\Psi(R')} \langle R|\exp(-\tau H)|R'\rangle \qquad (5)$$

In actual simulation the solution of (3) is found iteratively because the Green's function $G(R, R', \tau)$ is known in an analytical form only for small τ [7]. The fermion problem is treated by using the fixed-node approximation which enforces the nodes of the resulting wave function $\Phi(R)$ to be identical to the nodes of $\Psi(R)$ so that the product $\Phi(R)\Psi(R)$ is non-negative everywhere (the node is a subset of R-space defined by $\Psi(R) = 0$) [9]. In most cases the impact of the fixed-node approximation on the calculated quantities is rather small (of the order of 5% of the correlation energy). However, there are cases where the fixed-node error can be significant if an adequate trial function is not used, e.g., when there is a strong near-degeneracy effect so that the trial function has a multi-configuration character [9].

The result of the stochastic solution of (5) is a set of samples of the product $\Phi(R)\Psi(R)$ and the energy of the fixed node solution can be evaluated by the so-called mixed estimator,

$$E_{DMC} = \frac{\int \Phi(R) H \Psi(R) dR}{\int \Phi(R) \Psi(R) dR} = \frac{\int \Phi(R) \Psi(R) [H\Psi(R)/\Psi(R)] dR}{\int \Phi(R) \Psi(R) dR} \qquad (6)$$

Further details can be found in review [9] and references therein.

The trial function is one of the most important factors in obtaining accurate results by the VMC and DMC methods in an efficient way. There has been a remarkable development of the functional forms for the trial function which has brought a significant improvement in efficiency and accuracy of QMC calculations in recent years. One of the most common forms is given by

$$\Psi = \sum_n d_n D_n^\uparrow D_n^\downarrow \exp\left[\sum_l \sum_{i<j} u(r_{il}, r_{jl}, r_{ij})\right] \quad (7)$$

where $D_n^\uparrow$ and $D_n^\downarrow$ are Slater determinants of spin-up and spin-down one-electron orbitals, d_n are expansion coefficients and $u(r_{il}, r_{jl}, r_{ij})$ is the function which depends on electron-electron and electron-ion distances and therefore captures electron-electron and electron-electron-ion correlations. Orbitals both from LDA and Hartree-Fock (HF) calculations have been used for building the Slater determinants and also several functional forms and parametrizations of the correlating function have been proposed [9].

The variational parameters which enter into the function $u(...)$ are found by minimization of the variational energy or by minimization of the variance of the local energy given by

$$\sigma_H^2 = \frac{\int |\Psi(R)|^2 \left[H\Psi(R)/\Psi(R) - E_V\right]^2 dR}{\int |\Psi(R)|^2 dR} \quad (8)$$

or a combination of both. The number of variational parameters used has varied between a few and several tens [9].

EFFECTIVE CORE POTENTIALS.

From a number of available effective core potentials or pseudopotentials [20, 21, 22] in the first place we were interested in accuracy. Previous tests and calculations have shown that semicore states are important for d elements ([23, 24]). This eliminated essentially the whole class of "large core" $nd, (n+1)sp$ type of pseudopotentials. Based on our previous calculations of Fe atom [25] the pseudopotentials from the Stuttgart group of M. Dolg and others [26] proved to be quite accurate, leading to the agreement with experiment for excited states within 0.1 eV or so. These pseudopotentials were developed in the Dirac-Fock formalism with adjustments to reproduce not only single-particle quantities such as single-particle eigenvalues and charge/norm conservation but also total energy differences such as excitation energies in subsequent correlated calculations. We used these pseudopotentials for Ba and other heavy elements. Very recently, the group of W. Lester, Jr., has developed improved ECPs [27] for the first two rows which have finite amplitudes in the origin (ie, position of the ion). This leads to a much smoother behavior of the local energy in the vicinity of the ion and increases the efficiency of the DMC method significantly. A similar set of pseudopotentials have been developed very recently of the third row by Y. Lee and coworkers [28]. These are "small core", implying that 3s and 3p semicore states are in the valence space and according to our tests they are among the most accurate available so far. We estimate that they would be able to get the energy differences within 0.05 eV, *i.e.*, essentially the limit of accuracy within the scalar-relativistic approximation in the third row.

VARIATIONAL WAVE FUNCTIONS: TESTS ON TMO MOLECULES.

As we have mentioned the systems with transition metals are quite challenging for electronic structure calculations. The key difficulties can be listed as follows:

- large Coulomb, exchange and correlation energies of d states with a tendency of exchange and correlation to partially compensate each other
- near-degeneracy of nd, $(n+1)s$ and $(n+1)p$ states which have very different spatial ranges but are energetically rather close (for example, the states $4s^2 3d^8$ and $4s^1 3d^9$ of the neutral Ni atom are within 0.1 eV)
- easy (partial) ionization of TM element which creates a mixture of ionic and covalent bonding and leads to many possibilities for hybridizations with orbitals of neighboring atoms
- "semicore" states, mentioned above, which lead to large increase of the total energy when included into valence space

All these effects imply that the wave function can be complicated, especially due to the near-degeneracy effects. Therefore one can expect difficulties in constructing adequate trial functions since it is not clear how to come up with a good mean-field theory for getting the one-particle orbitals and correspoding wave function. Therefore we have decided to test various wavefunctions on transition metal oxygen molecules which are a good laboratory for the effects mentioned above since they exhibit charge transfer, spin polarization, mixture of ionic and covalent bonding. In addition, multi-reference wavefunctions are often necessary to describe the correct spin/space symmetries. In fact, these molecules are still quite challenging for the most accurate quantum chemistry methods. The advantage for our purposes is that the system is sufficiently small to enable us to test various types of wave functions and analyze the impact on the fixed-node errors. We were able to carry out a systematic study for a series of wave function types. In particular, we have tested trial functions based on single determinant of HF orbitals, multi-determinant wave function with multi-configuration HF optimized orbitals, single determinant with B3LYP orbitals and a combined type using multi-determinats and B3LYP orbitals. Surprisingly, we have found that B3LYP orbitals have led to the best wave functions with the smallest fixed-node errors. This prompted us to look at the B3LYP functional in more detail. B3LYP is essentially a sum of correlation functional of Lee-Yang-Parr plus a weighted average of HF (nonlocal) exchange and a local approximation for the exchange. By varying the relative weight of these two exchange terms we tried to find optimized one-particle orbitals. In Fig.1 is a plot of total energy of MnO vs. the weight of the HF component in the B3LYP functional.

One can observe a shallow minimum with the value of the weight of about 0.17 which is actually quite close to the suggested correction of B3LYP weight of about 0.15 in studies molecular systems with transition metal atoms [29]. Although the improvement

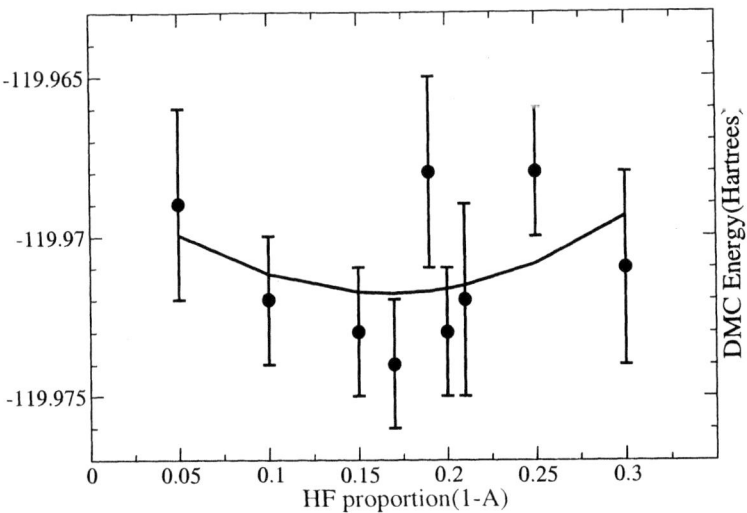

FIGURE 1. MnO total energy as a function of the weight of exact HF exchange in the B3LYP-type of DFT functional.

TABLE 1. Binding energies (eV) by various methods. The DMC(MCSCF) was a multi-reference wave function with ≈ 20 configurations with the largest weights.

	TiO	MnO
State	$^3\Delta$	$^6\Sigma$
Hartree-Fock	2.64	-0.92
LSDA	9.12	5.88
PW91	7.45	4.79
B3LYP	6.62	3.39
VMC(HF orbitals)	6.0(1)	3.1(1)
DMC(HF orbitals)	6.3(1)	2.9(1)
DMC(B3LYP orbitals)	6.9(1)	3.4(2)
DMC(MCSCF orbitals)	6.7(2)	3.4(2)
Experiment	6.98/6.87 (7)	3.70

is not very pronounced in this case it is clear that this of sufficient interest to be tested in the solids calculations. Table 1 compares the obtained binding energies from various wave functions. More details can be found in the upcoming paper [30].

CORRELATED SAMPLING FOR ENERGY DIFFERENCES IN SOLIDS.

The key QMC development, which is currently the focus of our effort, is the efficient calculation of the energy differences. Indeed this is very important for calculating quan-

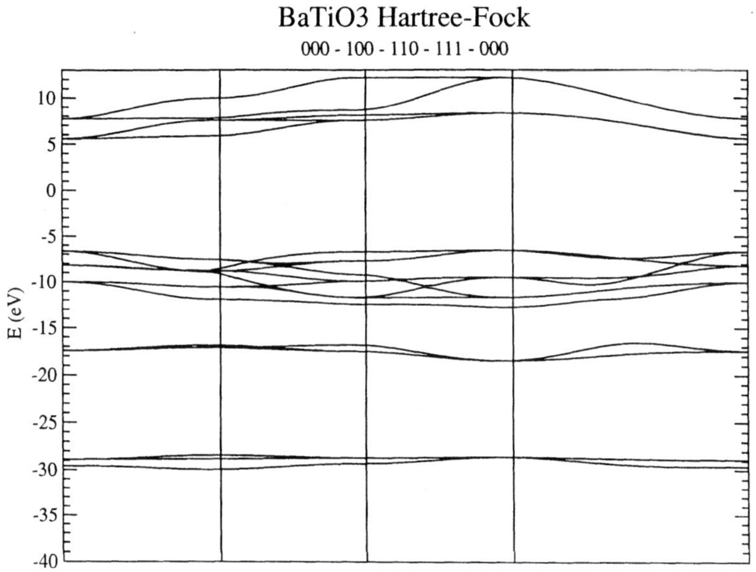

FIGURE 2. BaTiO$_3$ band structure in Hartree-Fock method.

tities such as band gaps and also energy differences for various phases effciently. In variational Monte Carlo we were able to estimate the band gap for BaTiO$_3$ system as $E_g = 3.9(9)$ eV from two separate calculations, one for the ground state and the other for the excited state. The excited state is created by promoting electron which occupies a state in the valence band into a state in the conduction band. The band structures from HF, B3LYP and PW91 calculations are in Fig2-4 and the excitation was from $\Gamma \rightarrow X$ point.

While it is straightforward to push the runs further to descrease the error bar the demands on computer time would be substantial. Such calculations would necessarily fall into a category of "unique" projects while we are interested in developemnt of computational tools and methods which could be used on a routine basis. A well-known strategy for VMC is to employ correlated sampling which enables cancelation of (to the first order) random fluctuations from the difference and obtain an error bar which is smaller than the error bars of the separate estimators. In VMC we implemented the correlated sampling for the excitation energy (gap) as a difference

$$E_{diff} = \sum_m (w_m^{exc} E_m^{exc} - w_m^{gr} E_m^{gr}) \tag{9}$$

where

$$w_m^* = \frac{|\Psi^*(R_m)|^2/\Psi_G(R_m)}{\sum_m |\Psi^*(R_m)|^2/\Psi_G(R_m)} \tag{10}$$

and $*$ denotes either the excited or the ground state, ie, $* = exc$ or gr, respectively, while Ψ^* is the corresponding variational wavefunction and E_m^* is the corresponding local

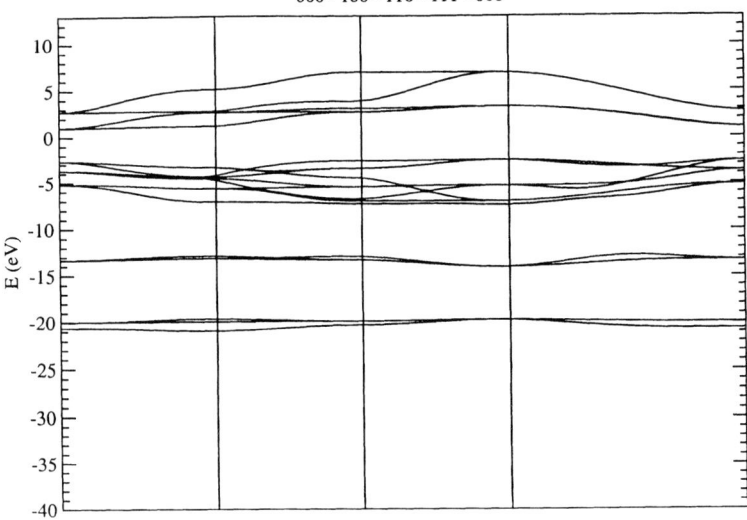

FIGURE 3. BaTiO$_3$ band structure with B3LYP functional.

energy. The configurations $\{R_m\}$ are distributed according to $\Psi_G(R)$. This sampling distribution can be chosen arbitrarily since it cancels out from the averages and therefore can be used to enhance the noise cancellation in (9). We have tried two possibilities

$$\Psi_G = |\Psi^{exc}|^2 + |\Psi^{gr}|^2$$

and

$$\Psi_G = |\Psi^{exc}||\Psi^{gr}|$$

Our current tests indicate that the first choice leads to a significant improvement of the error bars for the difference, approximately by an order of magnitude. This result is quite encouraging since its implementation is quite straightforward.

Currently, similar development is being pursued for the DMC method which will be more involved since instead of the ratio of the trial and sampling functions we will need to evaluate corresponding weights given by the ratio of Green's functions (5). This will open exciting possibilities for efficient evaluation of excitations in solids within the high accuracy DMC method.

The same approach can be easily applied to evaluation of energy differences of close points on the energy surface such as for the case of ferroelectric distortion and it would be possible to estimate the constants of the effective Hamiltonians for example. The noise cancellation in this case will be even more dramatic since the wave functions for two states will be almost identical and one can expect error bar improvements roughly by two orders of magnitude.

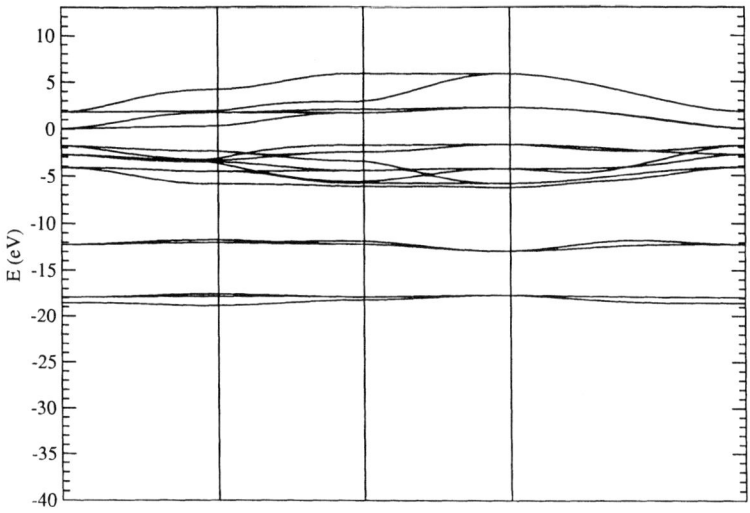

FIGURE 4. BaTiO$_3$ band structure with PW91 functional.

ACKNOWLEDGMENTS

Support by ONR-N0014-01-1-0408 and DARPA/ONR-N00014-01-1-1062 grants is gratefully acknowledged. We are also grateful to Dr. Y. Lee for providing us with his recently developed Dirac-Fock ECPs for the 3d atom series.

REFERENCES

1. King-Smith, R. D., and Vandebilt, D., *Phys. Rev. B*, **49**, 5828–5844 (1994).
2. Zhong, W., Vanderbilt, D., and Rabe, K. M., *Phys. Rev. Lett.*, **73**, 1861–1864 (1994).
3. Zhong, W., Vanderbilt, D., and Rabe, K. M., *Phys. Rev. B*, **52**, 6301–6312 (1995).
4. Cohen, R. E., and Krakauer, H., *Phys. Reb. B*, **42**, 6416–6423 (1990).
5. Singh, D. J., *Phys. Rev. B*, **52**, 12559–12563 (1995).
6. Cohen, R. E., and Krakauer, H., *Ferroelectrics*, **136**, 65–83 (1992).
7. Hohenberg, P., and Kohn, W., *Phys. ReV.* (1964).
8. Kohn, W., and Sham, L. J., *Phys. Rev.* (1965).
9. W. M. C. Foulkes, R. N., L. Mitas, and Rajagopal, G., *Rev. Mod. Phys.*, **73**, 33–83 (2001).
10. Therrien, J., et al., *Appl. Phys. Lett.*, **80**, 841 (2002).
11. Therrien, J., et al., *Appl. Phys. Lett.*, **78**, 1918 (2001).
12. Grossman, J. C., et al., *Phys. Rev. Lett.*, **86**, 472 (2001).
13. Torelli, T., and Mitas, L., *Phys. Rev. Lett.*, **85**, 1702 (2000).
14. Mitas, L., Grossman, J. C., Stich, I., and Tobik, J., *Phys. Rev. Lett.*, **84**, 1479 (2000).
15. Grossman, J. C., and Mitas, L., *Phys. Rev. Lett.*, **79**, 4353 (1997).
16. Grossman, J. C., Raghavachari, K., and Mitas, L., *Phys. Rev. Lett.*, **75**, 3870 (1995).
17. Grossman, J. C., and Mitas, L., *Phys. Rev. Lett.*, **74**, 1323 (1995).
18. Mitas, L., and Martin, R. M., *Phys. Rev. Lett.*, **72**, 2438 (1994).

19. Kato, T., *Commun. Pure Appl. Math.*, **10**, 151 (1957).
20. Vanderbilt, D., *Phys. Rev. B*, **41**, 7892 (1990).
21. Troullier, N., and Martins, J. L., *Phys. Rev. B*, **43**, 1993 (1991).
22. Bachelet, G. B., Hamann, D. R., and Schlütter, M., *Phys. Rev. B*, **26**, 4199 (1982).
23. Bernholc, J., and Holzwarth, N. A. W., *Phys. Rev. Lett.*, **50**, 1451 (1983).
24. Ceperley, D. M., and Mitas, L., *Advances in Chemical Physics*, Wiley, New York, 1996, p. 1.
25. Mitas, L., *Phys. Rev. A*, **49**, 4411 (1994).
26. Stuttgart-Köln Group: H. Preuss, M. D. e. a., H. Stoll (2002), URL `http://www.theochem.uni-stuttgart.de/pseudopotentials/`.
27. Ovcharenko, I., Aspuru-Guzik, A., and Lester Jr., W. A., *J. Chem. Phys.*, **114**, 7790 (2001).
28. Lee, Y., Kent, P., Towler, M., Needs, R., and Rajagopal, G., *Phys. Rev. B*, **62**, 13347 (2000).
29. O. Salomon, B. H., M. Reiher, *J. Chem. Phys.*, **117**, 4729 (2002).
30. Wagner, L., and Mitas, L., *Chem. Phys. Lett., in press.* (2003).

Phase-free quantum Monte Carlo method: random walks using general basis sets

Henry Krakauer and Shiwei Zhang

Department of Physics, College of William and Mary, Williamsburg, VA 23187-8795

Abstract. Fermion quantum Monte Carlo (QMC) methods that work in a general basis space may be more effective for some problems than the traditional diffusion QMC method. Projection of the ground state energy is achieved by random walks in the space of Slater determinants whose one-particle orbitals are expressed in terms of the chosen basis set. A complication is that the determinants will in general acquire complex phases. This a consequence of ground state Monte Carlo projection using the Hubbard-Stratonovich transformation of the two-body interaction. To control the resulting "sign" decay, we describe a method we have recently introduced for the propagation of phaseless determinants. The approximation relies on importance sampling with a trial wave function. The approximation has features in common with diffusion MC fixed node and lattice-model constrained path methods. Using a planewave basis and non-local pseudopotentials, we apply the method to Si atom, dimer, and 2, 16, 54 atom (216 electrons) bulk supercells. Single Slater determinant wave functions from density functional theory calculations were used as $|\Psi_T\rangle$ with no additional optimization. The calculated binding energy of Si_2 and cohesive energy of bulk Si are in excellent agreement with experiments and are comparable to the best existing theoretical results.

INTRODUCTION

Quantum Monte Carlo (QMC) methods based on auxiliary fields (AF) are used in areas spanning condensed matter physics, nuclear physics, and quantum chemistry. These methods [1, 2] allow essentially exact calculations of ground-state and finite-temperature equilibrium properties of interacting many fermion systems. The required CPU time scales in principle as a power law with system size, and the methods have been applied to study a variety of problems including the Hubbard model, nuclear shell models, and molecular electronic structure. The central idea of these methods is to write the imaginary-time propagator of a many-body system with two-body interactions in terms of propagators for independent particles interacting with external auxiliary fields. The independent particle problems are solved for configurations of the AF and averaging over different AF configurations is then performed by Monte Carlo (MC) techniques.

QMC methods with auxiliary fields have several appealing features. For example, they allow one to choose *any* one-particle basis suitable for the problem, and to fully take advantage of well-established techniques to treat independent particles. Given the remarkable development and success of the latter [3], it is clearly very desirable to have a QMC method that can use exactly the same machinery and systematically include correlation effects by simply building stochastic ensembles of the independent particle solutions. Vigorous attempts have been made from several fields to explore this possibility [4, 5, 6, 7].

A significant hurdle exists, however: except for special cases (e.g., Hubbard), the two-body interactions will require auxiliary fields that are *complex*. As a result, the single-particle orbitals become complex, and the MC averaging over AF configurations becomes an integration over complex variables in many dimensions. A phase problem thus occurs which ultimately defeats the algebraic scaling of MC and makes the method scale exponentially. This is analogous to but more severe than the fermion sign problem with real AF [8, 9] or in real-space methods [10]. No satisfactory, general approach exists to control the phase problem. As a result, only small systems or special forms of interactions can be treated.

In this paper we describe a new method for many-fermions that allows the use of any one-particle basis [11]. It projects out the ground state by random walks in the space of Slater determinants. The phase problem is eliminated with an approximation that relies on a trial wave function $|\Psi_T\rangle$. We demonstrate the method by applying it to electronic systems using a planewave basis and non-local pseudopotentials, which can be implemented straightforwardly in this method. We calculate the binding energy of Si_2 and the cohesive energy of bulk Si using fcc supercells consisting of up to 54 atoms (216 electrons). These calculations represent the first application of AF-based QMC to solids. The results are in excellent agreement with experiments and are comparable to the best existing theoretical results. Particularly worth noting is that our results were obtained with a trial wave function which is a single Slater determinant formed by orbitals from density functional theory (DFT) calculations (with the local density approximation (LDA)), with no additional parameters or optimization.

METHODOLOGY

The Hamiltonian for any many-fermion system with two-body interactions can be written in any one-particle basis in the general form

$$\hat{H} = \hat{H}_1 + \hat{H}_2 = \sum_{i,j}^{M} T_{ij} c_i^\dagger c_j + \frac{1}{2} \sum_{i,j,k,l}^{M} V_{ijkl} c_i^\dagger c_j^\dagger c_k c_l, \tag{1}$$

where M is the size of the chosen one-particle basis, and $c_i^\dagger$ and c_i are the corresponding creation and annihilation operators. Both the one-body (T_{ij}) and two-body matrix elements (V_{ijkl}) are known.

To obtain the ground state $|\Psi_G\rangle$ of $\hat{H}$, QMC methods use the imaginary time propagator $e^{-\tau \hat{H}}$ acting on a trial wave function $|\Psi_T\rangle$: $\lim_{n\to\infty} (e^{-\tau \hat{H}})^n |\Psi_T\rangle \propto |\Psi_G\rangle$. $|\Psi_T\rangle$ must not be orthogonal to $|\Psi_G\rangle$, and we will assume that it is of the form of a single Slater determinant or a linear combination of Slater determinants. The time step τ is chosen to be small enough so that $\hat{H}_1$ and $\hat{H}_2$ in the propagator can be accurately separated with the Trotter decomposition.

The propagator $e^{-\tau \hat{H}_1}$ is the exponential of a one-body operator. A propagator of this form acting on a Slater determinant is straightforward to calculate, and it simply yields another determinant. The two-body propagator $e^{-\tau \hat{H}_2}$ can be expressed as an integral of propagators of this form, as follows. Any two-body operator can be written as a quadratic form of one-body operators:

$$\hat{H}_2 = -\frac{1}{2} \sum_\alpha \lambda_\alpha \hat{v}_\alpha^2, \qquad (2)$$

where λ_α is a real number and $\hat{v}_\alpha$ is a one-body operator. The Hubbard-Stratonovich (HS) transformation [12] then allows us to write

$$e^{-\tau \hat{H}_2} = \prod_\alpha \left(\frac{1}{\sqrt{2\pi}} \int_{-\infty}^\infty e^{-\frac{1}{2}\sigma_\alpha^2} e^{\sqrt{\tau}\sigma_\alpha \sqrt{\lambda_\alpha} \hat{v}_\alpha} d\sigma_\alpha \right). \qquad (3)$$

Introducing vector representations $\sigma \equiv \{\sigma_1, \sigma_2, \cdots\}$ and $\hat{\mathbf{v}} = \{\sqrt{\lambda_1}\hat{v}_1, \sqrt{\lambda_2}\hat{v}_2, \cdots\}$, we have the desired form

$$e^{-\tau \hat{H}} = \int P(\sigma) B(\sigma) d\sigma, \qquad (4)$$

where $P(\sigma)$ is the normal distribution in Eq. (3) and

$$B(\sigma) \equiv e^{-\tau \hat{H}_1/2} e^{\sqrt{\tau}\sigma \cdot \hat{\mathbf{v}}} e^{-\tau \hat{H}_1/2} \qquad (5)$$

is a one-body propagator.

The imaginary-time propagation thus requires evaluating the multidimensional integral in Eq. (4) over time slices n and the corresponding auxiliary fields. MC techniques are the only way to evaluate such integrals efficiently. We use a random walk approach [9]. In each step, a walker $|\phi\rangle$, which is a single Slater determinant, is propagated to a new position $|\phi'\rangle$: $|\phi'(\sigma)\rangle = B(\sigma)|\phi\rangle$, where σ is a random variable sampled from $P(\sigma)$. After a sufficient number of steps (iterations), the ensemble of random walkers is a MC representation of the ground-state wave function: $|\Psi_G\rangle \doteq \sum_{\phi'} |\phi'\rangle$.

FORMULATION FOR A PLANEWAVE BASIS

In this paper, all the simulations employ a normalized planewave basis using periodic boundary conditions with N electrons in a simulation cell of volume Ω: $|\psi\rangle = \frac{1}{\sqrt{\Omega}} \exp(i\mathbf{G} \cdot \mathbf{r})$, where $\mathbf{G}$ is a reciprocal lattice vector. Only planewaves with $G^2/2 \leq E_{cut}$, where E_{cut} is a fixed cutoff energy are retained in the basis, as in conventional planewave-based density functional calculations. In this basis, the one-body kinetic energy, local and non-local pseudopotential terms in the Hamiltonian are given by (constant terms in the Hamiltonian including ion-ion interactions and those arising from self-interaction Ewald-like terms are not shown):

$$K = \frac{1}{2} \sum_i G_i^2 c_i^\dagger c_i$$

$$V_{ei,L} = \frac{1}{2} \sum_{\mathbf{Q} \neq 0} V_L(\mathbf{Q}) \left[\rho(\mathbf{Q}) + \rho^\dagger(\mathbf{Q}) \right].$$

$$V_{ei,NL} = \sum_{i,j} V_{NL}(\mathbf{G}_i, \mathbf{G}_j) c_i^\dagger c_j \qquad (6)$$

where the one-body density operator ρ is given by

$$\rho(\mathbf{Q}) = \sum_{\mathbf{G},\lambda} c^\dagger_{\mathbf{G}+\mathbf{Q},\lambda} c_{\mathbf{G},\lambda} \theta\left(E_{cut} - |\mathbf{G}+\mathbf{Q}|^2/2\right), \quad (7)$$

and the step function ensures that $(\mathbf{G}+\mathbf{Q})$ lies within the planewave basis.

The two-body terms in the Hamiltonian can be expressed after manipulation using the Fermion commutation relations as (additional one-body terms arising from the use of the commutation relations are not shown):

$$\begin{aligned}\hat{H}_2 &= \frac{1}{2\Omega} \sum_{\mathbf{Q}\neq 0} \frac{4\pi}{Q^2} \rho^\dagger(\mathbf{Q})\rho(\mathbf{Q}) \\ &= \frac{1}{4} \sum_{\mathbf{Q}\neq 0} [A^2(\mathbf{Q}) + B^2(\mathbf{Q})], \end{aligned} \quad (8)$$

where we have defined Hermitian operators $A(\mathbf{Q})$ and $B(\mathbf{Q})$ as

$$\begin{aligned} A(\mathbf{Q}) &\equiv \sqrt{\tfrac{2\pi}{\Omega Q^2}} [\rho(\mathbf{Q}) + \rho^\dagger(\mathbf{Q})] \\ B(\mathbf{Q}) &\equiv i\sqrt{\tfrac{2\pi}{\Omega Q^2}} [\rho(\mathbf{Q}) - \rho^\dagger(\mathbf{Q})] \end{aligned}. \quad (9)$$

Equation (8) is now clearly in the form of a sum of squares of (Hermitian) one-body operators, as in Eq. (2) (with all the $\lambda_\alpha = -1/2$).

Although the formalism of our method is independent of details of $\hat{v}_\alpha$ and the transformation, it is important to emphasize that the quadratic form for $\hat{H}_2$ is not unique. The number of terms in the sum, which is in general of $\mathcal{O}(M^2)$, depends on the particular break-up, as does the actual form of $\hat{v}_\alpha$. Different choices lead to different HS decompositions, which in turn can affect the quality of the results (e.g., severity of the sign or phase problem). Using a planewave basis and the break-up given in Eq. (8), the number of terms in the sum is $\mathcal{O}(M)$, i.e. the number of unique Q vectors.

PHASE PROBLEM

In general λ_α cannot be made all positive in Eq. (3). As mentioned, in the planewave formulation above all the $\lambda_\alpha = -1/2$. Moreover, we have found, as have others[4], that shifting the potential to make λ_α real causes large fluctuations and does not work well in general. The one-body operators $\hat{v}$ are therefore complex. As the projection proceeds, the orbitals in the random walkers will become complex. As a result, the statistical fluctuations in the MC representation of $|\Psi_G\rangle$ increase exponentially with projection time $\beta \equiv n\tau$. This is the phase problem referred to earlier. It is of the same origin as the sign problem that occurs when $B(\sigma)$ is real. The phase problem is more severe, however, because for each $|\phi\rangle$, instead of a $+|\phi\rangle$ and $-|\phi\rangle$ symmetry [9], there is now an infinite set $\{e^{i\theta}|\phi\rangle\}$ ($\theta \in [0,2\pi)$) from which the random walk cannot distinguish. At large β, the phase of each $|\phi\rangle$ becomes random, and the MC representation of $|\Psi_G\rangle$ becomes

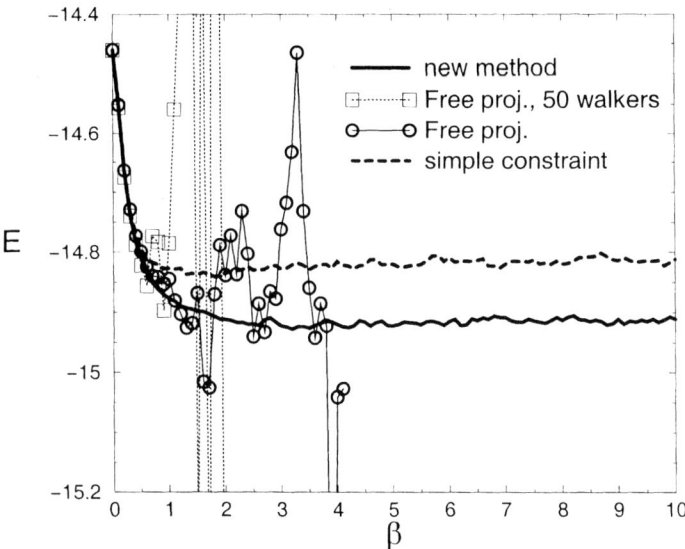

FIGURE 1. Illustration of the phase problem and constraints to control it. The total valence energy (in Ry) of an fcc Si primitive cell (2 atoms) is shown as a function of projection time $\beta = n\tau$, with $\tau = 0.05$ Ry^{-1}. Unless otherwise indicated, 10,000 walkers are used. Increasing the number of walkers from 50 to 10,000 only slightly delays the onset of the phase problem. Simple generalization of the constraint that worked well for real determinants leads to poor results. The new method gives accurate results (note the agreement with the solid free projection curve, which is exact, until the latter becomes too noisy at $\beta \sim 1.5$).

dominated by noise. This problem is generic, and the same analysis would apply if we had chosen, instead of the random walk, the standard AF QMC sampling approach [2]. In Fig. 1, the curves labeled "free projection" (i.e. no constraint is applied to the random walk) illustrate the phase problem.

Existing fixed-node type approximations have often worked very well to control the sign/phase problem in real space [13, 14] or in Slater determinant space when the propagator is real [9]. The phase problem here is unique because not only do the determinants acquire overall phases, but the internal structures of their orbitals become complex. The real-space analogy would be to have walkers whose coordinates become complex. This makes straightforward generalization of existing approaches ineffective. For example, similar to the constrained path approximation [9] we could impose the condition $\text{Re}\langle \Psi_T | \phi \rangle > 0$. Or, in the spirit of the fixed-phase approximation in real space [14] we could project the walker by including a factor $\cos(\Delta\theta)$ in the weight, where $\Delta\theta$ is the phase of $\langle \Psi_T | \phi' \rangle / \langle \Psi_T | \phi \rangle$. They give similar results and do not work well [6]. The former is shown in Fig. 1 ("simple constraint"). Importance sampling with $\text{Re}\langle \Psi_T | \phi \rangle$ or $|\langle \Psi_T | \phi \rangle|$ does not change the results [15].

RANDOM WALKS WITH PHASELESS DETERMINANTS

To formulate a new method that can better separate the overall phase from the determinant, we first borrow from the idea of importance sampling [16], although our choice of the so-called importance function, $\langle \Psi_T | \phi \rangle$, is actually *complex*. We modify Eq. (4) to obtain the following new propagator for $|\phi\rangle$:

$$\int \langle \Psi_T | \phi'(\sigma - \bar{\sigma}) \rangle P(\sigma - \bar{\sigma}) B(\sigma - \bar{\sigma}) \frac{1}{\langle \Psi_T | \phi \rangle} d\sigma, \tag{10}$$

where we have included a constant shift [5] $\bar{\sigma}$ in the integral in Eq. (4), which does not affect the equality. Eq. (10) can be re-written as

$$\int P(\sigma) W(\sigma, \phi) B(\sigma - \bar{\sigma}) d\sigma, \tag{11}$$

where

$$W(\sigma, \phi) \equiv \frac{\langle \Psi_T | \phi'(\sigma - \bar{\sigma}) \rangle}{\langle \Psi_T | \phi \rangle} e^{\sigma \cdot \bar{\sigma} - \frac{\bar{\sigma} \cdot \bar{\sigma}}{2}}. \tag{12}$$

The new propagator in Eq. (11) defines a new random walk. In each step the walker $|\phi\rangle$ is propagated to $|\phi'\rangle$ by $B(\sigma - \bar{\sigma})$: $|\phi'(\sigma - \bar{\sigma})\rangle = B(\sigma - \bar{\sigma})|\phi\rangle$, where σ is again sampled from $P(\sigma)$. $W(\sigma, \phi)$ is a c-number which can be accounted for by having every walker carry an overall weight factor and updating them according to: $w_{\phi'} = W(\sigma, \phi) w_\phi$. Formally the MC representation of $|\Psi_G\rangle$ in the new random walk is:

$$|\Psi_G\rangle \doteq \sum_{\phi'} w_{\phi'} \frac{|\phi'\rangle}{\langle \Psi_T | \phi' \rangle}. \tag{13}$$

For any choice of the shift $\bar{\sigma}$, the new random walk is an exact procedure to realize the imaginary time propagation, in the sense of Eq. (13). The optimal choice of $\bar{\sigma}$ is determined by minimizing the fluctuation of $W(\sigma, \phi)$ with respect to σ. To $\mathcal{O}(\sqrt{\tau})$ this yields

$$\bar{\sigma} = -\sqrt{\tau} \frac{\langle \Psi_T | \hat{v} | \phi \rangle}{\langle \Psi_T | \phi \rangle}. \tag{14}$$

With this choice the leading σ-dependent term in W is reduced to $\mathcal{O}(\tau)$ and, by expanding $B(\sigma - \bar{\sigma})$ in $|\phi'\rangle$ in Eq. (12), we can manipulate W into the following form:

$$W(\sigma, \phi) \doteq \exp\left[-\tau \frac{\langle \Psi_T | \hat{H} | \phi \rangle}{\langle \Psi_T | \phi \rangle}\right] \equiv \exp[-\tau E_L(\phi)], \tag{15}$$

where the term E_L parallels the local energy in real-space QMC methods. Both E_L and the shift $\bar{\sigma}$ in Eq. (14) are independent of any overall phase factor of $|\phi\rangle$.

The weight of the walker in the new random walk is determined by E_L. In the limit of an exact $|\Psi_T\rangle$, E_L is a real constant, and the weight of each walker remains real. The so-called mixed estimate for the energy is phaseless:

$$E_G = \frac{\langle \Psi_T | \hat{H} | \Psi_G \rangle}{\langle \Psi_T | \Psi_G \rangle} \doteq \frac{\sum_{\phi'} w_{\phi'} E_L(\phi')}{\sum_{\phi'} w_{\phi'}}. \tag{16}$$

With a general $|\Psi_T\rangle$ which is not exact, a natural approximation is to replace E_L in Eq.'s (15) and (16) by its real part, $\text{Re}(E_L)$. We have thus obtained a phaseless formalism for the random walk, with real and positive weights in Eq.'s (13) and (16).

Despite this, an additional constraint is still required. To illustrate the problem we consider the overlap $\langle\Psi_T|\phi'\rangle$ during the random walk. Let us denote the phase of $\langle\Psi_T|\phi'(\sigma-\bar{\sigma})\rangle/\langle\Psi_T|\phi\rangle$ by $\Delta\theta$, which is in general non-zero (of order $-\sigma\text{Im}\bar{\sigma}$). This means that, the walkers will undergo a random walk in the complex plane defined by $\langle\Psi_T|\phi'\rangle$. At large β they will therefore populate the complex plane symmetrically, independent of their initial positions. It is useful to contrast the situation with the special case of a *real* $\hat{\mathbf{v}}$. For any $\hat{\mathbf{v}}$ the shift $\bar{\sigma}$ diverges as a walker approaches the origin in the complex plane, i.e., as $\langle\Psi_T|\phi'\rangle \to 0$. The effect of the divergence is to move the walker away from the origin. With a *real* $\hat{\mathbf{v}}$, $\Delta\theta = 0$ and the random walkers move only on the real axis. If they are initialized to have positive overlaps with $|\Psi_T\rangle$, $\bar{\sigma}$ will ensure that the overlaps remain positive throughout the random walk, much like in fixed-node diffusion Monte Carlo (DMC) in real space. Thus in this case the phaseless formalism reduces to the constrained path Monte Carlo method of Ref. [9], and it alone is sufficient to control the sign problem. For a *complex* $\hat{\mathbf{v}}$, however, the random walk is "rotationally invariant" in the complex plane, and the divergence of $\bar{\sigma}$ is not enough to prevent the build-up of a finite density at the origin. Near the origin the local energy E_L diverges, which causes diverging fluctuations in the weights of walkers. To address this we make an additional approximation. We project the random walk to "one-dimension" and multiply the weight of each walker in each step by $\max\{0,\cos(\Delta\theta)\}$. Imposing instead $\text{Re}\langle\Psi_T|\phi'\rangle > 0$ gave similar results, but with somewhat larger variance.

APPLICATION TO SILICON ATOM, DIMER, AND BULK

We apply the new method to Si atom, molecule, and bulk. The Si^{4+} ions are represented by a norm-conserving LDA Kleinman-Bylander (KB) non-local pseudopotential [17]. We use periodic boundary conditions, and a planewave basis with a kinetic energy cut-off $E_{cut} = 12.25$ Ry. The error resulting from E_{cut} was estimated through LDA calculations and is smaller than the MC statistical errors. The pseudopotential can be applied in essentially the same way as in planewave-based LDA calculations [15]. Calculations involving $\hat{\mathbf{v}}$ and the local part of the pseudopotential are efficiently handled using fast Fourier transforms. The separable KB form of the non-local pseudopotential makes its application as efficient as in LDA planewave codes. Our $|\Psi_T\rangle$ is a single Slater determinant consisting of LDA orbitals generated using ABINIT [18].

In Table 1, we show results for the atom and molecule. Additional calculations with $a = 22a_B$ supercells show that finite-size errors at $a = 19a_B$ were smaller than the MC statistical errors. Our calculated Si_2 binding energy is in excellent agreement with the experimental value [19].

In the bulk calculations, we use fcc supercells consisting of 2, 16, 54 atoms (5209 plane waves). As Fig. 1 shows, the new method leads to a large improvement. Results for 16 and 54 atoms are shown in Table 2. Our calculation for 54 atoms took several days on 20 Compaq Alpha 667 MHz processors. For the bulk cohesive energy, we first included

TABLE 1. Total valence energies of Si and Si_2, and binding energy of Si_2. The Si_2 ground state is $^3\Sigma_g^-$ (electronic configuration 5 ↑ 3 ↓). Calculations were done at the experimental equilibrium bond length of $4.244 a_B$, in a cubic supercell with $a = 19 a_B$ (4945 plane waves). Energies are in eV. Error bars are in the last digit and are in parentheses.

	Si	Si_2	$Si_2\ E_B$
LDA	−102.648	−209.175	3.879
QMC	−103.45(2)	−210.03(7)	3.12(8)
Experiment			3.21(13)

TABLE 2. Cohesive energy of bulk Si. Calculations are done for fcc supercells with 16 and 54 atoms, at $a_{\exp} = 5.43$ Å. QMC result at ∞ is from 54 atoms and includes two finite-size corrections: (i) an independent-particle correction of 0.311 eV from LDA and (ii) an additional Couloumb correction of −0.174 eV from Ref. [20, 22]. A zero-point energy correction of −0.061 eV was also added to the calculated results at ∞. Energies are in eV per atom. Error bars are in the last digit and are in parentheses.

	16	54	∞
LDA	3.836	4.836	5.086
QMC	3.79(4)	4.51(3)	4.59(3)
Experiment			4.62(8)

a correction for the independent-particle finite-size error from the LDA results. We then corrected for the remaining Coulomb finite-size error using the results of Kent *et al.* [20]. [In Ref. [20], the DMC calculations used L-point sampling, but it was pointed out that, after the independent-particle correction, the remaining Coloumb finite-size error is approximately the same for Γ-point sampling.] Our result is again in excellent agreement with the experimental value (from Ref. [13]). It also compares very well with the result of a recent fixed-node DMC calculation [21], which also used a 54-atom supercell and gave 4.63(2) eV per atom after similar finite-size and zero-point energy corrections.

DISCUSSION

Without an exact solution to the sign/phase problem, reducing the reliance on trial wave functions is clearly of key importance to increasing the predictive power of QMC. For continuum electronic systems such as our test cases above, fixed-node DMC has often been the most accurate theoretical method [13]. It is encouraging that the new method, using simple LDA trial wave functions, gave comparable results to DMC. For similar supercells DMC often uses trial wave functions with 30-100 additional parameters [13].

Obtaining a good enough $|\Psi_T\rangle$ is instrumental to a successful DMC calculation, and often constitutes a substantial effort. The quality of $|\Psi_T\rangle$ controls the systematic errors from the fixed-node approximation and the variance. It also affects errors due to the locality approximation [23], which has been employed by most DMC calculations with non-local pseudopotentials. In the new method the latter approximation is eliminated. It remains to be seen whether the present method could lead to more accurate results than fixed-node DMC for continuum systems. This has been possible in some cases [24] with real AF in the Hubbard model.

We have presented a general framework. Various possibilities exist for further improvement of the method. An improved $|\Psi_T\rangle$ will give improved results. The freedom to choose the one-particle basis and the form of HS transformation, both of which can impact the quality of the results, offers significant opportunities. For periodic systems it should be possible to generalize the formalism to allow **k**-point sampling.

CONCLUSIONS

In conclusion, we have described a method for ground-state QMC calculations that allows the use of any one-particle basis. The method is general and applies to any Hamiltonian of the form in Eq. (1). It provides an approximate way to control the phase problem in all AF-based QMC methods, while allowing many of their advantages to be retained that lead to their applications spanning several areas. We have shown that the method gave accurate results for systems from an atom to a large supercell, using a simple trial wave function.

ACKNOWLEDGEMENTS

We are very grateful to E. J. Walter for help with LDA calculations and with several programming issues. We thank P. Kent for sending us his DMC data and J. Carlson for useful conversations. This work is supported by the NSF under Grant No. DMR-9734041 and the Research Corporation, and by ONR Grant DMR-N000149710049. We acknowledge computing support by the Center for Piezoelectrics by Design.

REFERENCES

1. Blankenbecler, R., Scalapino, D. J., and Sugar, R. L., *Phys. Rev. D*, **24**, 2278 (1981).
2. Sugiyama, G., and Koonin, S. E., *Ann. Phys. (NY)*, **168**, 1 (1986).
3. Kohn, W., *Rev. Mod. Phys.*, **71**, 1253 (1999).
4. Silvestrelli, P. L., Baroni, S., and Car, R., *Phys. Rev. Lett.*, **71**, 1148 (1993).
5. Rom, N., Charutz, D. M., and Neuhauser, D., *Chem. Phys. Lett.*, **270**, 382 (1997).
6. Wilson, M. T., and Gyorffy, B. L., *J. Phys. Condens. Matter*, **7**, 371 (1995).
7. Baer, R., and Neuhauser, D., *J. Chem. Phys.*, **112**, 1679 (2000), also see e.g., Y. Alhassid *et al. Phys. Rev. Lett.* **72**, 613 (1994); C. W. Johnson and D. J. Dean, *Phys. Rev. C* **61** 044327 (2000); and references therein.

8. Loh Jr., E. Y., Gubernatis, J. E., Scalettar, R. T., White, S. R., Scalapino, D. J., and Sugar, R., *Phys. Rev. B*, **41**, 9301 (1990).
9. Shiwei Zhang, Carlson, J., and Gubernatis, J. E., *Phys. Rev. B*, **55**, 7464 (1997).
10. Schmidt, K. E., and Kalos, M. H., "Few- and Many-Fermion Problems," in *Applications of the Monte Carlo Method in Statistical Physics*, edited by K. Binder, Springer Verlag, Heidelberg, 1984.
11. Zhang, S., and Krakauer, H., *Phys. Rev. Lett.*, **xx**, xx (2003), in press.
12. Stratonovich, R. L., *Sov. Phys. Dokl.*, **2**, 416 (1958).
13. Foulkes, W. M. C., Mitas, L., Needs, R. J., and Rajagopal, G., *Rev. Mod. Phys.*, **73**, 33 (2001), and references therein.
14. Ortiz, G., Ceperley, D. M., and Martin, R. M., *Phys. Rev. Lett.*, **71**, 2777 (1993).
15. Zhang, S., and Krakauer, H. (2003), unpublished.
16. Kalos, M. H., Levesque, D., and Verlet, L., *Phys. Rev. A*, **9**, 2178 (1974).
17. Rappe, A. M., Rabe, K. M., Kaxiras, E., and Joannopoulos, J. D., *Phys. Rev. B*, **41**, 1227 (1990).
18. Gonze, X., Beuken, J.-M., Caracas, R., Detraux, F., Fuchs, M., Rignanese, G.-M., Sindic, L., Verstraete, M., Zerah, G., Jollet, F., Torrent, M., Roy, A., Mikami, M., Ghosez, P., Raty, J.-Y., and Allan, D., *Comp. Mat. Sci.*, **25**, 478 (2002), URL http://www.abinit.org.
19. Schmude, R. W., Jr., Ran, Q., and Gingerich, K. A., *J. Chem. Phys.*, **99**, 7998 (1993).
20. Kent, P. R. C., Hood, R. Q., Williamson, A. J., Needs, R. J., Foulkes, W. M. C., and Rajagopal, G., *Phys. Rev. B*, **59**, 1917 (1999).
21. Leung, W.-K., Needs, R. J., Rajagopal, G., Itoh, S., and Ihara, S., *Phys. Rev. Lett.*, **83**, 2351 (1999).
22. Kent, P. R. C. (2002), private communication.
23. Mitas, L., Shirley, E. L., and Ceperley, D. M., *J. Chem. Phys.*, **95**, 3467 (1991).
24. Zhang, S., "Constrained Path Monte Carlo for Fermions," in *Quantum Monte Carlo Methods in Physics and Chemistry*, edited by M. P. Nightingale and C. J. Umrigar, Kluwer Academic Publishers, 1999, URL http://lanl.arxiv.org/cond-mat/9909090.

Site-Specific X-ray Photoelectron Spectroscopy: A New Method To Measure Partial Density of Valence States

Joseph C. Woicik

National Institute of Standards and Technology, Gaithersburg, Maryland 20899

Abstract. X-ray photoelectron spectroscopy has emerged as a premiere method for examining the electronic structure of solids and films. However, as the intensity of a monochromatic photon beam is constant over the dimensions of the crystalline unit cell, standard photoemission measurements are unable to produce *direct*, site-specific valence information. Here we demonstrate that by utilizing the sinusoidal variation of the electric-field intensity that occurs within the vicinity of a crystal x-ray Bragg reflection, photoelectron partial density of states of the individual atoms within the crystalline unit cell may be directly obtained. The technique is demonstrated for Rutile TiO_2 and compared to state of the art density functional calculations.

One of the most powerful experimental tools for examining the electronic structure of a solid or film is photoelectron spectroscopy. Due to the conservation of energy between the incident photon and the ejected photoelectron, much direct and important information pertaining to the occupied valence-band density of states has been obtained for many materials. This information has been used to establish the validity of complicated band-structure calculations for metals, semiconductors, insulators, and alloys [1].

Traditionally, photoemission measurements have been performed with excitation sources that are monochromatic plane waves. As the intensity of a plane wave is constant over the dimensions of the crystalline unit cell, standard photoemission measurements are unable to produce *direct*, site-specific valence information. Such information is important to further advance our understanding of how chemical bonding results in the solid-state electronic structure. Here we demonstrate an experimental method for obtaining site-specific valence-electronic structure. Unlike previous photoemission methods that rely on the detailed atomic properties of the individual atoms within the unit cell, it utilizes only the sinusoidal variation of the electric-field intensity that occurs within the vicinity of a crystal x-ray Bragg reflection.

Under the condition of x-ray Bragg reflection, the spatial dependence of the electric field is given by the superposition of the incident $\mathbf{E}_0$ and reflected $\mathbf{E}_h$ x-ray beams that travel with wave vectors $\mathbf{k}_0$ and $\mathbf{k}_h$, polarization vectors $\mathbf{e}_0$ and $\mathbf{e}_h$, and frequency ω:

$$\mathbf{E}(\mathbf{r},t) = [\ \mathbf{e}_0 E_0 e^{i\mathbf{k}_0 \cdot \mathbf{r}} + \mathbf{e}_h E_h e^{i\mathbf{k}_h \cdot \mathbf{r}}\ ] e^{-i\omega t}. \tag{1}$$

k_0 and k_h are connected by the Bragg condition $h = k_h - k_0$, where h is a reciprocal-lattice vector of the crystal. For the σ-polarization geometry of a symmetric reflection, this field squares to give the electric-field intensity at an arbitrary point r in space:

$$I(r) = |E_0|^2 [1 + R + 2\sqrt{R} \cos(v + h \cdot r)]. \qquad (2)$$

v is the phase of the complex-field amplitude ratio $E_h/E_0 = \sqrt{R}e^{iv}$ and R is the reflectivity function $R = |E_h/E_0|^2$ that is given by the dynamical theory of x-ray diffraction [2]. Germane to the method is the unique ability to position the maxima (or minima) of the electric-field intensity at any location within the crystalline unit cell by experimentally varying the phase of the complex reflectivity function between $0 < v < \pi$. This is achieved by slightly varying either the sample angle or the photon energy within the natural width of the crystal x-ray Bragg reflection.

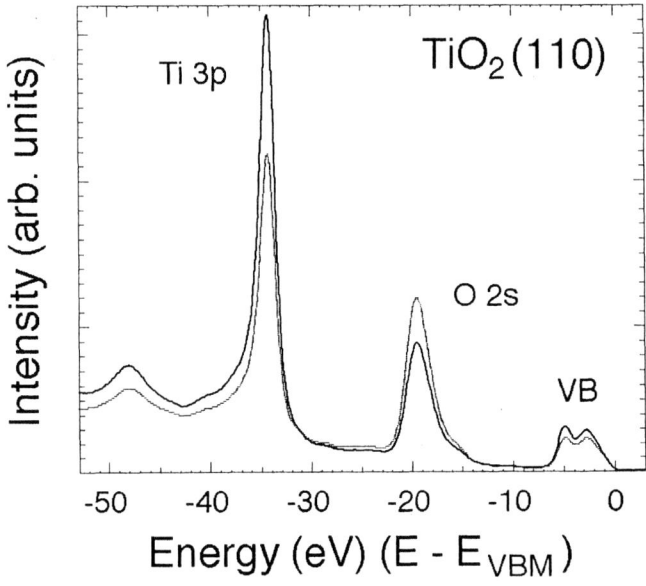

FIGURE 1. X-ray photoelectron spectra from the Rutile $TiO_2(110)$ surface recorded within the photon-energy width of the $TiO_2(200)$ Bragg back-reflection condition. The photon energies were chosen to maximize the electric-field intensity on either the Ti (solid curve) or O (dotted curve) atomic plane.

In the dipole approximation, the probability of emission of an electron from an atom in an external electric field is proportional to the electric-field intensity at the location of its atomic core. Despite the large spatial extent of the valence electrons, it has been verified experimentally that the dipole approximation is a good approximation for valence-electron emission in angle-integrated photoelectron measurements at the low x-ray energies [3]. This situation arises due to the lack of overlap between the plane-wave part of the initial-state valence-wave function and the rapidly oscillating photoelectron final state. Therefore, only the core part of the valence-wave function contributes to the x-ray photocurrent. Consequently, by selectively positioning the electric-field intensity within the crystalline unit cell and recording high-resolution valence-photoelectron spectra, spatially resolved components to the photoemission valence-band density of states may be directly obtained.

To illustrate this methodology, Fig. 1 shows the photoelectron spectra from a clean $TiO_2(110)$ surface recorded within the photon-energy width ($\Delta E \sim 0.38$ eV) of the $TiO_2(200)$ Bragg back-reflection condition (hv $\sim$ 2700 eV). The well known x-ray standing-wave effect [4] is observed for the Ti $3p$ and O $2s$ core lines, as well as for the crystal valence band.

To extract the photoelectron partial density of states, the photoelectron spectra were aligned relative to the energy position of the Ti $3p$ core line and normalized to the electric-field intensity at either the Ti or O atomic position. The electric-field intensity was taken to be proportional to the intensity of the Ti $3p$ or O $2s$ core line, respectively. After removal of an integrated background from the valence-band region, linear combinations of the two spectra yielded the *experimental* photoemission partial density of states curves around the Ti and O sites, as shown in Fig. 2. Clearly, the large contribution of Ti to the valence-band spectrum indicates significant covalent bonding between the Ti and O atoms, despite the formal Ti^{4+} charge state of Ti in Rutile. These density of states curves are compared to theoretical partial density of states curves calculated by Professor Chelikowsky's group at the University of Minnesota using *ab initio* Troullier-Martins pseudopotentials within the local density approximation using a plane-wave basis [5]. Details of the theoretical calculations as well as the experiment have been given previously [6]. The theoretical partial density of states curves were obtained by projecting the obtained wave functions over the Ti and O valence atomic orbitals within spheres centered around the Ti and O atoms. The use of finite sphere radii produces the theoretical site specificity, and the sphere radii were chosen to be equal to the known covalent radii of each species, ~ 1.3 Å for Ti and ~ 0.75 Å for O.

Clearly, the agreement between theory and experiment is much less than satisfactory. In particular, the second lobe of the Ti valence band is nearly absent in the theory, and the triply peaked structure of the O valence band is poorly reproduced. Because the photoemission process conserves angular momentum, the theoretical Ti and O partial density of states curves were further decomposed into their angular-momentum resolved components. These curves are shown in Fig. 3. It is clear that the second lobe of the experimental Ti spectrum cannot be reproduced without inclusion of the Ti $4p$ component, and the intermediate structure of the experimental O spectrum cannot be reproduced without inclusion of the O $2s$ component.

The photoelectron spectra were then modeled from the weighted sums of the different orbital components of Fig. 3 using the tabulated, theoretical, angular-momentum dependent atomic-photoelectron cross sections (Ti $\sigma_{4s}/\sigma_{3d} \sim 9.9$ and O $\sigma_{2s}/\sigma_{2p} \sim 29$) [7]. Agreement between the theoretical and experimental partial density of states curves was much improved; however, even better agreement was obtained for both the Ti and O components if the relative atomic cross sections were scaled by an additional factor of 2. This discrepancy could arise from the choice of theoretical deconvolution radii used in the calculation, as well as "solid-state" effects on the theoretical cross sections in going from the atomic to the solid state [8]. The resulting theoretical corrected partial density of states are shown in Fig. 4 together with the experimental data. The agreement now between theory and experiment is startling.

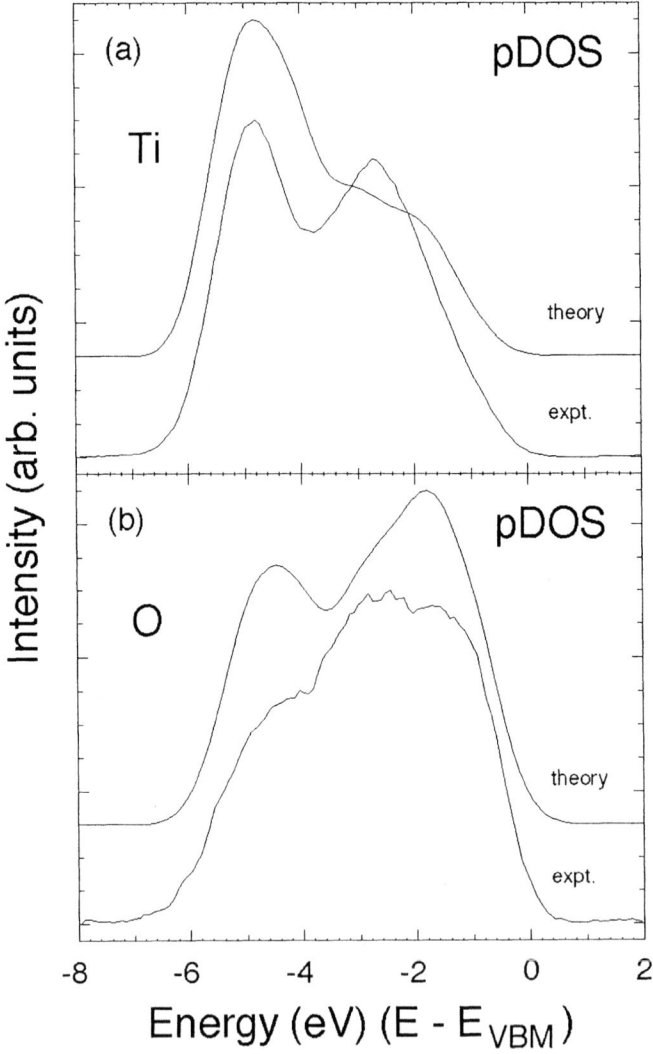

FIGURE 2. Theoretical partial density of states and the experimental site-specific valence-photoelectron spectrum: (a) Ti; (b) O. Theoretical and experimental curves have been scaled to equal peak height. The O component has been scaled by a factor of 4 relative to the Ti component.

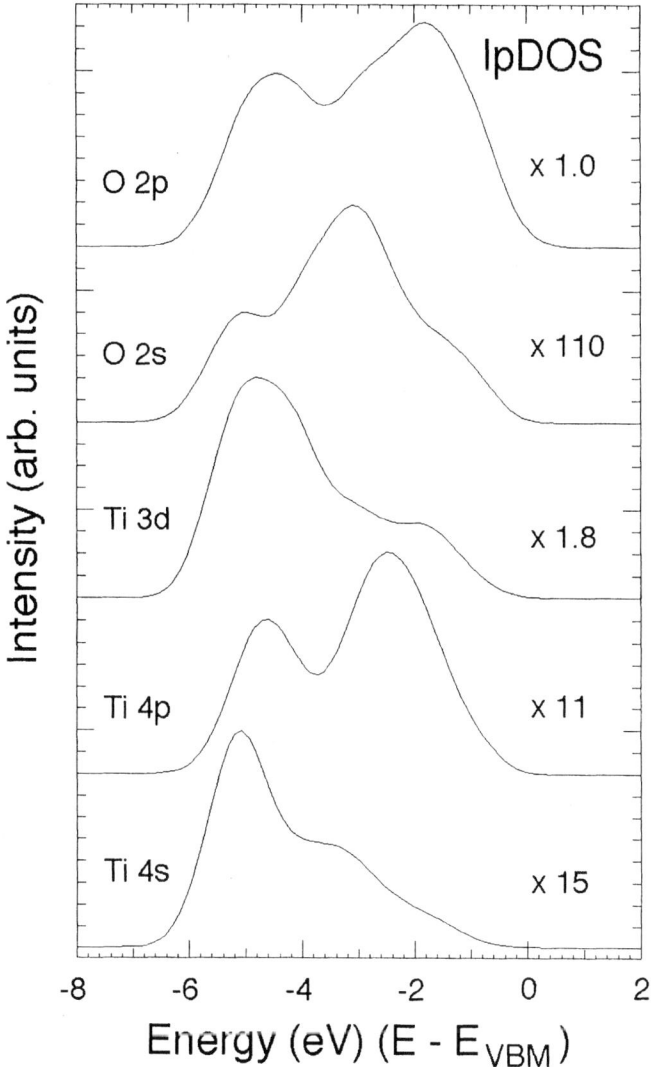

FIGURE 3. The different *angular-momentum* components of the theoretical Ti and O partial density of states. Multipliers relative to the O 2p component are indicated in each case.

Having accurately reproduced the photoelectron partial density of states, it is now instructive to consider the molecular orbitals for an octahedral complex containing a first-row transition-metal ion within octahedral (O_h) symmetry [9]. In this depiction, the metal 4s orbitals bond with the ligand σ orbitals to form the $a_{1g}(\sigma^b)$ level, the metal $3d_{x^2-y^2}$ and $3d_{z^2}$ orbitals bond with the ligand σ orbitals to form the $e_g(\sigma^b)$ level, the metal 4p orbitals bond with both the ligand σ and π orbitals to form the $t_{1u}(\sigma^b)$ and $t_{1u}(\pi^b)$ levels, and the metal $3d_{xy}$, $3d_{xz}$, and $3d_{yz}$ orbitals bond with the ligand π orbitals to form the $t_{2g}(\pi^b)$ level. Additionally, there are ligand π orbitals [$t_{1g}(\pi)$ and $t_{2u}(\pi)$] that are left over and are rigorously non-bonding in O_h symmetry.

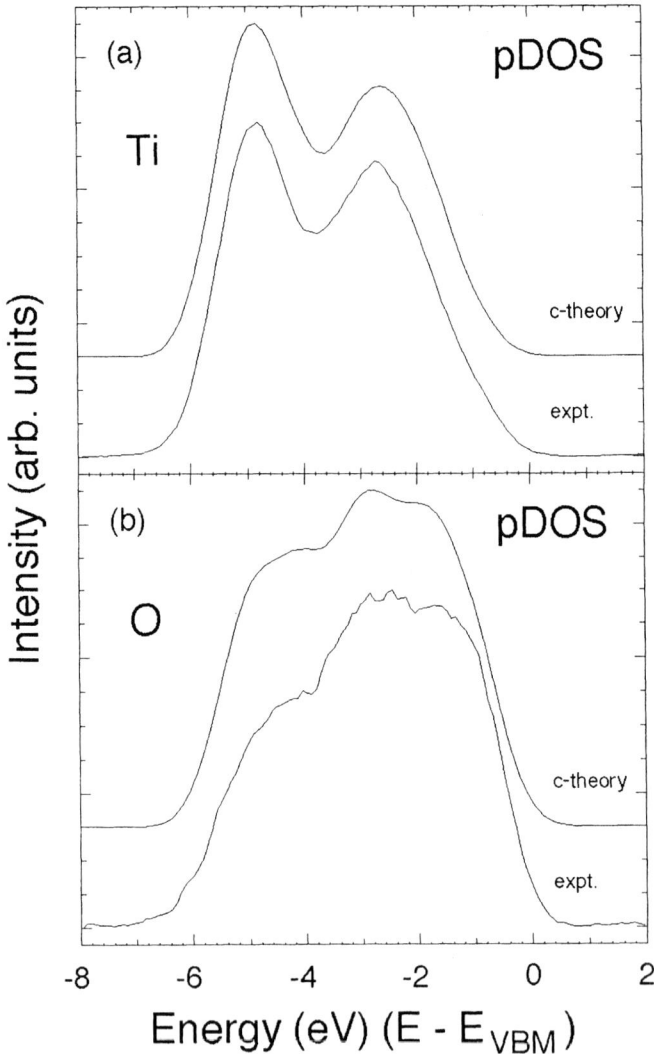

FIGURE 4. Theoretical partial density of states corrected for individual angular-momentum dependent photoelectron cross sections and the site-specific experimental valence-photoelectron spectrum: (a) Ti; (b) O. The curves have been scaled to equal peak height.

Examination of both the experimental partial density of states and the theoretical corrected partial density of states of Fig. 4 leads to an attractive interpretation of the electronic structure within this σ and π bonding scheme. The doubly lobed Ti structure of the valence band may be attributed to the energy splitting between the σ and π groupings of the valence states, with the σ bonds lying at the lower energy. These states are mirrored in the triply lobed O structure, indicative of the sharing of electrons in a covalent bond. The O non-bonding π states then naturally compose the third lobe of the O spectrum; they occur at higher energy than the O bonding π states and reveal little or no electron density on the metal atoms, as expected. Inspection of the angular-momentum resolved components of Fig. 3 supports these conclusions, although the solid-state electronic structure is much more complicated than that of an isolated TiO_6 molecule. In particular, the groupings of the orbitals into discrete σ and π states is not so transparent, and the effect of translational symmetry spreads the states into bands.

As both the O 2s and O $2p_z$ atomic orbitals belong to the same symmetry representation of the O_h point group, the ligand σ orbitals will contain a mixture of these states [9]: $|\psi_{L,\sigma}\rangle = \alpha |2s\rangle + \sqrt{(1 - \alpha^2)} |2p_z\rangle$. This hybridization orients the ligand-charge density towards the metal atoms, leading to a larger charge-density overlap within the chemical bond. It has been stated by Mulliken that "a little bit of hybridization goes a long way" to stabilize a chemical bond [9], and, from the theoretical calculations of Fig. 3, we see that α, the mixing coefficient, is only ~ 10 %, even though the O 2s hybrid component accounts for as much as ~ 30 % of the experimental O valence spectrum. This relatively small value of α results from the relatively large energy separation between the O 2s and O 2p atomic orbitals that, as seen from the data in Fig. 1, is ~ 17 eV. The much larger amount of Ti 4s, 4p, and 3d hybridization observed on the Ti sites is due to the much smaller energy separation (~ 8 eV) between the Ti atomic orbitals [9]. Note that it is the added complexity of the photoemission process, i.e., the angular-momentum dependence of the photoelectron cross sections, that allows this experimental observation of chemical hybridization in the solid-state electronic structure.

In conclusion, we have demonstrated that x-ray photoelectron spectroscopy performed in conjunction with crystal x-ray Bragg diffraction can reveal site-specific valence information that is directly related to the partial density of occupied valence states of the individual atoms within the crystalline unit cell. This technique is ideally suited for testing both the predictive power and the accuracy of theoretical calculations that model solid-state electronic structure, and it has determined that the local density approximation accurately predicts the chemical bonding and bond hybridization in Rutile TiO_2. Unlike traditional XPS measurements, the site-specific XPS technique uniquely reveals the genealogy of the solid-state chemical bond.

ACKNOWLEDGMENTS

This work was performed at the National Synchrotron Light Source, which is supported by the U.S. Department of Energy. Additional support was provided by the Computational Materials Science Network of the Department of Energy, the National Science Foundation, and the Minnesota Superconducting Institute.

REFERENCES

1. Hufner, S., Photoelectron Spectroscopy: Principles and Applications, 2nd ed., Berlin: Springer-Verlag, 1996.
2. Batterman, B.W., and Cole H., Rev. Mod. Phys. **36**, 681 (1964).
3. Woicik, J.C., Nelson, E.J., and Pianetta, P., Phys. Rev. Lett. **84**, 773 (2000).
4. Zegenhagen, J., Surf. Sci. Rep. **18**, 199 (1993).
5. Glassford, K.M., and Chelikowsky, J.R., Phys. Rev. B **46**, 1284 (1992).
6. Woicik, J.C., Nelson, E.J., Kronik, L., Jain, M., Chelikowsky, J.R., Heskett, D., Berman, L.E., and Herman, G.S., Phys. Rev. Lett. **89**, 77401 (2002).
7. Trzhaskovskaya, M.B., Nefedov, V.I., and Yarzhemsky, V.G., At. Data Nucl. Data Tables **77**, 97 (2001).
8. We note that the Ti 4p atomic cross section is not tabulated because it is unoccupied in the free Ti atom. The Ti 4p occupation observed here arises strictly from the bonding between the Ti and O atoms in the solid. Note as well that the O 2s solid-state hybrid component differs in energy by ~ 17 eV from its atomic counterpart. Consequently, it should not be expected that free-atom calculations would accurately predict either of these cross sections in the solid state.
9. Ballhausen, C.J., and Gray, H.B., Molecular Orbital Theory, New York: Benjamin 1964.

Car-Parrinello Molecular Dynamics in a Finite Homogeneous Electric Field

P. Umari*† and Alfredo Pasquarello*†

*Ecole Polytechnique Fédérale de Lausanne (EPFL), CH-1015 Lausanne, Switzerland
†Institut Romand de Recherche Numérique en Physique des Matériaux (IRRMA), CH-1015 Lausanne, Switzerland

Abstract. We introduce a variational total-energy functional to treat finite homogeneous electric fields with periodic boundary conditions and show that this functional can be implemented within a Car-Parrinello molecular dynamics scheme. The coupling to an electric field is achieved through the Berry-phase expression of the polarization. The minimization of this extended functional gives a ground state which describes the polarized state in an electric field. For a crystalline system, the ground state of this extended functional preserves the Bloch symmetry. The reliability of the method is demonstrated in the case of bulk MgO for the Born effective charges, and the high- and low-frequency dielectric constants. In the latter case, we evaluated the static dielectric constant by performing a damped molecular dynamics in the presence of a finite electric field, completely avoiding the calculation of the dynamical matrix.

INTRODUCTION

The Car-Parrinello method [1] is routinely used to perform molecular dynamics simulations in which the electronic structure is described within density functional theory. This method has successfully been applied to a large variety of systems, giving access to detailed electronic and structural properties [2]. It would be highly desirable to extend the Car-Parrinello molecular dynamics scheme to the study of dynamical transformation processes induced by an electric field. However, the treatment of a finite electric field within periodic density functional calculations is severely hindered by the nonperiodic and unbound nature of the electric potential [3, 4, 5]. Nonetheless, it is of interest to describe the long-living metastable physical state which arises in response to the field [4, 5]. A few studies in this direction have been reported [5, 6], but appear at this stage not suitable for routine applications. This situation should be contrasted with the successful description given by density-functional perturbational approaches in the limit of a vanishing electric field [7, 8, 9, 10, 11, 12].

For a periodic infinitely-extended electron system, it is tempting to formulate a variational energy functional expressing the coupling to an electric field $\mathcal{E}$ as:

$$E^{\mathcal{E}}[\{\psi_i\}] = E^{(0)}[\{\psi_i\}] - \Omega \, \mathcal{E} \cdot P[\{\psi_i\}], \quad (1)$$

where $E^{(0)}[\{\psi_i\}]$ is the energy functional in the absence of an electric field, $P[\{\psi_i\}]$ is the polarization (along the direction of the electric field) as given by the modern theory of polarization [14, 15, 16], and Ω is the volume of the adopted unit cell [13].

However, for a finite value of $\mathscr{E}$, functional (1) presents a pathological behavior [5, 17]. Nunes and Vanderbilt addressed the finite-field problem expressing the polarization in terms of truncated Wannier functions [5]. This formulation was later extended to realistic density functional calculations by Fernandez, Dal Corso, and Baldereschi [6], but did not appear sufficiently practical for implementation in a first-principles molecular dynamics scheme. Using a simple tight-binding model, Nunes and Gonze recognized that functional (1) could be regularized when the Berry-phase polarization is defined for a discrete mesh of k points. In this case, minimization of functional (1) yields a well-defined ground state.

Functional (1) has also been used in the context of density-functional perturbational approaches. Putrino, Sebastiani and Parrinello [11] developed a variational linear response method based on functional (1), in which the polarization is given by the single k-point expression proposed by Resta [18]. For doubly occupied wave functions:

$$P[\{\psi_i\}] = -\frac{1}{\Omega}\frac{L}{\pi}\text{Im}\left(\ln\det S[\{\psi_i\}]\right), \tag{2}$$

where L is the periodicity of the unit cell, here taken to be cubic, and $S[\{\psi_i\}]$ is obtained from the electronic wave functions ψ_i:

$$S_{ij} = \langle\psi_i|e^{2\pi ix/L}|\psi_j\rangle, \tag{3}$$

for an electric field in the x direction. For a crystal, it can be shown that the polarization defined by (2) is equivalent to the polarization defined for a discrete mesh of equally spaced k points [18]. Nunes and Gonze also addressed functional (1) within a perturbational scheme and found that it yields derivatives with respect to the electric field identical to those obtained within more conventional perturbational methods [8, 9, 10].

We recently introduced a new method to treat a finite homogeneous electric field within periodic density-functional calculations [19]. Our method consists in using the variational functional (1) with the definition of the polarization as given in (2). In parallel to our work, a method based on similar concepts was proposed independently by Souza, Íñiguez, and Vanderbilt [20].

METHOD

We here show how our method for treating finite electric fields [19] can be combined with the Car-Parrinello molecular dynamics scheme. We considered the following extended Car-Parrinello Lagrangian, which allows us to study the simultaneous evolution of the ionic and electronic degrees of freedom in the presence of an electric field:

$$\mathscr{L} = \mu\sum_i\int d\mathbf{r}\left|\frac{d\psi_i(\mathbf{r})}{dt}\right|^2 + \frac{1}{2}\sum_I M_I\left(\frac{d\mathbf{R}_I}{dt}\right)^2 - E^{\mathscr{E}}[\{\psi_i\}] + \Omega\,\mathscr{E}\cdot P_{\text{ion}}, \tag{4}$$

where μ is the fictitious electron mass, $\mathbf{R}_I$ and M_I are the position and the mass of the Ith ion, respectively, and P_{ion} is the ionic polarization defined as [13]:

$$P_{\text{ion}} = \frac{1}{\Omega} \sum_{I=1}^{N_{\text{ion}}} Z_I \cdot X_I, \qquad (5)$$

where X_I and Z_I are the position coordinate of the Ith ion in the direction of the applied field and its core charge, respectively. Consistently with the definition of the polarization (2), the wave functions ψ_i are taken at the Γ point of the Brillouin zone. The Lagrangian (4) is subjected to the orthonormality constraints for the wave functions:

$$\langle \psi_i | \psi_j \rangle = \delta_{ij}. \qquad (6)$$

The Euler equations of motion describing the evolution of the electronic and ionic degrees of freedom become:

$$\mu \frac{d^2 \psi_i}{dt^2} = -\frac{\delta E^{(0)}}{\delta \psi_i^*} + \mathcal{E} \cdot \frac{L}{\pi} \text{Im} \sum_j (S^{-1})_{ij} \psi_j + \sum_j \Lambda_{ij} \psi_j \qquad (7)$$

$$M_I \frac{d^2 \mathbf{R}_I}{dt^2} = -\frac{\partial E^{(0)}}{\partial \mathbf{R}_I} + \mathcal{E} Z_I \hat{x}, \qquad (8)$$

where $\hat{x}$ is a unitary vector along the direction of the electric field, and Λ_{ij} are Lagrange multipliers used to ensure the orthonormality constraints (6). To derive the Euler equations (7), we used the relation:

$$d(\ln \det S) = \sum_{ij} (S^{-1})_{ij} dS_{ji}. \qquad (9)$$

The computational overhead of our implementation with respect to a conventional Car-Parrinello scheme is limited to the inversion of the matrix S (3).

The pathological behavior of functional (1), which arises in the limit of a dense k-point sampling in the work of Ref. [17], corresponds in our formulation to the limit of infinite cell size L. For a given cell size L, our formulation therefore guarantees the occurrence of a well defined minimum. Once the Riemann plane defining the logarithmic function in (2) is chosen, this minimum is uniquely defined. We note that, for a fixed cell size L, an instability develops when the electric field exceeds a given critical value $\mathcal{E}_c$. This occurs when the difference in electric potential across the cell size L is of the order of the electronic gap.

In our formulation, we take advantage of the time-reversal symmetry to use *real* wave functions as in a conventional Car-Parrinello scheme. Indeed, it can be shown that, taking real wave functions as a starting point, the evolution as defined by the Euler equation (7) preserves the real character of the wave functions.

The formulation presented here applies to norm-conserving pseudopotentials but can readily be extended to the case of ultrasoft ones [21] in a similar way as in Ref. [12]. We note that this extension is considerably less involved than the treatment of ultrasoft pseudopotentials within a perturbational approach [12, 22].

The ground state in our formulation corresponds to a long-living resonance in the real physical system. In the case of a crystal, Nunes and Vanderbilt argued that the density matrix describing this state should be periodic [5]. Souza, Íñiguez, and Vanderbilt proved this statement for a real physical system using an electric field potential described in terms of the true position operator $\vec{r}$ [20]. Because of this reasoning, these authors considered in their approach only wave functions with Bloch symmetry.

The question remains whether the use of the Berry-phase polarization in functional (1) *intrinsically* leads to ground-state wave functions carrying the Bloch symmetry. This question is not trivial because the non-Hamiltonian character of the Berry-phase polarization prevents the study of commutation properties. This question cannot be addressed with the Berry-phase polarization as expressed for a discrete mesh of k points since this formulation already implies the Bloch symmetry from the outset [17, 20]. Instead, a supercell formulation with the polarization as given in (2) should be used. In this formulation, the electronic wave functions are allowed to vary in a Hilbert space in which the Bloch symmetry is not enforced *a priori*. In Ref. [19], we proved that, despite this freedom, the ground-state wave functions which minimize functional (1) do carry the Bloch symmetry. The validity of this property also sustains the work of Nunes and Gonze [17], who assumed the Bloch symmetry of the wave functions from the outset. On the basis of their work, we conclude that the present formulation for treating electric fields gives the same derivatives with respect to the field as obtained with more conventional perturbational methods [8, 10, 9].

APPLICATION

To examine the reliability of this approach, we focused on the dielectric properties of bulk MgO. We modeled MgO by a periodic cubic cell containing 64 atoms at the experimental lattice constant. Only valence wave functions were treated explicitly, while core-valence interactions were described by ultrasoft pseudopotentials [23]. The exchange and correlation energy was described by the local density approximation.

First, we investigated the stability of the energy functional (1) as a function of the electric field for *fixed* atomic positions. A variational minimum was found for field intensities ranging over several orders of magnitude up to 0.034 atomic units, corresponding to 1.7 V/Å. Beyond this value of the field, the functional could not be minimized. The instability occurs for fields of the same order as the critical field $E_{\text{gap}}/L = 0.5$ V/Å, where L is the periodicity of our simulation cell. In Fig. 1, the variation of the polarization $\Delta P^{\mathscr{E}} = P^{\mathscr{E}} - P^0$ shows a close to linear behavior for fields $\mathscr{E}$ in the range of stability of the energy functional. Within the linear regime, the high-frequency dielectric constant can be estimated from [13]:

$$\varepsilon_\infty = 4\pi \frac{\Delta P^{\mathscr{E}}}{\mathscr{E}} + 1, \qquad (10)$$

valid for cubic crystals. For a field of 10^{-3} a.u., we found $\varepsilon_\infty = 2.79$. To check the validity of this result, we generalized the linear response approach of Ref. [11] to treat ultrasoft pseudopotentials [24], and found a numerically equivalent result (within 10^{-4}).

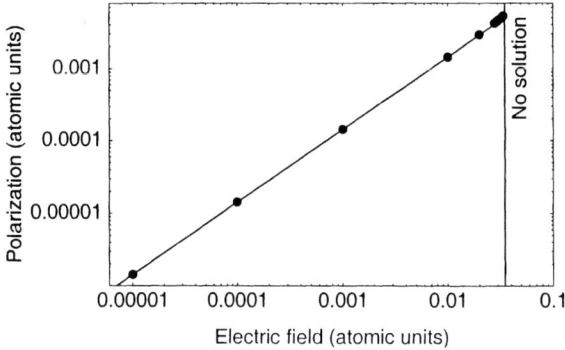

FIGURE 1. Calculated polarization per unit volume versus electric field for bulk MgO (discs). The solid line corresponds to the result from linear response [11]. The polarization is given as a difference with respect to the case of vanishing field. For electric fields higher than ~0.035 a.u. functional (1) could not be minimized.

Our theoretical result is in good agreement with the experimental one ($\varepsilon_\infty^{\text{expt.}} = 2.96$ [26]). This application illustrates that functional (1) with the polarization as given in (2) also provides reliable results for *finite* $\mathcal{E}$.

The Born effective charges Z^* are defined as the derivatives of the atomic forces with respect to the applied field [16]. The Z^* can then be obtained in a straightforward way by using the atomic force $F^{\mathcal{E}}$ acting on the ions in the presence of a field $\mathcal{E}$:

$$Z^* = \frac{F^{\mathcal{E}}}{\mathcal{E}}, \qquad (11)$$

where we used $F^{\mathcal{E}=0} = 0$ and the cubic symmetry of the crystal. For a field $\mathcal{E} = 10^{-3}$ a.u., we obtained $Z^* = 1.96$. The charges Z^* can alternatively be defined as derivatives of the polarization with respect to atomic displacements at vanishing electric field [16]. Using a finite difference approach in which the Mg and O sublattices are relatively displaced by 0.2 bohr, we derived a value of $Z^* = 1.96$, numerically coincident with the value obtained with the finite field method, and in excellent agreement with the experimental result ($Z^* = 1.98$ [26]).

To check the validity of the method for describing *atomic relaxations* in a finite electric field, we investigated the static dielectric constant. The difference between the low- and high-frequency dielectric constants is generally expressed in terms of the Born effective charges, the vibrational frequencies, and their associated eigenmodes [25]. For bulk MgO:

$$\Delta \varepsilon = \varepsilon_0 - \varepsilon_\infty = \frac{4\pi}{\Omega_0} \frac{(Z^*)^2}{\mu \omega^2}, \qquad (12)$$

where ω is the frequency of the zone-center optical phonon, μ the reduced mass of Mg and O atoms, and Ω_0 the volume of the fcc primitive cell. We obtained the zone-center frequency ω by a frozen phonon calculation. Direct application of (12) then gives $\Delta \varepsilon = 5.09$.

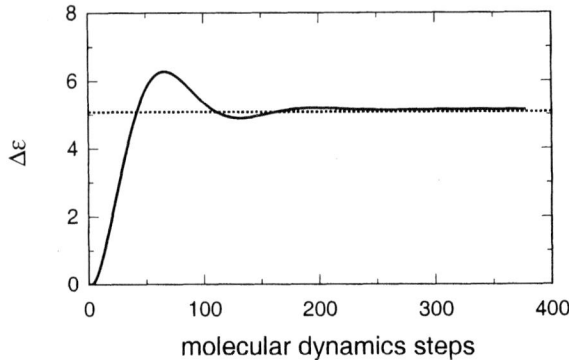

FIGURE 2. Evolution of $\Delta\varepsilon$ for a model of MgO during a damped molecular dynamics relaxation in a finite electric field of 10^{-3} a.u. The dotted line corresponds to the result (12) from linear response theory.

The functional (1) offers an alternative procedure for obtaining static dielectric constants, which entirely avoids the calculation of normal modes. We performed a damped molecular dynamics relaxation, in which the atomic positions are allowed to relax fully in the presence of a finite electric field $\mathscr{E}$. In this case, $\Delta\varepsilon$ can be expressed in terms of a difference between polarizations associated with atomic configurations before ($P^{\mathscr{E}}_{\text{nonrelaxed}}$) and after ($P^{\mathscr{E}}_{\text{relaxed}}$) atomic relaxation [13]:

$$\Delta\varepsilon = 4\pi \frac{P^{\mathscr{E}}_{\text{relaxed}} - P^{\mathscr{E}}_{\text{nonrelaxed}}}{\mathscr{E}}, \qquad (13)$$

where both polarizations, $P^{\mathscr{E}}_{\text{nonrelaxed}}$ and $P^{\mathscr{E}}_{\text{relaxed}}$, are calculated in the presence of a field $\mathscr{E}$. In Fig. 2, we plot the evolution of $\Delta\varepsilon$ during the relaxation of the atomic positions in an electric field of 10^{-3} a.u. After a few oscillations, $\Delta\varepsilon$ converges rapidly to 5.14, differing by less than the numerical accuracy from the estimate of 5.09, derived from (12). The experimental result is $\Delta\varepsilon = 6.67$ [26].

CONCLUSIONS

We introduced a method for treating homogeneous electric fields in density functional calculations with periodic boundary conditions. The method has been implemented within a Car-Parrinello molecular dynamics scheme. Results for bulk MgO illustrate the potential of this method. The present implementation is suitable to treat systems of relatively large size containing a few hundred atoms. This method offers the possibility of investigating properties which cannot be accessed with any other current approach. For instance, the stage is set for the study of structural and dielectric properties in an electric field at finite temperature.

ACKNOWLEDGMENTS

We acknowledge support from the Swiss National Science Foundation (Grant No. 620-57850.99) and the Swiss Center for Scientific Computing.

REFERENCES

1. R. Car and M. Parrinello, Phys. Rev. Lett. **55**, 2471 (1985).
2. For a review see G. Galli and A. Pasquarello in *"Computer Simulation in Chemical Physics"*, p.261, edited by M. P. Allen and D. J. Tildesley (Kluwer, 1993).
3. G. H. Wannier, Phys. Rev. **117**, 432 (1960).
4. G. Nenciu, Rev. Mod. Phys. **63**, 91 (1991).
5. R. W. Nunes and D. Vanderbilt, Phys. Rev. Lett. **73**, 712 (1994).
6. P. Fernandez, A. Dal Corso, and A. Baldereschi, Phys. Rev. B **58**, R7480 (1998).
7. R. Resta and A. Baldereschi, Phys. Rev. B **23**, 6615 (1981).
8. S. Baroni, P. Giannozzi, and A. Testa, Phys. Rev. Lett. **78**, 1861 (1987).
9. A. Dal Corso, F. Mauri, and A. Rubio, Phys. Rev. B **53**, 15638 (1996).
10. X. Gonze, Phys. Rev. B **55**, 10337 (1997).
11. A. Putrino, D. Sebastiani, and M. Parrinello, J. Chem. Phys. **113**, 7102 (2000).
12. P. Umari, X. Gonze, and A. Pasquarello. Phys. Rev. Lett. **90**, 027401 (2003).
13. The definition for P adopted here differs from that in Ref. [19], where P was defined as the electric dipole of the system.
14. R. D. King-Smith and D. Vanderbilt, Phys. Rev. B **47**, 1651 (1993).
15. R. Resta, Ferroelectrics **111**, 15 (1992).
16. R. Resta, Rev. Mod. Phys. **66**, 899 (1994).
17. R. W. Nunes and X. Gonze, Phys. Rev. B **63**, 155107 (2001).
18. R. Resta, Phys. Rev. Lett. **80**, 1800 (1998).
19. P. Umari and A. Pasquarello, Phys. Rev. Lett. **89**, 157602 (2002).
20. I. Souza, J. Íñiguez, and D. Vanderbilt, Phys. Rev. Lett. **89**, 117602 (2002).
21. A. Pasquarello, K. Laasonen, R. Car, C. Lee, and D. Vanderbilt, Phys. Rev. Lett. **69**, 1982 (1992); K. Laasonen, A. Pasquarello, R. Car, C. Lee, and D. Vanderbilt, Phys. Rev. B **47**, 10142 (1993).
22. A. Dal Corso, A. Pasquarello, and A. Baldereschi, Phys. Rev. B **56**, R11369 (1997).
23. D. Vanderbilt, Phys. Rev. B **41**, 7892 (1990).
24. P. Umari and A. Pasquarello (unpublished).
25. X. Gonze and C. Lee, Phys. Rev. B **55**, 10355 (1997).
26. M. E. Lines, Phys. Rev. B **41**, 3372 (1990); *Handbook of Chemistry and Physics*, edited by D. R. Lide (CRC Press, New York, 1998), p. 12-52.

First-Principles WDA Calculations for Ferroelectric Materials

Zhigang Wu*, Ronald E. Cohen* and David J. Singh[†]

Carnegie Institution of Washington, Washington DC 20015
[†]*Naval Research Laboratory, Washington DC 20375*

Abstract. First-principles calculations within the weighted density approximation (WDA) were performed for some common ferroelectric materials. We used a plane-wave basis, the Perdew-Wang pair-distribution function, and shell partitioning. Compared with the local density approximation (LDA), the WDA significantly improves the equilibrium volume of these materials, but it overestimates the ferroelectric instability in $BaTiO_3$ and $PbTiO_3$. We fixed this failure by imposing a better sum rule based on shell partitioning. Because an orbital-dependent xc potential is introduced, shell partitioning complicates calculation of atomic forces. We also investigated a hybrid WDA method, in which WDA is mixed with LDA.

I. INTRODUCTION

Since the early 1980s, first-principles calculations based on the density functional theory (DFT) [1] have been implemented to compute diverse properties of piezoelectric and ferroelectric materials. The main difficulty within DFT is how to treat the exchange-correlation (xc) energy, because the exact form of it remains unknown. The local density approximation (LDA) [1], in which the xc energy density depends only on local charge density, dominated these calculations due to its simplicity and surprising success. When performed at the experimental volume, the LDA predicts many properties of ferroelectric materials, such as phonon frequencies, ferroelectric transitions, polarization, elasticity, etc., with extraordinary accuracy [2]. The amazing success of the LDA is attributed to the partial cancellation of the errors of the exchange and correlation parts. But the LDA gives equilibrium volumes a few percent less than experiment, and ferroelectric properties are extremely sensitive to volume. For example, the ferroelectric instability in $BaTiO_3$ [3] and $KNbO_3$ [4, 5, 6, 7, 8, 9] are severely reduced, if not totally eliminated, in calculations done at the LDA zero pressure volume. Furthermore, even at the experimental volume, the LDA may incorrectly predict certain properties, e.g., overestimation of dielectric constants ε_∞ [10, 11, 12], and underestimation of band gaps [13, 14].

The generalized gradient approximation (GGA) [15, 16], which uses both local the density and the local density gradients, is among several approaches beyond LDA. Generally speaking, the semi-local GGA tends to improve upon the LDA in many aspects, such as atomic energies and structural energy differences [16, 19, 20]. On the other hand, the GGA often over-corrects the LDA [21, 8]. Here we did a full relaxation of tetragonal $PbTiO_3$ within both the LDA (Hedin-Lundqvist [22]) and the GGA (PBE [16]) using the LAPW+LO method [23]. As seen in Table 1, at the experimental volume,

TABLE 1. Fully relaxed structure of tetragonal P4mm PbTiO$_3$ with the LDA and the GGA. Volumes are in Å^3, and the numbers in parentheses are the deviations of strain (6.35%) and volume from experiment.

	volume (Expt.)	c/a	volume (Relaxed)	c/a
LDA	63.28	1.11 (+80%)	60.36 (-4.6%)	1.051(-20%)
GGA	63.28	1.068 (+7%)	70.3 (+11%)	1.22 (+250%)

the GGA predicts a better strain than the LDA; however, the fully relaxed GGA structure seems even worse than that of the LDA. But the LDA also has problems because the LDA optimized volume is about 4.6 percent less than experiment. The failures of LDA and GGA calculations indicate that more complicated non-local functionals are needed to include the contributions beyond the semi-local approximation.

The weighted density approximation (WDA) [17, 18, 24], within which the xc energy density depends on charge density over a finite region, was advanced in the late 1970's. The WDA assumes that any inhomogeneous electron gas can be regarded as continuously being deformed from a homogeneous electron gas, so the real pair-distribution function of the inhomogeneous gas is replaced by that of a homogeneous gas with the weighted density. By constructing a model xc hole, the weighted density at every point is determined; then the xc energy and xc potential are determined. Because it is complicated, computationally demanding, and also because of the success of the simpler LDA and GGA, the WDA has not attracted much attention until recently. However, the WDA is promising for predicting very accurate volumes so that true first-principles calculation can be done without relying on experiment. The first ferroelectric material calculated with WDA is KNbO$_3$ [9, 25]. An equilibrium volume in very good agreement with experiment was obtained in that study.

In this paper, a brief overview of WDA formalism is presented first. Then we report the successes and failures based on our first-principles WDA calculations of ground state properties of some of the most common ferroelectric materials in section III. In section IV we suggest improvements to overcome these failures. In section V, an ad hoc WDA hybrid approach is proposed and used to optimize the structure of tetragonal PbTiO$_3$.

II. FORMALISM

The general form of the xc energy of DFT scheme can be expressed in Rydberg units as

$$E_{xc}[n] = \int n(\mathbf{r})d\mathbf{r} \int \frac{n_{xc}(\mathbf{r},\mathbf{r}')}{|\mathbf{r}-\mathbf{r}'|} d\mathbf{r}', \quad (1)$$

where the xc hole density $n_{xc}(\mathbf{r},\mathbf{r}')$ is defined by the pair-distribution function $g_{xc}(\mathbf{r},\mathbf{r}')$,

$$n_{xc}(\mathbf{r},\mathbf{r}') = n(\mathbf{r}')[g_{xc}(\mathbf{r},\mathbf{r}') - 1], \quad (2)$$

and the xc hole satisfies

$$\int n_{xc}(\mathbf{r},\mathbf{r}')d\mathbf{r}' = -1. \quad (3)$$

Unfortunately the form of $g_{xc}(\mathbf{r},\mathbf{r}')$ is in general unknown. However, based on Monte Carlo simulations it is known with high accuracy for the uniform gas [26]. In the WDA, as mentioned, the unknown general $g_{xc}(\mathbf{r},\mathbf{r}')$ is replaced by G of a homogeneous gas with the weighted density $\bar{n}$,

$$g_{xc}(\mathbf{r},\mathbf{r}') - 1 = G[|\mathbf{r}-\mathbf{r}'|,\bar{n}(\mathbf{r})], \tag{4}$$

where $\bar{n}$ is fixed by the sum rule:

$$\int n(\mathbf{r}') G[|\mathbf{r}-\mathbf{r}'|,\bar{n}(\mathbf{r})] d\mathbf{r}' = -1. \tag{5}$$

The corresponding WDA xc energy is

$$E_{xc}[n] = \int\int \frac{n(\mathbf{r})n(\mathbf{r}')}{|\mathbf{r}-\mathbf{r}'|} G[|\mathbf{r}-\mathbf{r}'|,\bar{n}(\mathbf{r})] d\mathbf{r} d\mathbf{r}'. \tag{6}$$

Equation 5 assures lack of self-interaction in the one-electron limit. This corrects what is believed to be one of the key problems of the LDA.

The function G of homogeneous gas was parametrized by Perdew and Wang (P-W) [26]. As a matter of fact, one may speculate that other choices of G could be better in actual materials [28]. For example, the Gunnarsson-Jones (G-J [28]) and Gritsenko et. al. (GRBA [29]) ansatz are:

$$G^{G-J}(r,n) = c_1(n)\{1 - \exp(-[\frac{r}{c_2(n)}]^5)\}, \tag{7}$$

$$G^{GRBA}(r,n) = c_1(n)\exp(-[\frac{r}{c_2(n)}]^{1.5}). \tag{8}$$

The parameters c_1 and c_2 can be determined from the following conditions:

$$n\int G(r,n) d^3r = -1 \tag{9}$$

$$n\int \frac{G(r,n)}{r} d^3r = \varepsilon_{xc}(n). \tag{10}$$

$\varepsilon_{xc}(n)$ is the xc energy density of a uniform gas with density n.

The xc potential $v_{xc}(\mathbf{r})$ is the functional derivative of E_{xc},

$$v_{xc}(\mathbf{r}) = v_1(\mathbf{r}) \quad v_2(\mathbf{r}) \quad v_3(\mathbf{r}), \tag{11}$$

where

$$v_1(\mathbf{r}) = \int \frac{n(\mathbf{r}')}{|\mathbf{r}-\mathbf{r}'|} G[|\mathbf{r}-\mathbf{r}'|,\bar{n}(\mathbf{r})] d\mathbf{r}', \tag{12}$$

$$v_2(\mathbf{r}) = \int \frac{n(\mathbf{r}')}{|\mathbf{r}-\mathbf{r}'|} G[|\mathbf{r}-\mathbf{r}'|,\bar{n}(\mathbf{r}')] d\mathbf{r}', \tag{13}$$

$$v_3(\mathbf{r}) = \int\int \frac{n(\mathbf{r}')n(\mathbf{r}'')}{|\mathbf{r}'-\mathbf{r}''|} \frac{\delta G[|\mathbf{r}'-\mathbf{r}''|,\bar{n}(\mathbf{r}')]}{\delta n(\mathbf{r})} d\mathbf{r}' d\mathbf{r}''. \tag{14}$$

Examination of the above suggests that implementation of the WDA xc energy density and potential could be cumbersome. However, in a plane-wave representation, by using the convolution theorem, these terms can be evaluated efficiently, as detailed in Ref. [27].

One subtle issue in WDA implementations is shell partitioning [24]. In the WDA the range of integration of G is similar to the size of atoms, and there is no distinction between core and valence electrons, so core and valence electrons dynamically screen valence electrons equally, which is unphysical. As a result, the inter-shell contribution to xc energy would be exaggerated. On the other hand, the LDA can give correct inter-shell contributions, since the LDA depends only on the local density, and outside the core region, the core density vanishes. Based on this observation, a shell partitioning approach was proposed [24, 27], in which the valence-valence interactions are treated with the WDA, while core-core and core-valence interactions are based on the LDA. The total xc energy and potential are

$$E_{xc}[n] = E_{xc}^{LDA}[n] \quad E_{xc}^{WDA}[n_v] - E_{xc}^{LDA}[n_v], \tag{15}$$

$$v_{xc}(\mathbf{r}) = v_{xc}^{LDA}[n(\mathbf{r})] \quad v_{xc}^{WDA}[n_v(\mathbf{r})] - v_{xc}^{LDA}[n_v(\mathbf{r})], \tag{16}$$

where n_v is the valence density and n is the total density. For simplicity, the WDA sum rule of this scheme becomes

$$\int n_v(\mathbf{r}')G[|\mathbf{r}-\mathbf{r}'|,\bar{n}(\mathbf{r})]d\mathbf{r}' = -1, \tag{17}$$

as if the valence states were separated from the core states. In this paper, the semi-core states of metal ions and the O 2s states are treated with LDA, the higher states with the WDA, and the lower states are pseudized.

III. GROUND-STATE CALCULATIONS

The WDA was implemented [27] within a plane-wave basis pseudopotential method using hard Troullier-Martins pseudopotentials [30]. The semi-core states of metal ions include 3s and 3p states of K and Ti, 4s and 4p states of Nb and Sr, 5s and 5p states of Ta and Ba, and 5d states of Pb. Plane-wave basis sets with a cut-off of 132 Ry were used to assure convergence. A standard 3-point interpolation was employed to obtain $\bar{n}(\mathbf{r})$ with a logarithmic grid of increment $\bar{n}_{i+1} = 1.25\bar{n}_i$.

We first calculated the lattice constant of five ferroelectric materials, namely $KNbO_3$, $KTaO_3$, $SrTiO_3$, $BaTiO_3$, and $PbTiO_3$, constrained with cubic symmetry. As mentioned before, the LDA lattice constant is about 1-2% less than experiment, and this small error makes many ferroelectric properties incorrect. As shown in Table 2, the WDA with the P-W form of G dramatically improves the lattice constant over the LDA. Actually all these WDA lattice parameters are very close to experimental data except that of $PbTiO_3$.

TABLE 2. Calculated LDA and WDA lattice constants in Å within for some ferroelectric materials in cubic state, compared with experimental data. The Perdew-Wang form of G was used in WDA. Numbers in parentheses are the percentage deviations from experiment.

material	LDA	WDA	Expt.
$KNbO_3$	3.96 (-1.6)	4.02 (-0.0)	4.016
$KTaO_3$	3.92 (-1.6)	3.98 (-0.1)	3.983
$SrTiO_3$	3.86 (-1.2)	3.92 (+0.4)	3.905
$BaTiO_3$	3.95 (-1.2)	4.01 (+0.3)	4.000
$PbTiO_3$	3.98 (-2.1)	3.93 (-1.0)	3.969

Then we checked the ferroelectric instability in rhombohedral $KNbO_3$ and $BaTiO_3$, and tetragonal $PbTiO_3$. We used a $6 \times 6 \times 6$ special k-point sampling since the energy difference in $KNbO_3$ and $BaTiO_3$ is small, only several mRy/cell. We displaced atom positions according to the experimental distortion patterns, and we performed frozen-phonon calculations at the experimental structures within both LDA (plane-wave and LAPW) and WDA (P-W). We used cubic cells with lattice constants of 4.000 Å and 4.016 Å for $KNbO_3$ and $BaTiO_3$, respectively. The rhombohedral distortion of the lattice parameters from the pseudo-cubic structure in these materials is very small, and was neglected. For tetragonal $PbTiO_3$, $c/a = 1.0635$ and $V = 63.28$ Å^3.

As mentioned, the LDA describes the ferroelectric instability very well at the experimental volume, so one may hope that the WDA retains this desired feature. Fig. 1 shows the calculated LDA and WDA energy versus ferroelectric displacement curves. For $KNbO_3$, the LDA and WDA curves match pretty well, although the WDA energy difference is a little too small compared with the LDA. Our present $KNbO_3$ results agree with previous calculations [25]. For $BaTiO_3$ and $PbTiO_3$, the soft-mode amplitudes of the WDA are about 20% larger than experiment, while those of LDA agree with experiment quite well, within about 5%. Meanwhile, the WDA energy diference is much bigger than in the LDA. This indicates an overestimation of the ferroelectric instability in $BaTiO_3$ and $PbTiO_3$. We also tried the G-J and GRBA forms of G, although they predict slightly better lattice constants of $PbTiO_3$ (3.94 Å for G-J and 3.95 Å for GRBA), their frozen-phonon results are even worse. This implies some difficulty in the present WDA scheme. Fortunately, this can be fixed.

IV. A NEW WDA SUM RULE

As mentioned in the formalism section, the present WDA method includes shell partitioning, and the sum rule of Eq. 17 is used. This simple sum rule assumes valence states are separated from core states. But in the core region, this is not true. Considering that the core-valence xc interaction is treated with the LDA and valence-valence with WDA, one may write a new and more reasonable sum rule as:

$$\int {}^{\}n_v(\mathbf{r}')G[|\mathbf{r}-\mathbf{r}'|,\bar{n}(\mathbf{r})] \quad n_c(\mathbf{r})G[|\mathbf{r}-\mathbf{r}'|,n(\mathbf{r})] \ d\mathbf{r}' = -1, \quad (18)$$

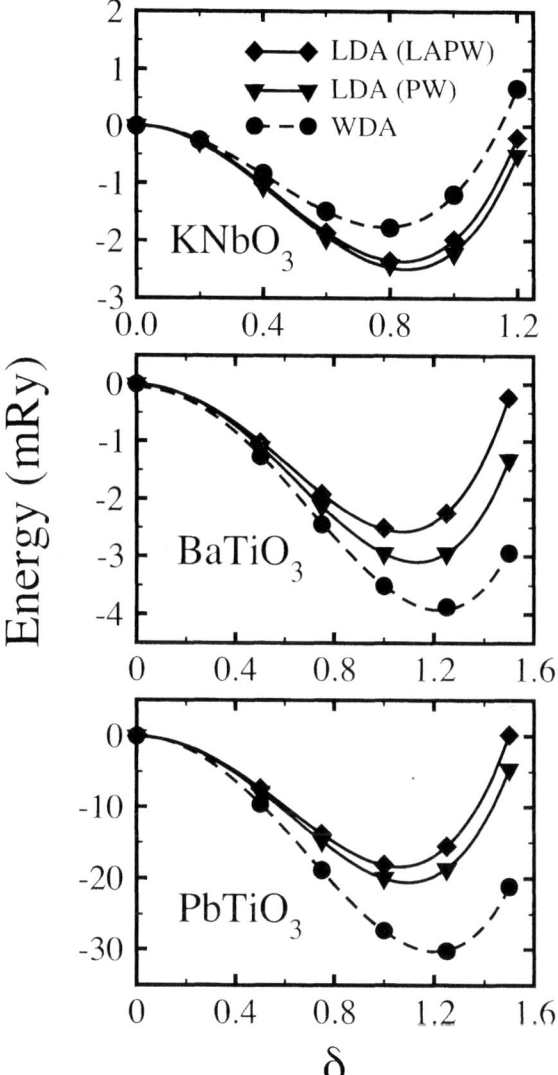

FIGURE 1. Total energy as a function of soft-mode displacement in $KNbO_3$, $BaTiO_3$ and $PbTiO_3$, with the LDA and the WDA. In the WDA the Perdew-Wang form of G was used. Energy is in mRy, and δ is the displacement relative to experiment. Calculations were done based on experimental structures.

which is similar to that proposed by Gunnarsson et. al. in Ref. [24]. The corresponding total xc energy is:

$$E_{xc}[n] = E_{xc}^{LDA}[n] + \int\int \left\{ \frac{n_v(\mathbf{r})n_v(\mathbf{r}')}{|\mathbf{r}-\mathbf{r}'|} G[|\mathbf{r}-\mathbf{r}'|,\bar{n}(\mathbf{r})] - \frac{n_v^2(\mathbf{r})}{|\mathbf{r}-\mathbf{r}'|} G[|\mathbf{r}-\mathbf{r}'|,n(\mathbf{r})] \right\} d\mathbf{r}d\mathbf{r}'. \quad (19)$$

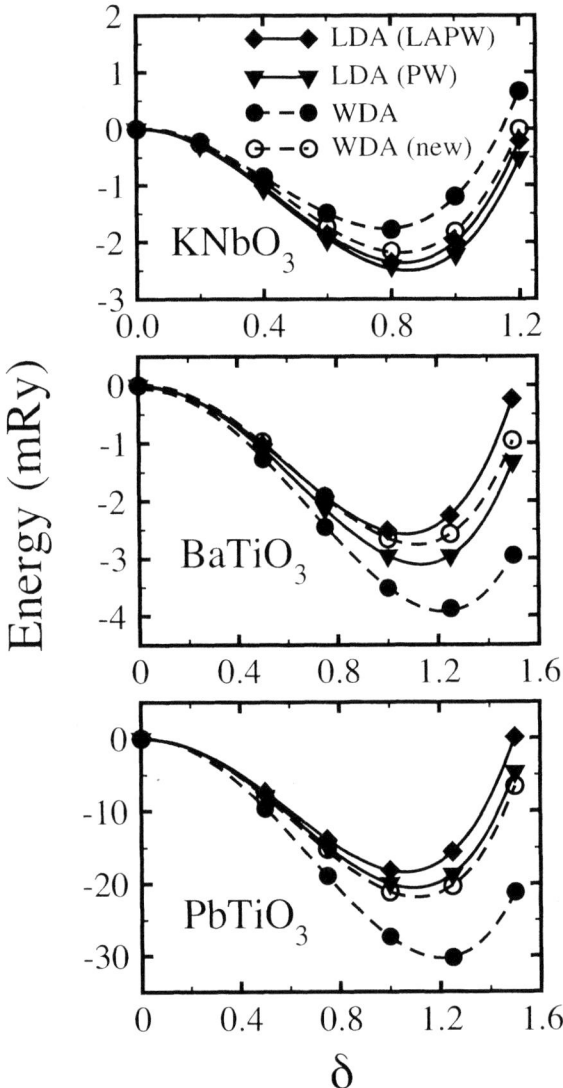

FIGURE 2. Total energy as in Fig. 1, but the results of the WDA with the new sum rule are added.

Enforcing this new sum rule, we recalculated the lattice constants and frozen-phonon curves, and the results are shown in Table 3 and Fig. 2. The new WDA method still gives very good equilibrium volumes for these materials. Comparing with the previous WDA results, we found that the new WDA method reduces lattice constants a little bit. Now the new WDA frozen-phonon curves of all the three materials match with the LDA curves very well. This suggests that it is the sum rule of Eq. 17 that causes overestimation of the ferroelectric instability in $BaTiO_3$ and $PbTiO_3$.

TABLE 3. Calculated lattice constants as in Table 2 within the WDA with previous and the new sum rules. The Perdew-Wang form of G was used.

material	WDA	WDA (new)	Expt.
$KNbO_3$	4.02 (-0.0)	4.01 (-0.1)	4.016
$KTaO_3$	3.98 (-0.1)	3.97 (-0.3)	3.983
$SrTiO_3$	3.92 (+0.4)	3.91 (+0.2)	3.905
$BaTiO_3$	4.01 (+0.3)	4.00 (+0.0)	4.000
$PbTiO_3$	3.93 (-1.0)	3.93 (-1.1)	3.969

V. A HYBRID WDA METHOD

In order to optimize structures, it is very helpful to have the ability to efficiently calculate accurate atomic forces. Unfortunately shell partitioning complicates atomic force calculations in that it introduces an orbital-dependent xc potential, as seen in Eq. 16. As a result, the xc potential is not the xc energy derivative with respect to the total charge density, and the Hellmann-Feynman theory [31] does not directly apply [32].

On the other hand, it is straightforward to obtain atomic forces in a one-window (no shell partitioning) version of the WDA. But as mentioned before, one-window WDA is incorrect, and it predicts lattice constants way too large. However, we can regard one-window WDA as overusing non-local information, while the LDA does not use any non-local information, so the real system should be something in between. The simplest way is to put WDA and LDA together, and so we propose a hybrid WDA method consisting of direct mixing of the WDA and the LDA,

$$E_{xc}[n] = (1.0 - \alpha)E_{xc}^{LDA}[n] \quad \alpha E_{xc}^{WDA}[n], \tag{20}$$

where the only parameter α can be determined by fitting to the lattice constant calculated from shell partitioning WDA. This scheme is very similar to Becke's hybrid method [33], in which the xc energy is a direct mixture of the exact exchange energy and the GGA xc energy. In table 4 parameters α computed from fitting to data in Table 3 of the new WDA method are given. Interestingly, all α are about 28% except for $PbTiO_3$, and Becke's universal mixing parameter for the exact exchange part is also about 28% [33].

We used the hybrid WDA method with $\alpha = 0.19$ to relax tetragonal $PbTiO_3$. The optimized atom positions for the experimental structure are in good agreement with experiment, as seen in Table 5. The optimized $c/a = 1.09$ (50% larger) at the experimental volume. For the fully relaxed structure, we obtained a volume of about 65.8 Å^3 (4% larger), and $c/a = 1.12$ (90% larger). Compared with the GGA data in Table 1, the current hybrid WDA method predicts a better fully optimized structure. The partial success of the ad hoc hybrid WDA method implies that the WDA could be improved by adding a discontinuous δ function to the normally continuous function G, and the corresponding xc energy will have an ordinary WDA part and an additional LDA-like contribution.

TABLE 4. Fitting parameters α in the hybrid WDA scheme.

material	KNbO$_3$	KTaO$_3$	SrTiO$_3$	BaTiO$_3$	PbTiO$_3$
α	0.28	0.28	0.30	0.27	0.19

TABLE 5. Internal coordinates of tetragonal $P4mm$ PbTiO$_3$ for the experimental volume and strain. u_z are given in terms of the lattice constant c. In hybrid WDA, the parameter $\alpha = 0.19$.

	LDA (LAPW)	LDA (PW)	WDA (hybrid)	Expt.
u_z(Pb)	0.000	0.000	0.000	0.000
u_z(Pb)	0.538	0.545	0.543	0.538
u_z(O$_1$,O$_2$)	0.613	0.628	0.618	0.612
u_z(O$_2$)	0.105	0.122	0.114	0.117

VI. CONCLUSIONS

We have investigated the WDA method by ground state calculations for some of the most common ferroelectric materials. In short, with the P-W uniform electron gas form of G and shell partitioning, the WDA yields much better equilibrium volumes than LDA for these materials. But because of the sum rule of Eq. 17, the WDA overestimates the ferroelectric instability in BaTiO$_3$ and PbTiO$_3$, i.e., we found that the WDA double wells are wider and deeper than the LDA ones. We present a better sum rule in Eq. 18, and with it, the corresponding frozen-phonon curves match very well with the LDA curves in KNbO$_3$, BaTiO$_3$ and PbTiO$_3$. At the same time, the new WDA still yields very good lattice constants. In order to do structural relaxation, we proposed an ad hoc hybrid WDA method, in which WDA is directly mixed with LDA, to avoid the difficulty of atomic force calculations incurred by shell partitioning. We obtained the mixing parameter α from fitting, and finally we relaxed tetragonal PbTiO$_3$ using the hybrid WDA method. The optimized structure agrees fairly with experiment. This suggests a possible way to get rid of shell partitioning by adding a discontinuous δ function to the function G.

ACKNOWLEDGMENTS

This work was supported by the Office of Naval Research under ONR Grants N00014-02-1-0506 and N0001403WX20028. Calculations were done on the Center for Piezoelectrics by Design (CPD) computer facility, and on the Cray SV1 supported by NSF and the Keck Foundation.

REFERENCES

1. P. Hohenberg, and W. Kohn, Phys. Rev. **136**, B864 (1964). W. Kohn, and L.J. Sham, Phys. Rev. **140**, A1133 (1965).

2. See e.g., *Theory of ferroelectrics: A vision for the next decade and beyond*, R.E. Cohen, J. Phys. Chem. Solids **61**, 139-146 (1999).
3. R.E. Cohen, and H. Krakauer, Phys. Rev. B **42**, 6416 (1990). R.E. Cohen, Nature (London) **358**, 136 (1992). R.E. Cohen, and H. Krakauer, Ferroelectrics **136**, 65 (1992).
4. D.J. Singh, and L.L. Boyer, Ferroelectrics **136**, 95 (1992).
5. A.V. Postnikov, T. Neumann, G. Borstel, and M. Methfessel, Phys. Rev. B **48**, 5910 (1993) A.V. Postnikov, T. Neumann, and G. Borstel, Ferroelectrics **164**, 101 (1995).
6. R. Yu, and H. Krakauer, Phys. Rev. Lett. **74**, 4067 (1995).
7. C.-Z. Wang, R. Yu, and H. Krakauer, Ferroelectrics **194**, 97 (1997).
8. D.J. Singh, Ferroelectrics **164**, 143 (1995).
9. D.J. Singh, Ferroelectrics **194**, 299 (1997).
10. A. Dal Corso, S. Baroni, and R. Resta, Phys. Rev. B **49**, 5323 (1994).
11. R. Resta, Ferroelectrics **194**, 1 (1997).
12. S. Dallolio, R. Dovesi, and R. Resta, Phys. Rev. B **56**, 10105 (1997).
13. D.R. Hamann, Phys. Rev. Lett. **42**, 662 (1979).
14. M.T. Yin, and M.L. Cohen, **26**, 5668 (1982).
15. D.C. Langreth, and M.J. Mehl, Phys. Rev. B **28**, 1809 (1983). A.D. Becke, Phys. Rev. A **38**, 3098 (1988).
16. J.P. Perdew, K. Burke, and M. Ernzerhof, Phys. Rev. Lett. **77**, 3865 (1996).
17. O. Gunnarsson, M. Jonson, and B.I. Lundquist, Phys. Lett. **59A**, 177 (1976). O. Gunnarsson, M. Jonson, and B.I. Lundquist, Solid State Commun. **24**, 765 (1977).
18. J.A. Alonso, and L.A. Girifalco, Phys. Rev. B **17**, 3735 (1978).
19. J.P. Perdew, J.A. Chevary, S.H. Vosko, K.A. Jackson, M.R. Pederson, D.J. Singh, and C. Fiolhais, Phys. Rev. B **46**, 6671 (1992); **48**, 4978(E) (1993).
20. B. Hammer, K.W. Jacobsen, and J.K. Norskov, Phys. Rev. Lett. **70**, 3971 (1993); B. Hammer, and M. Scheffler, Phys. Rev. Lett. **74**, 3487 (1995).
21. C. Filippi, D.J. Singh, and C. Umrigar, Phys. Rev. B **50**, 14947 (1993).
22. L. Hedin, and B.I. Lundqvist, I. Phys. C **4**, 2064 (1971).
23. D.J. Singh, *Planewaves, Pseudopotentials and LAPW method* (Kluwer Academic publishers, Boston, 1994).
24. O. Gunnarsson, M. Jonson, and B.I. Lundqvist, Phys. Rev. B **20**, 3136 (1979).
25. I.I. Mazin, and D.J. Singh, arXiv:cond-mat/9801301.
26. J.P. Perdew, and Yue Wang, Phys. Rev. B **46**, 12947 (1992).
27. D.J. Singh, Phys. Rev. B **48**, 14099 (1993).
28. O. Gunnarsson, and R.O. Jones, Phys. Scr. **21**, 394 (1980).
29. O.V. Gritsenko, A. Rubio, L.C. Balbás, and J.A. Alonso, Chem. Phys. Lett. **205**, 348 (1993).
30. N. Troullier, and J.L. Martins, Phys. Rev. B **43**, 1993 (1991).
31. H. Hellmann, *Einfuhrung in die Quantenchemie* (Deuieke, Leipzig, 1973), p. 285; R.P. Feynman, Phys. Rev. **56**, 340 (1939).
32. J.C. Slater, J. Chem. Phys. **57**, 2389 (1972).
33. A.D. Becke, J. Chem. Phys. **104**, 1040 (1996).

AUTHOR INDEX

B

Bellaiche, L., 139
Blinc, R., 20
Borstel, G., 204, 210, 231
Bunina, O. A., 118
Burton, B. P., 146

C

Car, R., 168
Caracas, R., 186
Chan, W.-H., 10
Chao, L. K., 33
Chen, H., 10, 55
Chien, R., 152, 160
Cockayne, E., 146
Cohen, R. E., 65, 276
Colla, E. V., 10, 33
Cooper, V. R., 220

D

Davies, P. K., 108
Dec, J., 26
Dkhil, B., 74
Dmowski, W., 48, 108
Dorfman, S., 231
Dorner, B., 74

E

Egami, T., 48, 108
Eglitis, R. I., 204, 210, 231
Emelyanov, S. M., 118
Eremkin, V. V., 118

F

Fornari, M., 124
Fu, H., 139
Fuks, D., 231

G

Gagarina, E. S., 118
Gebauer, R., 168
Goddard III, W. A., 210
Gonze, X., 186
Grachev, V., 196
Grinberg, I., 108, 130, 220

H

Halilov, S. V., 124
Heifets, E., 210
Hellwig, H., 65
Hemley, R. J., 65
Huang, L.-W., 152

I

Itoh, M., 1, 20, 26
Ivanov, A., 74

J

Jin, Y. M., 84
Juhás, P., 108

K

Khachaturyan, A. G., 84
Kholkin, A., 74
Kiat, J.-M., 74
Kleemann, W., 1, 26
Kohanoff, J., 176
Kotomin, E. A., 204, 210, 231
Koval, S., 176
Krakauer, H., 251

L

La-Orauttapong, D., 98
Lebar, A., 20
Li, J. F., 84

M

Malovichko, G., 196
Mamontov, E., 48
Migoni, R. L., 176
Mitas, L., 241

N

Naberezhnov, A. A., 74

P

Pasquarello, A., 269
Prosandeev, S. A., 41, 118, 146

R

Raevski, I. P., 41, 118
Rappe, A. M., 108, 130, 220
Resta, R., 168

S

Sahkar, E. V., 118
Samara, G. A., 1
Savenko, F. I., 118
Schirmer, O., 196
Schmidt, V. H., 152, 160
Sen, P., 241
Shih, I.-C., 160
Shwartsman, V., 74
Singh, D. J., 124, 276
Smotrakov, V. G., 118
Srinivasan, V., 168
Svitelskiy, O., 98

T

Tkachuk, A., 55
Tosatti, E., 176

Toulouse, J., 98
Trepakov, V. A., 231
Tu, C.-S., 152, 160

U

Umari, P., 269

V

Vakhrushev, S. B., 48, 74
Venturini, E. L., 1
Viehland, D., 84
Vikhnin, V. S., 204

W

Waghmare, U. V., 41
Wagner, L. K., 241
Wang, R., 26
Wang, Y., 84
Weissman, M. B., 33
Woicik, J. C., 261
Wu, Z., 276

X

Xu, Z. K., 10

Z

Zakharchenko, I. N., 118
Zalar, B., 20
Zhai, J., 10
Zhang, S., 251